Waterbirds i[n t]h[e UK 20]'06
The Wetl[a]nd[s]

GW01454178

Andy Musgrove, [Alex Banks,]
Neil Calbrade, Richard Hearn and Graham Austin

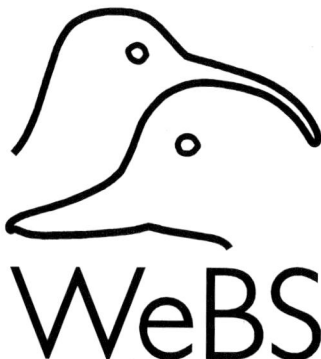

WeBS

Published by

British Trust for Ornithology, Wildfowl & Wetlands Trust,
Royal Society for the Protection of Birds and Joint Nature Conservation Committee

November 2007

BTO   **WWT**   **RSPB**   **JOINT NATURE CONSERVATION COMMITTEE**

ISBN 978-1-906204-27-3
ISSN 1755-6384

This publication should be cited as:
Musgrove, A.J., Collier, M.P., Banks, A.N., Calbrade, N.A., Hearn, R.D. & Austin, G.E. 2007. *Waterbirds in the UK 2005/06: The Wetland Bird Survey.* BTO/WWT/RSPB/JNCC, Thetford.

Published by: BTO/WWT/RSPB/JNCC

Cover: Whooper Swans and Pink-footed Geese by James McCallum. James McCallum is a wildlife artist living and working in North Norfolk. He is best known for his watercolour paintings and sketches of natural history, particularly birds, made outdoors from life at the time of watching. Many of James' paintings are reproduced in a series of books, the latest, '*Arctic Flight*', is a collection of sketches and painting from trips to northern and Arctic regions. For more details of James' work visit *www.jamesmccallum.co.uk*.

Photographs: John Bowers, Paul Doherty, John Hardy, Rob Robinson.

Produced by the British Trust for Ornithology.

Printed by Crowes Complete Print, 50 Hurricane Way, Norwich, NR6 6JB.

Typeset in Times New Roman, Arial and Trebuchet MS fonts.

Available from: BTO, The Nunnery, Thetford, Norfolk IP24 2PU, UK.

This report is provided free to all WeBS counters and those who participate in the other national waterbird surveys, none of whom receive financial reward for their invaluable work. Further feedback is provided to counters through the annual WeBS Newsletter. For further information please contact the WeBS Office at the BTO.

ACKNOWLEDGEMENTS

This book represents the twenty-fifth report of the Wetland Bird Survey and comprises information from WeBS and complementary national and local surveys, *e.g.* goose censuses. It is entirely dependent on the many thousands of dedicated volunteer ornithologists who supply the data and to whom we are extremely grateful. The Local Organisers who coordinate these counts deserve special thanks for their contribution.

We are also grateful to the following people for providing technical assistance, supplementary information and additional data, and comments on draft texts:

Helen Baker, Niall Burton, Dave Butterfield, Peter Cranswick, Olivia Crowe, Colette Hall, Martin Heubeck, Baz Hughes, Rowena Langston, Ilya Maclean, Heidi Mellan, Margaret Morris, Mark Rehfisch, Lucy Smith, David Stroud, Chris Waltho and Jenny Worden. Many amateur observers also provide reports of their studies; these are acknowledged within the text.

Grateful thanks to all and apologies to anyone who has inadvertently been missed.

Any maps partially based on Ordnance Survey products have been produced under the following license agreement. © Crown copyright. All rights reserved JNCC Licence Number 100017955.2007.

The WETLAND BIRD SURVEY

Organised and funded by

British Trust for Ornithology
The Nunnery, Thetford, Norfolk IP24 2PU
www.bto.org

Wildfowl & Wetlands Trust
Slimbridge, Gloucestershire GL2 7BT
www.wwt.org.uk

Royal Society for the Protection of Birds
The Lodge, Sandy, Bedfordshire SG19 2DL
www.rspb.org.uk

Joint Nature Conservation Committee
Monkstone House, City Road, Peterborough
PE1 1JY
www.jncc.org.uk

CONTACTS

WeBS National Coordinator: Andy Musgrove
WeBS Core Counts: Mark Collier
WeBS Low Tide Counts: Neil Calbrade

General queries: webs@bto.org

WeBS Office
British Trust for Ornithology
The Nunnery
Thetford
Norfolk IP24 2PU
UK
Tel: 01842 750050
Fax: 01842 750030
E-mail: firstname.surname@bto.org
or webs@bto.org
www.bto.org/webs

NATIONAL GOOSE CENSUSES

Organised and funded by the Wildfowl &
Wetlands Trust and the Joint Nature
Conservation Committee.
Contact: Richard Hearn
E-mail: Richard.Hearn@wwt.org.uk
or monitoring@wwt.org.uk

Wildfowl & Wetlands Trust
Slimbridge
Glos GL2 7BT
UK
Tel: 01453 891225
Fax: 01453 891901
www.wwt.org.uk/research/monitoring

OTHER NATIONAL WATERBIRD SURVEYS

Details of and contacts for many of the other
waterbird surveys used in this report, and of
forthcoming surveys, can be obtained via the
web sites of the four WeBS partner
organisations.

ERRATUM TO 2004/05 REPORT

Please note the following corrections to data
presented in the 2004/05 WeBS annual report:

p142 The 2000/01 peak counts of 268 Black-
tailed Godwit on the Ouse Washes was
from a supplementary daytime count and
should have be indicated by a method 13.

CONTENTS

Summary

The Wetland Bird Survey and Waterbirds in the UK

The Wetland Bird Survey (WeBS) is a joint scheme of the British Trust for Ornithology (BTO), the Wildfowl & Wetlands Trust (WWT), Royal Society for the Protection of Birds (RSPB) and Joint Nature Conservation Committee (JNCC) to monitor non-breeding waterbirds in the UK. The principal aims of the scheme are to identify population sizes, to determine trends in numbers and distribution, and to identify important sites for waterbirds. WeBS Core Counts are made annually at around 2,000 wetland sites of all habitats; estuaries and large still waters predominate. Monthly coordinated counts are made mostly by volunteers, principally from September to March, with fewer observations during summer months. Data from other sources, *e.g.* roost counts of grey geese, are included in this report where relevant.

This report presents total numbers counted for all species in the most recent year in Great Britain and Northern Ireland. Annual indices are provided for the more numerous species, as are monthly indices showing relative abundance during the winter.

Species accounts provide yearly maxima for all sites supporting internationally and nationally important numbers. Sites with changed status are highlighted and significant counts are discussed. Counts are placed in an international context where possible, and relevant research is summarised. Waterbird totals are provided for all sites meeting criteria for international importance and species occurring in internationally important numbers on each are identified.

WeBS Low Tide Counts are made on selected estuaries to determine the distribution of birds during low tide and to identify important feeding areas that may not be recognised during Core Counts, which are made mostly at high tide. A summary of results for these estuaries, and distribution maps for selected species, are provided.

Waterbird totals recorded by the Irish Wetland Bird Survey, a similar scheme operating in the Republic of Ireland, are also included.

The 2005/06 year

This report summarises counts during 2005/06 and previous years (since 1960 for wildfowl, 1969 for waders and the early 1980s or 1990s for other species). During 2005/06, WeBS counters covered 3,750 count sectors at around 2,100 count sites, during the crucial 'winter' period of September to March. At least 1,500 were counted in any one of these months and almost 1,150 were covered continually throughout this period. This, again, represents a fantastic effort all around and a huge thank you must go to all those involved.

In Britain, wintering numbers of both Bewick's and Whooper Swans showed a convincing increase during 2005/06 with the Whooper Swan index reaching its highest ever. The British index for Mute Swan also reached record levels, however, numbers in Northern Ireland fell to their lowest ever recorded.

The number of Pink-footed Geese fell slightly, although remain at a high level. Following the long-term decline in the number of European White-fronted Geese wintering in Britain numbers in 2005/06 remained at a similar level to that of the previous year. Greenland White-fronted Geese also showed further signs of decline, although the national index was similar to that of 2004/05. Unlike European Whitefronts, this decline has been observed across the entire population and is thought to be linked to poor breeding success.

Icelandic Greylag Geese wintered in similar numbers to the past few years. An increase in Greenland Barnacle Geese was noted at several key sites including at the principal site, Islay. There was a slight decline in Svalbard Barnacle Geese, which was attributed to lower numbers at the key site, Solway Estuary. Numbers of Dark-bellied Brent Geese rose for the second year running with increases witnessed at several of the key sites. Peak numbers of East Canadian High Arctic Light-bellied Brent Geese were similar to the previous few years, whilst numbers of Svalbard Light-bellied Brent Geese was almost double that of the previous year, albeit similar to figures two-years hence. As usual the majority of birds wintered on Lindisfarne, numbers at which were the third highest ever recorded during WeBS.

Shelduck remained at a similar level to 2004/05 in Britain and continued to show an increase in Northern Ireland. The Wigeon index rose to its highest ever level in Britain; however, the Northern Ireland wintering population continues to fall. Numbers of wintering Gadwall also reached their highest-ever level in Britain, whilst numbers in Northern Ireland were still low. As usual Teal numbers were subject to a high level of fluctuation, although this species too was recorded in highest ever numbers. Following a long-term decline the numbers of Mallard appear to be stable having reached record low levels in both Britain and Northern Ireland in 2004/05. Pintail numbers in both Britain and Northern Ireland have risen sharply since the beginning of the millennium and continue to do so. Shoveler too reached a new high level in the index in Britain numbers whilst numbers in Northern Ireland remained low.

The recent dramatic decline of some diving duck species in Northern Ireland appeared to be abating with some signs of a revival, whilst Scaup continued to increase, peaking at their highest ever level. Numbers of these three species have been less erratic in Britain with a slight decline for Pochard, a continued rise for Tufted Duck and little change for Scaup. Eider numbers in Britain rose slightly following a recent decline, whereas the long tern increase in Northern Ireland continued. The wintering population of Goldeneye in the UK remained low, though both the British and Northern Ireland indices appeared relatively stable following recent declines. Both Red-breasted Merganser and Goosander numbers fell slightly although numbers for the latter remain relatively stable.

Counted maxima for Red-throated Diver increased on those of 2004/05, while those for Black-throated and Great Northern Divers were at their highest levels ever. The British trend for Little Grebe continues to show a steady increase, whilst the trend for Great Crested Grebes is stable. Little Egret again continued to expand its population and range, particularly in the west. In Britain both Moorhen and Coot numbers continue to remain relatively stable. However, Coot numbers in Northern Ireland remained low despite notable increases at Loughs Neagh and Beg and Upper Lough Erne.

Oystercatcher numbers remained fairly stable, though the index did show a slight decrease with peak counts at many sites below average. Avocets have shown a steady increase over the past 15 years though the British index dropped slightly for the second consecutive year. Numbers of wintering Golden Plover continued to increase in both Britain and Northern Ireland, reaching a new peak in Britain, whilst Lapwing numbers also rose in Britain and remained stable in Northern Ireland. Ringed Plover numbers rose slightly in Britain and also rose more markedly in Northern Ireland. Conversely, Grey Plover have continued to decline in Britain, though Northern Ireland did see a small increase. Although the British Knot index showed a slight decline numbers on the key site, The Wash, were the highest for over twelve years. Wintering numbers of Dunlin continued to decline. There were mixed fortunes for the two godwit species, with a continued fall in Bar-tailed Godwit numbers yet a rise in Black-tailed Godwits to such an extent that the peak monthly count for Black-tailed Godwit outnumbered that of Bar-tailed for the first time. Both Redshank and Greenshank showed a slight decrease in Britain, whilst Curlew and Turnstone numbers remained similar to the past few years. Turnstone numbers in Northern Ireland have shown a rise during the past few years.

Declines were again recorded in most of the main gull species, the exception being Great Black-backed Gull, where the maximum count rose by almost 23% on the previous year. Kittiwake numbers too were higher than for 2004/05, with the maximum being more than doubled. Summed maxima of Little Tern and Arctic Tern were higher than in the previous year. However, Sandwich Tern and Common Tern numbers were lower than in 2005/06. As with all gulls and terns the optional coverage of these groups during WeBS counts means that numbers recorded will largely be dependent on coverage across sites.

Introduction

The UK is of outstanding international importance for waterbirds. Lying on some of the major flyways for Arctic-nesting species, large numbers of waterbirds are attracted, especially during winter, by the relatively mild climate and extensive areas of wetland, notably estuaries. The UK thus has both moral and legal obligations to conserve both these waterbirds and the wetlands upon which they depend.

As a signatory to a number of international conservation conventions, and as a member of the EU, the UK is bound by international law. In particular, the 'Ramsar' Convention on Wetlands of International Importance especially as Waterfowl Habitat, the EC Birds Directive and the EU Habitats and Species Directive, between them, require the UK to identify important examples of wetland and other habitats and sites important for birds and designate them for protection. Implicit in these obligations is the need for regular monitoring to identify and monitor such sites. These instruments also lay particular significance on the need to conserve migratory populations, and consequently most of the waterbird populations in the UK.

The UK has ratified the Agreement on the Conservation of African-Eurasian Migratory Waterbirds (AEWA) of the 'Bonn' Convention on the Conservation of Migratory Species of Wild Animals. AEWA entered into force in 1999. It is a specific Agreement requiring nations to take coordinated measures to conserve migratory waterbirds given their particular vulnerability due to their migration over long distances and their dependence on networks that are decreasing in extent and becoming degraded through non-sustainable human activities. Article three of the Agreement requires, among other things, that sites and habitats for migratory waterbirds are identified, protected and managed appropriately, that parties initiate or support research into the ecology of these species, and exchange information and results. Explicit in this Agreement is that adequate monitoring programmes are set in place to fulfil these objectives and the Action Plan to the Agreement specifically requires that nations endeavour to monitor waterbird populations.

AIMS AND OBJECTIVES OF WeBS

The Wetland Bird Survey (WeBS) aims to monitor all non-breeding waterbirds in the UK to provide the principal data on which the conservation of their populations and wetland habitats is based. To this end, WeBS has three main objectives:
- to assess the size of non-breeding waterbird populations in the UK;
- to assess trends in their numbers and distribution; and
- to assess the importance of individual sites for waterbirds.

These results also form the basis for informed decision-making by conservation bodies, planners and developers and contribute to the sustainable and wise use and management of wetlands and their dependent waterbirds. The data and the WeBS report also fulfil some of the objectives of the Conventions and Directives listed above. WeBS also provides UK data to Wetlands International to assist their function to coordinate and report upon waterbird monitoring at an international scale.

Structure and organisation of WeBS
WeBS is a partnership scheme of the British Trust for Ornithology (BTO), Wildfowl & Wetlands Trust (WWT), Royal Society for the Protection of Birds (RSPB) and the Joint Nature Conservation Committee (JNCC), the last on behalf of Natural England (NE), Scottish Natural Heritage (SNH) and the Countryside Council for Wales (CCW), and the Environment and Heritage Service in Northern Ireland (EHS).

WeBS continues the traditions of two, long-running count schemes which formed the mainstay of UK waterbird monitoring since 1947 (Cranswick et al. 1997). WeBS Core Counts are made at a wide variety of wetlands throughout the UK. Synchronised counts are conducted once per month, particularly from September to March, to fulfil all three main objectives. In addition, WeBS Low Tide Counts are undertaken on selected estuaries with the aim of identifying key areas used during the low tide period, principally by feeding birds; areas not otherwise noted for their importance by Core Counts which are normally conducted at high tide.

The success and growth of these count schemes accurately reflects the enthusiasm and dedication of the several thousands of volunteer ornithologists who participate. It is largely due to their efforts that waterbird monitoring in the UK is held in such international high regard.

Aim of this report

This report presents syntheses of data collected between July 2005 and June 2006 (see *The WeBS Year*), and in previous years, in line with the WeBS objectives. Data from other national and local waterbird monitoring schemes, notably annual goose censuses, are included where WeBS data alone are insufficient to fulfil this aim, so that the report provides a single, comprehensive source of information on waterbird status and distribution in the UK. All nationally and internationally important sites for which data exist are listed.

WEATHER IN 2005/06

This summary of UK weather is drawn from the Meteorological Office web site at www.metoffice.gov.uk. Bracketed figures following the month refer to the Core Count priority date for the month in question. Arctic breeding conditions for birds that winter in the UK are summarised from information collated by Soloviev & Tomkovich at the web site www.arcticbirds.ru.

United Kingdom

Mean monthly temperatures during **July** (24) were above average for the time of year. Levels of rainfall were below average in Scotland and Northern Ireland but above the norm in England and Wales.

Temperatures remained above average throughout **August** (21). The amount of rain was below average throughout most of the country except for in East Anglia and parts of western and northern Scotland, which experienced above average rainfall.

East Anglia and northern Scotland remained wet and temperatures were generally above average for most of the country with some areas experiencing their warmest **September** (18) for over five years. For much of southern England and parts of Wales September was the ninth out of the past 11 months in which

rainfall was below average. For these areas 2005 was the driest year since 1973.

October (16) brought mild conditions across the whole of the UK, in spite of this northeast England and southeast Scotland experienced lower than average sunshine. October was wetter than usual with many areas receiving rainfall double the monthly average.

November (6) began mild but became cold in the second half of the month. Sunshine levels were above average with some areas documenting record levels of sunshine. Rainfall was close to average except in the south and east of England, which remained dry.

A cold **December** (4) throughout many parts of southern England, however, temperatures were above average in much of Scotland and Northern Ireland. In general most of the country was drier and sunnier than expected for the time of year.

January (15) was the driest since 1997 for England, Wales and Northern Ireland. Temperatures were close to average for the southern half of the UK but above average in the north of Scotland.

During **February** (12) southwest England experienced cold conditions, while temperatures were above average in Scotland. Southeast England and East Anglia were wetter than average and Northern Ireland remained drier.

March (12) was predominantly cold throughout much of the UK. Average temperatures in Scotland were lower in March than in the three preceding months; the first time that this has been recorded for 30 years. Rainfall varied between regions and some areas experienced twice the monthly average.

April (9) saw temperatures return to nearer average in Scotland and Northern Ireland, with those in England and Wales being slightly above average. Northwest Scotland was predominantly wet while southeast England remained dry. Sunshine was above average throughout most of the UK, particularly eastern Scotland.

May (14) was very wet throughout most of the UK; some parts of England and Wales received double the monthly average. East Anglia was particularly warm.

June (11) was warm, dry and sunny for most of the UK. Temperatures were above average and rainfall below average.

Table 1. The percentage of inland count units (lakes, reservoirs, gravel pits, rivers and canals) in the UK with any ice and with 75% or more of their surface covered by ice during WeBS counts in winter 2005/06 (England divided by a line drawn roughly between the Humber and the Mersey Estuaries).

Region	Ice	S	O	N	D	J	F	M
Northern Ireland	>0%	0	0	0	0	0	0	0
	>74%	0	0	0	0	0	0	0
Scotland	>0%	0	0	2	3	6	6	7
	>74%	0	0	2	1	5	4	5
N England	>0%	0	0	0	2	4	5	4
	>74%	0	0	0	<1	<1	3	0
S England	>0%	0	0	0	<1	1	7	2
	>74%	0	0	0	<1	<1	2	<1
Wales	>0%	0	0	1	0	0	4	1
	>74%	0	0	1	0	0	2	0

Arctic Breeding Conditions 2005

Spring temperatures were above average throughout Fennoscandia, arctic Russia, Greenland and eastern Canada. Indications of phenology pointed towards an early spring in Greenland and across most of arctic Russia, while perceptions were that the season was later in Fennoscandia. Summer temperatures remained above average in eastern Greenland, Fennoscandia and western arctic Russia, south through from the Yamal Peninsula. However, there was considerable local variation between summer temperatures at sites in Fennoscandia and western arctic Russia with temperatures at many sites being warmer than average and at nearby sites colder than average. Temperatures on the Novaya Zemlya archipelago, west through the Taimyr Peninsula and west through arctic Russia were below average.

Rodent abundance was low in Greenland, Fennoscandia and arctic Russia east from the Yamal Peninsula. Higher abundances were recorded from the northern Yamal west through the Taimyr Peninsula.

SURVEY METHODS

The main source of data for this report is the WeBS scheme, providing regular monthly counts for most waterbird species at the majority of the UK's important wetlands. In order to fulfil the WeBS objectives, however, data from a number of additional schemes are included in this report. In particular, a number of species groups necessitate different counting methodologies in order to monitor numbers adequately, notably most geese and seaducks, and the results of other national and local schemes for these species are routinely included.

The methods for these survey types are outlined below and more detail can be found in Gilbert *et al.* (1998). It should be noted that site definition is likely to vary between these surveys (see *Interpretation of Waterbird Counts*).

WeBS Core Counts
WeBS Core Counts are made using so-called 'look-see' methodology (Bibby *et al.* 2000), whereby the observer, familiar with the species involved, surveys the whole of a predefined area. Counts are made at all wetland habitats, including lakes, lochs/loughs, ponds, reservoirs, gravel pits, rivers, freshwater marshes, canals, sections of open coast and estuaries. Numbers of all waterbird species, as defined by Wetlands International (Rose & Scott 1997), are recorded. In the UK, this includes swans, geese, ducks, divers, grebes, cormorants, herons, Spoonbill, rails, cranes, waders and Kingfisher. Counts of gulls and terns are optional.

In line with the recommendations of Vinicombe *et al.* (1993), records of all species recorded by WeBS, including escapes, have been published to contribute to the proper assessment of naturalised populations and escaped birds. Following Holmes & Stroud (1995), non-native species, which have become established are termed 'naturalised'. These species are categorised according to the process by which they became established: naturalised feral (domesticated species gone wild); naturalised introduction (introduced by man); naturalised re-establishment (species re-established in an area of former occurrence); or naturalised establishment (a species which occurs, but does not breed naturally, *e.g.* potentially Barnacle Goose in southern England). With the exception of vagrants, all other non-native species have been classed as 'escapes'. The native range is given in the species account for naturalised species, escapes and vagrants.

Most waterbirds are readily visible. Secretive species, such as snipes, are generally under-recorded. No allowance is made for these habits by the observer and only birds seen or heard are recorded. The species affected by such biases are well known and the problems of interpretation are highlighted individually in the Species accounts. Most species and many subspecies are readily identifiable during the counts. Categories may be used, *e.g.* unidentified scoter species, where it is not possible to be confident of identification, *e.g.* under poor light conditions.

Species present in relatively small numbers or dispersed widely may be counted singly. The number of birds in large flocks is generally estimated by mentally dividing the birds into groups, which may vary from five to 1,000 depending on the size of the flock, and counting the number of groups. Notebooks and tally counters may be used to aid counts.

Counts are made once per month, ideally on predetermined 'priority dates'. This enables counts across the whole country to be synchronised, thus reducing the likelihood of birds being double counted or missed. Such synchronisation is imperative at large sites, which are divided into sectors, each of which can be practicably counted by a single person in a reasonable amount of time. Local Organisers ensure coordination in these cases due to the high possibility of local movements affecting count totals. The priority dates are pre-selected with a view to optimising tidal conditions for counters covering coastal sites at high tide on a Sunday (see *Coverage*). The dates used for individual sites may vary due to differences in the tidal regime around the country. Coordination within a site takes priority over national synchronisation.

Counts suspected to be gross underestimates of the true number of non-secretive species present are specifically noted, *e.g.* a large flock of roosting waders only partially counted before being flushed by a predator, or a distant flock of seaduck in heavy swell. These counts may then be treated differently when calculating site totals (see *Analysis*).

Data are input by a professional data input company. Data are keyed twice by different people and discrepancies identified by computer for correction. Any particularly unusual counts are checked by the National Organisers and are confirmed with the counters if necessary.

WeBS Low Tide Counts
This survey aims to assess numbers of waterbirds present during low tide on estuaries, primarily to assess the distribution of feeding birds at that time (Musgrove *et al.* 2003; see the section *Low Tide Counts* for a full explanation of methods).

This survey occasionally provides higher counts for individual sites than Core Counts, for example, where birds feed on one estuary but roost on another. These data are validated before being used for site assessment against 1% thresholds.

Supplementary daytime and roost counts
Supplementary counts are made at some sites where WeBS counts are known to under-represent the true value of the site. In particular, some species occur in much larger sites when using the site as a nighttime roost, *e.g.* geese, Goosander and gulls, that are not present during WeBS daytime counts. Some sites are also counted more frequently than once monthly by some observers.

Supplementary counts are collected by counters familiar with the site for WeBS survey, thus employing the same site definition and, for daytime counts, the same counting methods, and are submitted on standardised recording forms adapted from those used for WeBS Core Counts.

Goose roost censuses
Many geese (*Anser and Branta spp*) spend daylight hours in agricultural landscapes, and are therefore missed during counts at wetlands by WeBS. These species are usually best counted as they fly to or from their roost sites at dawn or dusk since these are generally discrete wetlands and birds often follow

traditional flight lines approaching or leaving the site. Even in half-light, birds can generally be counted with relative ease against the sky, although they may not be specifically identifiable at mixed species roosts.

In order to produce population estimates, counts are synchronised nationally for particular species (see *National totals* below), though normally only one or two such counts are made each year. The priority count dates are determined according to the state of the moon, since large numbers of geese may remain on fields during moonlit nights. Additional counts are made by some observers, particularly during times of high turnover when large numbers may occur for just a few days.

In some areas, where roost sites are poorly known or difficult to access, counts of birds in fields are made during the daytime. As with WeBS Core Counts, the accuracy of the count is noted.

Additional counts
Additional, *ad hoc*, data are also sought for important sites not otherwise covered by regular monitoring, particularly open coast sections in Scotland, whilst the results of periodic, coordinated surveys - such as the non-estuarine coastal waterbird survey (NEWS), International Greenland Barnacle Goose Census and International Whooper & Bewick's Swan Census - are included where the data collected are compatible with the presentation formats used in this report.

The accuracy of counts of waterbirds on the sea is particularly dependent on prevailing weather conditions at the time of or directly preceding the count. Birds are often distant from land, and wind or rain can cause considerable difficulty with identifying and counting birds. Wind not only causes telescope shake, but even a moderate swell at sites without high vantage points can hamper counts considerably. The need to count other waterbirds in 'terrestrial' habitats at the site often precludes the time required for an accurate assessment of seaducks. Many sites may be best covered using aerial surveys, though this technique has been little used in the UK historically. Consequently, the best counts of most divers, grebes and seaduck at open coast and many estuarine sites are made simply when conditions allow; only rarely will such conditions occur by chance during WeBS counts. Synchronisation between different sites may be difficult or impossible to achieve, and thus coordination of most counts to date has occurred at a regional or site level, *e.g.* within the Moray Firth and within North Cardigan Bay.

The extensive use of aerial survey methods in nearshore marine waters in recent years means that data are available for a number of sites. However, the boundaries of such sites frequently do not correspond to those counted for WeBS Core Counts, and indeed the area surveyed from the air can vary between years. As a result, such aerial surveys are now tabulated separately within the relevant species accounts. These surveys employ a 'distance sampling' methodology (see Buckland *et al.* 2001, 2004), whereby only a proportion of birds is counted, and the missed proportion estimated by statistical means. Some published reports from these surveys provide only the counted number, whilst others include the calculated estimates also (which often have relatively wide confidence intervals).

Some data are provided directly by individuals (for example, reserve wardens), often undertaking counts for site survey purposes, but whose data are not formally published in a report.

A significant point is that these additional data are taken from published sources, from surveys with the specific aim of monitoring waterbirds, and where methods have been published - or where data have been collected by known individuals, usually undertaking site-based surveys, and are provided directly for use in *Waterbirds in the UK*. Casual records and data from, *e.g.* county bird reports, where the methods and/or site boundaries used are not documented, are not included. Reports and data for important sites from surveys that the authors know to have taken place in recent years are actively sought for inclusion in this report, but it is likely that other sources of suitable data are overlooked. The inclusion of additional data for some species and sites does not, thus, indicate that the tables in the Species accounts include all such suitable data.

Irish Wetland Bird Survey

The Irish Wetland Bird Survey (I-WeBS) monitors non-breeding waterbirds in the

Republic of Ireland (Crowe 2005). I-WeBS was launched in 1994 as a joint partnership between BirdWatch Ireland, National Parks and Wildlife Service of Dúchas, The Heritage Service of the Department of Environment and Local Government (Ireland), and WWT, with additional funding and support from the Heritage Council and WWF UK (World Wide Fund for Nature). I-WeBS is complementary to and compatible with the UK scheme. The main methodological difference from UK-WeBS is that counts are made only between September and March, inclusive.

Productivity monitoring

Changes in numbers of waterbirds counted in the UK between years are likely to result from a number of factors, including coverage and weather, particularly for European and Russian breeding species which may winter further east or west within Europe according to the severity of the winter. Genuine changes in population size will, however, result from differences in recruitment and mortality between years.

For several species of swans and geese, young of the year can be readily identified in the field and a measure of productivity can be obtained by recording the number of young birds in sampled flocks, expressed as a percentage of the total number of birds aged. Experienced fieldworkers, by observing the behaviour of and relationship between individuals in a flock, can record brood sizes as the number of young birds associating with, usually, two adults.

ANALYSIS AND PRESENTATION

In fulfilment of the WeBS objectives, results are presented in a number of different sections. An outline of the analyses undertaken for each is given here; further details can be provided upon request. A number of limitations of the data or these analytical techniques necessitate caution when interpreting the results presented in this report (see *Interpretation of Waterbird Counts*).

Count accuracy and completeness

Counts at individual sites may be hampered by poor conditions, or parts of the site may not be covered. This may result in counts missing a significant proportion of one or more species.

It is important to flag such counts since using them at face value would under-represent the importance of the site and give misleading results, *e.g.* when used for trend calculations and assessment of site importance.

Counts at sites - and at individual sectors of large sites that are counted using a series of sub-divisions (known as 'complex sites') - are flagged as 'OK' or 'Low' by the counter, where 'Low' indicates that the counter feels a significant proportion of the birds present at the time of the count may have been missed, *e.g.* because all of the site or sector was not visited, or because a large flock of birds flew before counts were complete. Such assessments may be provided for individual species, or for all species present.

Similarly, at complex sites, one or more sectors may be missed in a particular month, again rendering the total count for the site incomplete to a greater or lesser degree for one or more species.

For single sector sites, counts are assessed as incomplete based on the 'OK/Low' information provided by the counter. For complex sites, an algorithm is used to assess whether missed sectors and/or 'Low' counts in some sectors constitute an incomplete count at the site level. The mean count of each sector is calculated based on 'OK' counts from a window extending a month either side of the month of the count in question, and using earlier or subsequent years, such that within this window the 15 nearest counts are used to make the assessment. The total count for the site in any one month is considered incomplete if the sectors for which the count is missing or 'Low' in that month tend to hold, on the basis of their mean values, more than 25% of the sum of all sector means. The assessment is made on a species-by-species basis, recognising the fact that species distribution is not uniform across a site that and a missed sector may be particularly important for some species but not for others.

Completeness assessments are made for all WeBS Core Counts, and for most goose roost counts (which, as single-sector sites, are made on the basis of the 'OK/Low' assessment provided by the counter).

Because the completeness calculation for complex sites is based on a moving window of counts, and the use of different parts of the site by species may change, the addition of new data each year may result in counts flagged in previous *Waterbirds in the UK* (prior to 2004/05 published as *Wildfowl and Wader Counts*) as complete now being considered incomplete, and *vice versa*.

Counts are not flagged as 'Low' if a large number of the birds present is routinely missed, *e.g.* because they are cryptic, secretive, or hide in reeds - such as Snipe, Teal and Water Rail. 'Low' indicates that a significant proportion of the birds that could reasonably be expected to be counted under normal conditions was considered to have been missed. Similarly, many counts of waterbirds on the sea may be undercounts. Indeed, if the distribution of a flock stretches beyond the limits of visibility, the counter - as with birds hidden in reeds - can never know with confidence whether the count included all birds present. Counts flagged as incomplete are treated differently in trend analysis and site importance assessments.

The WeBS Year
Different waterbird species occur in the UK at different times of year. Most occur in largest numbers during winter, some are residents with numbers boosted during winter, while others occur primarily as passage migrants or even just as summer visitors.

Although WeBS counts concentrate primarily on winter months, survey is made year-round. Accordingly, different 12-month periods are used to define a year to report upon different species, in particular, to define the 'annual' maximum and to identify the peak 'annual' count for assessing site importance.

For most species, the year is defined as July to June, inclusive. Thus, for species present in largest numbers during winter, counts during autumn passage and spring passage the following calendar year are logically associated with the intervening winter. For species present as summer visitors - notably terns, Garganey and Little Ringed Plover - the calendar year is used to derive national and site maxima. The different format used for column headings (*e.g.* 05/06 or 2005) in the 'header' and tables in each species account identify whether a 'winter' or calendar year has been used.

Note that national totals (reported in Tables 3 and 4) present data for the period July 2005 to June 2006.

National totals and annual maxima

Total numbers of waterbirds recorded by WeBS and other schemes are presented (within Tables 3 and 4 and within individual species accounts). It is very important to appreciate that these national totals are not population estimates, as WeBS does not cover 100% of the population of any species. The totals are presented separately for Great Britain (including the Isle of Man but excluding the Channel Islands) and Northern Ireland in recognition of the different legislation that applies to each. Separate totals for England, Scotland, Wales, and the Channel Islands can be obtained from the BTO upon request. The count nearest the monthly priority date or, alternatively, the count coordinated with nearby sites if there is considered to be significant interchange, is chosen for use in this report if several accurate counts are available for the same month. A count from any date is used if it is the only one available.

Totals from different count methods are mostly not combined to produce national totals because the lack of synchronisation may result in errors, *e.g.* birds counted at roost by one method may be effectively double counted during the WeBS count at a different site in that month. Total counts from several national goose surveys are, however, used instead of WeBS Core Counts where the census total provides a better estimate of the total numbers, as follows:

- Pink-footed and Icelandic Greylag Geese in October, November and December;
- Greenland White-fronted Goose in December and March;
- Greenland Barnacle Geese in November and March;
- NW Scotland Greylag Geese in August and February;
- Canadian Light-bellied Brent Geese in October.

Additionally, counts of Svalbard Barnacle Geese from North Cumbria and Dumfries & Galloway are replaced by Solway-wide dedicated counts between October and March. Finally, the maximum British totals for both Bewick's and Whooper Swan do include roost counts from the Ouse and Nene Washes and Martin Mere in place of Core Counts at this site, given the particular concentration of these species feeding around and roosting at this site.

Counts from other site or regional-based surveys, for example of seaducks, are not included in national totals. Where a census total replaces standard Core Count data these are indicated by '*'.

Some of the goose populations are identified according to location (from research into movements of marked birds) as they cannot be separated in the field by appearance alone. In such cases, a standard region of the UK is used each year to assign individual birds to particular populations and thus to derive national totals. For full details please contact BTO but broadly, the breakdown is as follows:

- NW Scotland Greylag Goose - Inner and Outer Hebrides plus Southwest Highland.
- Icelandic Greylag Goose - all other areas of Scotland plus Northumberland and North Cumbria.
- Re-established Greylag Goose - other areas.
- Greenland Barnacle Goose - Scottish west coast plus Shetland and Orkney.
- Svalbard Barnacle Goose - other Scottish regions plus Northumberland and North Cumbria.
- Naturalised Barnacle Goose - other areas.
- Canadian Light-bellied Brent Goose - Northern Ireland, Wales, western and northern Scotland, Cornwall, Devon and Channel Islands.
- Svalbard Light-bellied Brent Goose - other areas.

(*Note that the separate populations overlap to some extent, and some birds are thus likely to be mis-assigned using these areas. This is particularly so in the case of Greylag Goose and future surveys are planned to help rectify this issue*).

Data from counts at all sites are used, irrespective of whether they are considered complete or not. Numbers presented in this report are not rounded. National and site totals calculated as the sum of counts from several sectors or sites may imply a false sense of accuracy if different methods for recording numbers have been used, *e.g.* 1,000 birds estimated on one sector and a count of seven individuals on another is presented as 1,007. It is safe to assume that any large count includes a proportion of estimated birds. Reproducing the submitted counts in this way is, however, deemed the most appropriate means of presentation and avoids the summation of 'rounding error'.

In the accounts of some scarcer species, including many escaped or introduced species, summed site maxima - calculated by summing the highest count at each site, irrespective of the month in which it occurred - have also been quoted. For some species, particularly more numerous ones, this is likely to result in double counting where birds have moved between sites.

Annual indices

Because the same sites are not necessarily covered by WeBS on every month in every year, relative changes in waterbird numbers cannot be determined simply by comparing the total number of birds counted each year (Tables 3 and 4). This issue is addressed by using indexing techniques that have been developed to track relative changes in numbers from incomplete data.

In summary, for occasions when a particular site has not been visited, an expected count for each species is calculated (imputed) based on the pattern of counts across months, years and other sites. This effectively means that a complete set of counts are available for all years and all months for a sample of sites. Only sites that have a good overall level of coverage are used (at least 50% of possible visits undertaken) and the underlying assumption is that the pattern of change in numbers across these sites (the index) is representative of the pattern of change in numbers at the country level (see *Interpretation of Waterbird Counts* below). Annual index values are expressed relative to the most recent year, which takes an arbitrary value of 100.

The 'Underhill index' was specifically developed for waterbird populations (see Underhill 1989, Prŷs-Jones *et al.* 1994, Underhill & Prŷs-Jones 1994 and Kirby *et al.* 1995 for a full explanation of this indexing process and its application for WeBS data). This report uses Generalized Additive Models (GAMs; Hastie & Tibshirani 1990) to fit both index values and a smoothed trend to the WeBS count data (see Maclean *et al.* 2005 for a full explanation of this process and its application for WeBS data) whilst retaining elements from the Underhill method that allows the assessment of whether or not counts flagged as incomplete should be treated as missing data. The generated smoothed trends

are less influenced by years of abnormally high or low numbers and sampling 'noise' than are the raw index values. This makes them especially useful when assessing changes through time (*e.g.* WeBS Alerts; Maclean *et al.* 2005). Following recent development work undertaken by WeBS, winter indices for waders are for the first time based on data from the months of November to March inclusive while those for other species additionally include September and October. Exceptions are made for the indices for Icelandic Greylag Goose, Pink-footed Goose, Greenland White-fronted Goose and Svalbard Barnacle Goose, for which annual census data are preferentially used to generate indices. Previously, the months used for indexing were assigned in a species-specific manner following established recommendations (Underhill & Prŷs-Jones 1994 and Kirby *et al.* 1995). The new approach, in addition to improving the robustness of the indices to changes in the timing of arrivals and departures with climate change and increasing comparability between species, brings WeBS indexing into line with other WeBS methodologies, specifically reporting of Alerts and computation of five-year mean of peaks.

Not all species are included in the indexing process. Gulls and terns are excluded because counting of these species is optional. Species that occur substantially on habitat not well monitored by WeBS (*e.g.* Moorhen and Snipe) are excluded as are species that occur at sites sporadically and/or in small numbers (*e.g.* Bean Goose and Smew).

The periods of years for which indices are calculated have been revised slightly in the light of recent analyses. Data for wildfowl continue to be presented for the period 1966/67 to the present. Data from 1974/75 onwards have been used for waders as a high proportion of counts before this winter were imputed. For species added later to the scheme, (*i.e.* Great Crested Grebe and Coot in 1982/83, Little Grebe in 1985/86, Cormorant in 1986/87 and gulls, terns, divers, rare grebes and other species from 1993/94), data from the first two years following their inclusion have been omitted from indices, as initial take-up by counters appears not to have been complete, resulting in apparent sharp increases in numbers during this time. For similar reasons the first two years of data have been excluded from Northern Ireland indices.

Index values, where calculated, are graphed within each account. The underlying trend, where calculated, is shown using a broken line. The actual index values used to produce the graphs in this report can be obtained on request from the British Trust for Ornithology (see *Contacts*).

Monthly indices

The abundance of different waterbird species varies during the winter due to a number of factors, most notably the timing of their movements along the flyway, whilst severe weather, particularly on the continent, may also affect numbers in the UK. However, due to differences in site coverage between months, such patterns cannot be reliably detected using count totals. Consequently, an index is calculated for each month to reflect changes in relative abundance during the season.

The imputing process used to derive missing data for generating annual trends also allows monthly indices to be calculated across the same suite of sites. This reveals patterns of seasonality for the species considered. These are presented as graphs in the species accounts, giving the value for the most recent winter and the average value and range over the five preceding winters. Monthly graphs are not presented for the goose species for which annual indices are based on censuses as data for these are available for a limited number of months only.

Broad differences in the monthly values between species reflect their status in the UK. Resident species, or those with large UK breeding populations, *e.g.* some grebes and Mallard, are present in large numbers early in the winter. Declines through the winter result in part from mortality of first year birds, but also birds returning to remote or small breeding sites that are not covered by WeBS. The majority of UK waterbirds either occur solely as winter visitors, or have small breeding populations that are swelled by winter immigrants, with peak abundance generally occurring in mid winter.

The vast majority of the wintering populations of many wader species are found on estuaries, and, since coverage of this habitat is relatively complete and more or less constant throughout winter, meaningful comparisons of total monthly counts can be made for many species.

Site importance

Criteria for assessing the international importance of wetlands have been agreed by the Contracting Parties to the Ramsar Convention on Wetlands of International Importance (Ramsar Convention Bureau 1988). Under criterion 6, a wetland is considered internationally important if it regularly supports 1% of the individuals in a population of one species or subspecies of waterbird, whilst any site regularly supporting 20,000 or more waterbirds qualifies under criterion 5. Similar criteria have been adopted for identification of SPAs under the EC Birds Directive in the UK legislation. A wetland in Britain is considered nationally important if it regularly holds 1% or more of the estimated British numbers of one species or subspecies of waterbird, and in Northern Ireland, important in an all-Ireland context if it holds 1% or more of the all-Ireland estimate. More detailed information about SPAs and Ramsar sites in the UK can be accessed via the JNCC website at http://www.jncc.gov.uk/page-4. There are currently 251 SPAs and 146 Ramsar sites in the UK.

Population estimates are revised once every three years, in keeping with internationally agreed timetables (Rose & Stroud 1994). International estimates used in this report follow recent revisions of international populations (Wetlands International 2006) and of estimates for Great Britain (Kershaw & Cranswick 2003, Rehfisch *et al.* 2003). The relevant 1% thresholds are given in Appendix 1. and are also listed at the start of each individual species account. (It should be noted that the estimates and thresholds for some species or populations which should be the same at an international and national level because all birds are found in Britain, *e.g.* for Pink-footed Goose, differ slightly because of the rounding conventions applied. In most species accounts, these differences have been rationalised and only one or other of the estimates used).

For some species (*e.g.* Lapwing) no national thresholds are available and arbitrary levels have been used to compile the table of sites, the chosen level being given in the subheading of the table. Passage thresholds, applied to counts of some wader species in Great Britain, are also listed.

'National threshold' is used as a generic term to imply the 1% British threshold for sites in Great Britain, and the all-Ireland threshold for sites in Northern Ireland. Similarly, the term 'national importance' implies sites in Great Britain and in Northern Ireland that meet the respective thresholds.

Tables in the Species accounts rank the principal sites for each species according to the mean of annual maxima for the last five years (the five-year peak mean), in line with recommendations of the Ramsar Convention, and identify those meeting national and international qualifying levels (see also *Interpretation of Waterbird Counts*). For each site, the maximum count in each of the five most recent years, the month of occurrence of the peak in the most recent year, and the five-year peak mean are given. Incomplete counts are bracketed.

In accounts for most wildfowl, divers, grebes, Cormorant, herons, gulls, terns and Kingfisher, annual maxima are derived from any month in the appropriate 12-month period (see *The WeBS Year*). Average maxima for sites listed in the wader accounts that are based on a 'winter' year are calculated using data from only the winter period, November to March. Data from other sources, often involving different methods, *e.g.* goose roost censuses, are used where these provide better, *i.e.* larger, counts for individual sites. The source of all counts, if not derived from WeBS Core Counts, is indicated using a superscripted number after the count (a list of sources is given at the beginning of the accounts).

In the first instance, five-year peak means are calculated using only complete counts; incomplete counts are not used if they depress the mean count. Incomplete counts are, however, included in the calculation of the mean if they raise the value of the mean. Where all annual maxima are incomplete, the five-year peak mean is the highest of these individual counts. Averages enclosed by brackets are based solely on incomplete counts.

Sites are selected for presentation using a strict interpretation of the 1% threshold (for convenience, sites in the Channel Islands and Isle of Man are identified using 1% thresholds

for Great Britain and included under the Great Britain section of the tables). For some species with very small national populations, and consequently very low 1% thresholds, an arbitrary, higher level has been chosen for the inclusion of sites. Where no thresholds are given, *e.g.* for introduced species, and where no or very few sites in the UK reach the relevant national qualifying levels, an arbitrary threshold has been chosen to select a list of sites for this report. These adopted thresholds are given in the sub-headings of the table. A blank line has been inserted in the table to separate sites that qualify as nationally important from those with five-year peak mean counts of less than 50 birds.

All sites that held numbers exceeding the relevant national threshold (or adopted qualifying level) in the most recent year, but with five-year peak means below this value, are listed separately. This serves to highlight important sites worthy of continued close attention.

For a number of wader species, where different thresholds exist for passage periods, the peak count during this period and month of occurrence are also listed. This list includes all those sites with counts above the relevant threshold, even if already listed in the main part of the table by virtue of the five-year winter peak mean attaining the national threshold.

Where the importance of a site has changed since the previous *Waterbirds in the UK* (prior to 2004/05 published as *Wildfowl and Wader Counts*) as a result of the data collected since then - *i.e.* it has become nationally or internationally important but was not following the previous year, or it has changed from international to national importance or *vice versa* - this is indicated in the table to the right of the five-year peak mean. Sites with elevated status have a black triangle pointing up (▲) to the right of the average, whilst those with lowered status are indicated using a triangle pointing down (▼). Sites for which the average fell below the threshold for national importance following 2003/04 are listed at the end of the table.

It should be noted that a site may appear to have been flagged erroneously as having elevated status if the most recent count was below the relevant threshold. However, a particularly low count six years previously will have depressed the mean in the previous report. The converse may be true for sites with lowered status and thus, in exceptional circumstances, a site may be listed in the relevant sections of the table as both no longer being of national importance yet also with a peak count in the most recent year exceeding the national threshold.

WeBS Alerts

WeBS Alerts have been developed to provide a standardised method of measuring and reporting on changes in wintering waterbird numbers at different temporal and spatial scales using WeBS data. Generalized Additive Models (GAMs) are used to fit smoothed trends to annual population indices (changes in population size calculated using these smoothed values are less susceptible to the effects of short-term fluctuations in population size or to errors when sampling than are results produced using raw data plots). Alerts are triggered for populations that have undergone major declines, and are intended to help identify where research into causes of decline may be needed and inform conservation management.

Proportional changes in the smoothed index value of a population over short- (5-year), medium- (10-year) and long- (25-year) term time frames are categorised according to their magnitude and direction. Population declines of between 25% and 50% trigger Medium Alerts and declines of greater than 50% trigger High Alerts. Increases of 33% and 100% (values chosen to be those necessary to return a population to its former size following declines of 25% and 50% respectively) are also identified, albeit that these are rarely of conservation concern.

National Alerts are generated for species (or specific populations of a species) using data from across the WeBS site network, for Great Britain and the constituent countries of the UK (Maclean *et al.* 2005). These Alerts provide some context for understanding finer scale changes in numbers. Alerts are calculated only for native species for which WeBS annual indices are calculated. Alerts are not available for some species over long time periods because there were only relatively recently included in WeBS Core Counts. Full results from the latest Alerts report are available for download from http://www.bto.org/webs/alerts/alerts/index.htm.

Principal sites

In addition to the assessment of sites against 1% thresholds in Species accounts, sites are identified for their importance in terms of overall waterbird numbers in the section *Principal Sites*. The peak count at each site is calculated by summing the individual species maxima during the season, irrespective of the month in which they occurred, or whether counts were complete or not. Data from all sources used for site assessment within the species accounts are used here, including wader numbers during passage periods. Non-native introduced or escaped species (*i.e.* those not in BOURC category A) are not included in these totals.

Counts made using methodologies that employ different site definitions to those used by WeBS (*e.g.* seaducks on the Moray Firth) are not incorporated into the calculations. Such sites are, however, listed at the end of the table.

INTERPRETATION OF WATERBIRD COUNTS

Caution is always necessary in the interpretation and application of waterbird counts given the limitations of these data. This is especially true of the summary form, which, by necessity, is used in this report. A primary aim here remains the rapid feedback of key results to the many participants in the WeBS scheme. More detailed information on how to make use of the data for research or site assessment purposes can be obtained from the British Trust for Ornithology (see *Contacts*).

Whilst the manner of presentation is consistent within this report, information collated by WeBS and other surveys can be held or used in a variety of ways. Data may also be summarised and analysed differently depending on the requirements of the user. Consequently, calculations used to interpret data and their presentation may vary between this and other publications, and indeed between organisations or individual users. The terminology used by different organisations may not always highlight these differences. This particularly applies to summary data. Such variations do not detract from the value of each different method, but offer greater choice to users according to the different questions being addressed. This should always be borne in mind when using data presented here.

For ease of reference, the caveats provided below are broadly categorised according to the presentation of results for each of the key objectives of WeBS. Several points, however, are general in nature and apply to a broad range of uses of the data.

National totals

The majority of count data are collected between September and March, when most species of waterbird are present in the UK in highest numbers. Data are collected during other months and have been presented where relevant. Caution is urged, however, regarding their interpretation both due to the relative sparsity of counts from this period and the different count effort for different sites. Data are presented for the months July to June inclusive (see *The WeBS Year*), matching the period for which data are provided *en masse* by counters.

A number of systematic biases of WeBS or other count methodology must be borne in mind when considering the data. Coverage of estuarine habitats and large, standing waters by WeBS is good or excellent. Consequently, counted totals of those species which occur wholly or primarily on these habitats during winter will approach a census. Those species dispersed widely over rivers, non-estuarine coast or small inland waters are, however, likely to be considerably under-represented, as will secretive or cryptic species, such as snipes, or those which occur on non-wetlands, *e.g.* grassland plovers. Species which occur in large numbers during passage are also likely to be under-represented, not only because of poorer coverage at this time, but due to the high turnover of birds in a short period. Furthermore, since counts of gulls and terns are optional, national totals are likely to be considerable underestimates of the number using the WeBS network of sites. Only for a handful of species, primarily geese, can count totals be considered as a census.

One instance of possible over-estimation may occur if using summed site maxima as a guide to the total number of scarcer species. For species with mobile flocks in an area well covered by WeBS, *e.g.* Snow Goose in southeast England, it is likely that a degree of double counting will occur, particularly if birds move between sites at different times of the year.

The publication of records of vagrants in this report does not imply acceptance by the British Birds Rarities Committee (*e.g.* Fraser *et al.* 2007).

Annual indices

For most species, the long-term trends in index values can be used to assess changes in overall wintering numbers with confidence. However, the comments above concerning the differential coverage of different habitats remain important. For some species, a substantial proportion of wintering birds occur away from those sites monitored by the WeBS Core Count scheme or use these sites at certain times of day that make them unlike to be encountered by WeBS counters. Consequently, this incomplete coverage needs to be borne in mind when interpreting the indices for some species. The proportion of some of these species being monitored by the WeBS Core Count scheme can be quantified and biases understood by comparison to other surveys. For example, from the Non-estuarine Coastal Waterbird Survey (NEWS) it is known that WeBS Core Counts monitor between one quarter and one half of wintering Ringed Plover, Purple Sandpiper, Sanderling and Turnstone and that the indices and trends reported will be biased towards changes occurring on estuaries. Similarly, trends reported for seaduck and grassland plovers will be biased towards changes occurring within estuaries although in these species the proportion of overall numbers monitored by WeBS Core counts is less well understood. In the case of winter swans, although the sites on which they occur are generally well monitored by WeBS Core counts they are mainly used as roost sites by the birds and therefore changes in the birds' daily routine with weather or local feeding opportunities may have considerable influence on whether they are present during the WeBS count and thus affect the reported indices and trends.

Indices and trends for Pink-footed Goose, Greenland White-fronted Goose, Icelandic Greylag Goose and Svalbard Barnacle Goose can be considered to be especially representative of national patterns. The numbers of these species are not well monitored by monthly WeBS Core Counts but rather are preferentially monitored by the annual coordinated censuses that cover the majority of British wintering birds. Indices for strictly or principally estuarine species (*e.g.* Wigeon and Knot) can also be considered especially representative as over 90% of British estuaries, including all major sites, are counted each month between September and March. Similarly, species that occur principally on larger inland waterbodies (*e.g.* Pochard and Goldeneye) are well monitored by WeBS Core Counts although the proportion of the numbers not being monitored is largely unquantified. For these species the indices and trends reported can be considered representative of the national pattern. For more widespread species (*e.g.* Mallard, Tufted Duck and Curlew) a large proportion of birds occur at small inland sites and habitats not well monitored by WeBS Core Counts. The selection of such sites follows no formal sampling pattern and therefore it is unclear as to whether these wetlands are a representative sample of the country as a whole.

Because short-term fluctuations provide a less rigorous indication of population changes, care should be taken in their interpretation. The underlying trend, denoted by the smoothed line in the annual index graphs, will give a better overall impression of trends for species with marked inter-annual variation, although it should be noted that unusually high or low index values in the most recent year will have a disproportionate effect on the trend at that point.

Caution should be used in interpreting figures for species that only occur in small numbers. Thus, numbers tend to fluctuate more widely for many species in Northern Ireland, largely as a result of the smaller numbers of birds involved but also, being at the western most limit of their range, due to variable use being made of Ireland by wintering waterbirds.

It should be borne in mind that the imputed values, used in place of missing and incomplete counts, are calculated anew each year, as in the completeness calculation for 'complex sites' which may cause the same count to change from complete to incomplete or *vice versa* with the addition of a new year's data. Because the index formula uses data from all years, each new year's counts will slightly alter the site, month and year factors. In turn, the assessment of missing counts may differ slightly and, as a result, the index values produced each year are likely to differ from those published in the previous *Waterbirds in the UK* (prior to 2004/05 published as *Wildfowl and Wader Counts*). Additionally, data submitted too late for inclusion are subsequently added to the dataset. The indices published here represent an improvement on previous figures as the additional year's data allow calculation of the site, month and year factors with greater confidence.

Monthly indices

As for annual indices, the reduced numbers of both sites and birds in Northern Ireland result in a greater degree of fluctuation in numbers used in the analyses of data from the province.

Site definition

To compare count data from year to year requires that the individual sites - in terms of the area surveyed - remain the same. The boundary of many wetlands are readily defined by the extent of habitat (*e.g.* for reservoirs and gravel pits), but are less obvious for other sites (*e.g.* some large estuaries) and here count boundaries have often been defined over time by a number of factors to a greater or lesser degree, including the distribution of birds at the time of the count, known movements of birds from roost to feeding areas, the extent of habitat, and even ease of access.

Sites are defined for a variety of purposes, and the precise boundary of sites describing ostensibly the same wetland may differ accordingly. For example, the boundaries used to define a large lake may differ for its definition as a wetland (based on habitat), as a waterbird count area (some birds may use adjacent non-wetland habitat), and as a statutorily designated site for nature conservation (which may be constrained by the need to follow boundaries easily demarcated in planning and legal terms). It should be recognised that the boundary of a site for counting may even differ between different waterbird surveys, particularly where different methodologies are employed, *e.g.* the 'Forth Estuary' comprises one large site for WeBS Core Counts, a slightly different area for Low Tide Counts, and two roost sites for Pink-footed Geese.

Data from different waterbird surveys have been used for assessment of site importance in this report if collected for ostensibly the same site, and are unlikely to cause significant discrepancies in the vast majority of cases (though see *Site importance*).

Particular caution is urged, however, in noting that, owing to possible boundary differences, totals given for WeBS or other sites in this report are not necessarily the same as totals for designated statutory sites (ASSIs/SSSIs, SPAs or Ramsar Sites) having the same or similar names.

It should also be borne in mind that whilst discrete wetlands may represent obvious sites for waterbirds, there is no strict definition of a site as an ecological unit for birds. Thus, some wetlands may provide all needs - feeding, loafing and roosting areas - for some species, but a 'site' for other species may comprise a variety of disparate areas, not all of which are counted for WeBS. Similarly, for some habitats, particularly linear areas such as rivers and rocky coasts, and marine areas, the definition of a site as used by waterbirds is not readily discerned without extensive survey or research that is usually beyond the scope of WeBS or other similar surveys. The definitions of such sites may thus evolve, and therefore change between *Waterbirds in the UK* (prior to 2004/05 published as *Wildfowl and Wader Counts*). Further, the number of birds recorded by WeBS at particular sites should not be taken to indicate the total number of birds in that local area.

In some cases, for example where feeding geese are recorded by daytime WeBS Core Counts over large sites, and again at discrete roosts within or adjacent to that same site, data are presented for both sites in the table of key sites given the very different nature or extent of the sites and often number of birds, even though the same birds will be counted at both.

Site importance

Sites are selected for presentation in this report using a strict interpretation of the 1% threshold. It should be noted, however that where 1% of the national population is less than 50 birds, 50 is normally used as a minimum qualifying threshold for the designation of sites of national importance. It should also be noted that the 'qualifying levels' used for introduced species are used purely as a guide for presentation of sites in this report and do not infer any conservation importance for the species or the sites concerned since protected sites would not be identified for these non-native birds.

It is necessary to bear in mind the distinction between sites that regularly hold wintering populations of national or international importance and those which may happen to exceed the appropriate qualifying levels only in occasional winters. This follows the Ramsar Convention, which states that key sites must be identified on the basis of demonstrated regular use (calculated as the mean winter maxima from the last five seasons for most species in this report), otherwise a large number of sites might qualify as a consequence of irregular visitation by one-off large numbers of waterbirds. However, the Convention also indicates that provisional assessments may be made on the basis of a minimum of three years' data. These rules of thumb are applied to SPAs and national assessments also. Sites with just one or two years' data are also included in the tables if the mean exceeds the relevant threshold for completeness but this does not, as such, imply qualification. This caveat applies also to sites that are counted in more than two years but, because one or more of the peak counts are incomplete, whose means surpass the 1% threshold based on counts from only one or two years.

Nevertheless, sites which irregularly support nationally or internationally important numbers may be extremely important at certain times, *e.g.* when the UK population is high, during the main migratory periods, or during cold weather, when they may act as refuges for birds away from traditionally used sites. For this reason also, the ranking of sites according to the total numbers of birds they support (particularly in *Principal Sites*) should not be taken as a rank order of the conservation importance of these sites, since certain sites, perhaps low down in terms of their total 'average' numbers, may nevertheless be of critical importance to certain species or populations at particular times.

Peak counts derived from a number of visits to a particular site in a given season will reflect more accurately the relative importance of the site for the species than do single visits. It is important to bear this in mind since, despite considerable improvements in coverage, data for a few sites presented in this report derive from single counts in some years. Similarly, in assessing the importance of a site, peak counts from several winters should ideally be used, as the peak count made in any one year may be unreliable due to gaps in coverage and disturbance- or weather-induced effects. The short-term movement of birds between closely adjacent sites may lead to altered assessments of a site's apparent importance for a particular species. More frequent counts than the once-monthly WeBS visits are necessary to assess more accurately the rapid turnover of waterbird populations that occurs during migration or cold weather movements.

It should also be borne in mind that because a count is considered complete for WeBS, it does not imply that it fully represents the importance of the site. A site of importance for a wintering species may have been counted only in autumn or spring, and thus while a valid complete count is available for that year, it under-represents the importance of the site for that species. This problem is overcome to some extent by the selection of counts from a limited winter window for wader species, although this will also tend to underestimation of the mean if it excludes large counts at other times of year. A similar issue arises for counts derived from different survey methods. For example, many sites important as gull roosts are identified on the basis of evening roost counts. Valid and complete counts may have been made by WeBS Core Counts during daytime over the course of a particular winter but, if no roost counts were made, the mean will be depressed by the much lower Core Count in that year. Thus, when counts appear to fluctuate greatly between years at individual sites on the basis of data from different sources - particularly for geese and gulls in the absence of roost counts, and for seaducks in the absence of dedicated survey - the five-year means and apparent trends over time should be viewed with caution.

Caution is also urged regarding the use of Low Tide Count data in site assessment. Whilst this survey serves to highlight the importance of some estuaries for feeding birds that, because they roost on other sites, are missed by Core Counts, the objectives of Low Tide Counts do not require strict synchronisation across the site and this may result in double counting of birds on some occasions. It should also be noted that count completeness assessments are not made for Low Tide Count totals at complex sites, and any undercounts from this scheme are not flagged in the tables, leading to under-estimation of the site's importance.

This list of potential sources of error in counting wetland birds, though not exhaustive, suggests that the net effect tends towards under- rather than over-estimation of numbers and provides justification for the use of maximum counts for the assessment of site importance or the size of a population. Factors causing under-estimation are normally constant at a given site in a given month, so that while under-estimates may occur, comparisons between sites and years remain valid.

It should be recognised that, in presenting sites supporting nationally or internationally important numbers of birds, this report provides just one means of identifying important sites and does not provide a definitive statement on the conservation value of individual sites for waterbirds, let alone other conservation interests. The national thresholds have been chosen to provide a reasonable amount of information in the context of this report only. Thus, for example, many sites of regional importance or those of importance because of the assemblage of species present are not included here. European Directives and conservation Conventions stress the need for a holistic approach to effect successful conservation, and lay great importance on maintaining the distribution and range of species, in addition to the conservation of networks of individual key sites.

For the above reasons of poor coverage, geographically or temporally, outlined above, it should be recognised that lists of sites supporting internationally and nationally important numbers of birds are limited by the availability of WeBS and other survey data. Whilst the counter network is likely to cover the vast majority of important sites, others may be missed and therefore will not be listed in the tables due to lack of appropriate data.

Some counts in this report differ from those presented previously; this results from the submission of late data and corrections, and in some cases, the use of different count seasons or changes to site structures. Additionally, some sites may have been omitted from tables previously due to oversight. It is likely that small changes will continue as definitions of sites are revised, in the light of new information from counters. Most changes are minor, but comment is made in the text where they are significant.

Note that sites listed under 'Sites no longer meeting table qualifying levels' represent those that would have been noted of national importance based on the preceding five years (*i.e.* 2000/01 to 2004/05) but which, following the 2005/06 counts, no longer met the relevant threshold. It is not an exhaustive list of sites, which at any time in the past have been of national or all-Ireland importance.

COVERAGE

WeBS Core Counts

Coordinated, synchronous counts are advocated to prevent double counting or birds being missed. Consequently, priority dates are recommended nationally. Due to differences in tidal regimes around the country, counts at a few estuaries were made on other dates to match the most suitable conditions. Weather and counter availability also result in some counts being made on alternative dates.

Table 2. WeBS Core Count priority dates in 2005/06

24 July	15 January
21 August	12 February
18 September	12 March
16 October	09 April
06 November	14 May
04 December	11 June

Standard Core Counts were received from 2,109 sites of all habitats for the period July 2005 to June 2006, comprising 3,801 count sectors (the sub-divisions of large sites for which separate counts are provided).

WeBS and I-WeBS coverage in 2005/06 is shown in Figure 1. The location of each count sector is shown using only its central grid reference. The grid reference of principal WeBS count sites mentioned in the Principal Sites table (Table 6.) are given in Table A2. in Appendix 2. and are shown in Figure A1. in Appendix 2.

As ever, areas with few wetlands (*e.g.* inland Essex/Suffolk) or small human populations (*e.g.* much of Scotland) are apparent on the map as areas with little coverage. Northwest Scotland is usually poorly covered, although in 2005/06 this area was covered by surveys by the RAF Ornithological Society, which are reported upon in this report. Northern Ireland remains relatively uncovered aware from the major sites and further volunteers from here, or indeed anywhere in the UK, are always welcome.

Goose censuses

In 2005/06, supplementary counts of Bean Geese were submitted by the Bean Goose Action Group (Slamannan Plateau) (Maciver 2006) and the RSPB (Middle Yare Marshes). National surveys of Pink-footed and Icelandic Greylag Geese (the Icelandic-breeding Goose Census) were undertaken at roosts in October, November and December 2005 (Rowell 2006). A census of the Northwest Scotland Greylag Goose population on the Uists was made in August 2005 and February 2006 (Uist Greylag Goose Management Committee), and counts of this population at other key sites (*e.g.* Tiree) were timed to coincide with these counts. Censuses of Greenland White-fronted Geese were carried out in autumn 2005 and spring 2006 by the Greenland White-fronted Goose Study (Fox & Francis 2006). Greenland Barnacle Geese were counted regularly by SNH and others on Islay and other key locations whilst the Svalbard Barnacle Geese on the Solway were counted regularly by WWT staff and volunteers. Data were also provided by the International Light-bellied Brent Goose census.

Seaduck surveys

Coastal counts of seaduck, divers and grebes were received from several sites. Aerial and/or shore-based counts from Aberdeen Bay, Outer Hebrides, Coll/Tiree/Mull, Moray/Dornoch Firths, Scapa Flow/Tankerness, Sound of Gigha & outer West Loch Tarbert, and the Firth of Clyde were provided by JNCC (Söhle *et al.* 2006). Continuing surveys of the Moray Firth were carried out between November 2005 and January 2006 (RSPB Scotland/Talisman Energy (UK) Ltd). Monthly aerial and/or land-based counts of Common Scoter in Carmarthen Bay were carried out between September 2005 and February 2006 (Banks *et al.* 2007). Results from aerial monitoring of Common Scoter and Red-throated Diver were provided by WWT. Continuing counts of key sites around Shetland were provided by SOTEAG (Heubeck and Mellor 2007). Continuing survey of the Eiders of the wider Firth of Clyde area were carried out in September 2005 (Waltho *pers. comm.*).

Figure 1. Position of all locations counted for standard WeBS and I-WeBS counts between July 2005 and June 2006.

TOTAL NUMBERS

The total numbers of waterbirds recorded by WeBS in 2005/06 are given in Tables 3 and 4 for Great Britain (including the Isle of Man, but excluding the Channel Islands) and Northern Ireland, respectively. Counts of waterbirds in the Republic of Ireland by I-WeBS are provided in Table 5.

Site coverage for gulls and terns is given separately since recording of these species was optional.

Introduced and escaped waterbirds

Many species of waterbird occur in the UK as a result of introductions, particularly through escapes from collections. Several have become established, such as Canada Goose and Ruddy Duck. The British Ornithologists' Union Records Committee categorises each species occurring in Britain according to its likely origin. The categories are explained more fully at www.bou.org.uk/reccats.html. Species that have been recorded as 'introductions, human-assisted transportees or escapes from captivity, and whose breeding populations (if any) are not thought to be self-sustaining' are included in the BOURC's category E. WeBS records of these species are included in this report both for the sake of completeness and in order to assess their status and monitor any changes in numbers, a key requirement given the need, under the African-Eurasian Waterbird Agreement of the Bonn convention '. . . to prevent the unintentional release of such species . . .' and once introduced, the need '. . . to prevent these species from becoming a threat to indigenous species' (Holmes *et al.* 1998).

Numbers of established populations (*e.g.* Canada Goose and Ruddy Duck, which are placed in category C) are excluded from Figure 2 below since the large numbers involved would swamp numbers of other species. Additionally, species that occur in both categories A and E (*e.g.* Pink-footed Goose) are also excluded since separation of escaped from wild birds is not readily possible using WeBS methods. However, Ruddy Shelduck (categories B/E) is included; the BOURC does not consider any recent records to have been of wild origin. Additionally, a small number of species not yet assigned to category by the BOURC (e.g. Coscoroba Swan) are also included.

A total of 21 category E species were recorded in 2005/06 at 179 sites, similar figures to those seen in 2004/05. The summed site maxima of 428 birds was the lowest for the last five years. As in previous years, over half of this total was attributable to Black Swan and Muscovy Duck, followed in abundance by Bar-headed Goose, Ruddy Shelduck, Emperor Goose, Chinese Goose and Wood Duck.

Although this figure will undoubtedly include some duplication of individual birds recorded at more than one site, plus occasional records of pinioned birds, the summed site maxima statistic probably provides a truer reflection of the numbers of introduced or escaped waterbirds frequenting WeBS sites than the peak monthly total of 158 birds in October 2005.

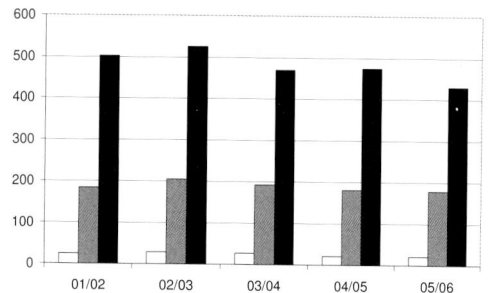

Figure 2. Number of species (white bars), number of sites at which birds were recorded (grey bars) and summed site maxima (black bars) for waterbirds in the BOURC's category E.

Table 3. Total numbers of waterbirds recorded by WeBS Core Counts in Great Britain in 2005/06. Census totals are indicated by '*'.

	Species	Jul	Aug	Sep	Oct	Nov
	Number of sites visited	*741*	*823*	*1442*	*1548*	*1614*
MS	Mute Swan	10364	11966	19182	21307	20611
AS	Black Swan	37	34	43	42	41
BS	Bewick's Swan	0	0	0	30	*1953
WS	Whooper Swan	16	20	67	*3095	*7256
OW	Coscoroba Swan	0	0	0	0	1
HN	Chinese Goose	7	5	5	8	3
BE	Bean Goose	1	3	*12	*287	*309
XF	Taiga Bean Goose	0	0	0	0	27
XR	Tundra Bean Goose	0	0	0	0	0
PG	Pink-footed Goose	9	29	40070	*234120	*258247
WG	White-fronted Goose	0	0	0	0	0
EW	European White-fronted Goose	0	1	1	53	153
NW	Greenland White-fronted Goose	0	0	0	1478	920
LC	Lesser White-fronted Goose	0	1	1	0	0
JI	Icelandic Greylag Goose	2077	2958	4776	*24390	*76490
JH	NW Scotland Greylag Goose	1543	*9530	2077	2180	1104
JE	Re-established Greylag Goose	11332	16436	26692	25179	22269
ZL	Greylag Goose (domestic)	156	208	364	303	409
HD	Bar-headed Goose	10	10	18	10	7
SJ	Snow Goose	3	41	58	48	52
RJ	Ross's Goose	1	2	1	0	0
EM	Emperor Goose	11	0	15	19	20
CG	Greater Canada Goose	24859	33862	53567	49383	49159
ZI	Canada Goose x domesticated Greylag	4	7	8	44	49
LQ	Lesser Canada Goose	0	0	0	1	0
NE	Hawaiian Goose	0	0	0	0	0
YN	Nearctic Barnacle Goose	6	4	29	2517	1459
YS	Svalbard Barnacle Goose	5	2	333	*16235	*28920
YE	Naturalised Barnacle Goose	242	377	958	415	475
ZH	Barnacle Goose x Canada Goose	0	0	1	2	0
BG	Brent Goose	0	0	0	2	40
DB	Dark-bellied Brent Goose	46	36	249	27688	56858
QN	Nearctic Light-bellied Brent Goose	0	0	102	50	63
QS	Svalbard Light-bellied Brent Goose	0	1	2798	1775	3756
BB	Black Brant	0	0	0	0	1
EB	Red-breasted Goose	2	0	2	0	1
UO	Unidentified goose	0	0	5	0	0
ZM	Hybrid goose	8	8	10	7	13
EG	Egyptian Goose	232	300	384	363	168
OJ	Orinoco Goose	0	0	0	1	0
UE	Cape Shelduck	0	0	0	3	0
UD	Ruddy Shelduck	5	1	5	22	4
SU	Shelduck	30388	23937	34948	38059	44233
MY	Muscovy Duck	15	21	42	44	65
DC	Wood Duck	0	1	1	6	4
MN	Mandarin	102	134	288	376	329
WN	Wigeon	459	1630	57317	240839	312082
AW	American Wigeon	0	0	0	0	0
HL	Chiloe Wigeon	2	0	2	0	0
GA	Gadwall	2137	4139	9583	13309	14298
T.	Teal	1392	14698	74488	114118	144902
TA	Green-winged Teal	0	0	0	1	2
KQ	Speckled Teal	0	0	0	0	2
MA	Mallard	45874	65596	110035	123238	123617
ZF	Feral/hybrid mallard type	399	361	737	445	568
QB	Chestnut Teal	0	2	0	0	0
PT	Pintail	24	106	7099	15674	22421
PN	Bahama Pintail	0	2	0	0	0
GY	Garganey	12	34	51	10	1
TB	Blue-winged Teal	0	0	0	0	0

Table 3. continued

	Dec	Jan	Feb	Mar	Apr	May	Jun
sites	1635	1683	1757	1561	879	760	742
MS	20131	19161	17412	15053	8213	8170	9274
AS	37	41	36	37	28	32	32
BS	*5102	*7663	*5627	*245	2	0	0
WS	*8051	*7977	*7397	*3897	215	12	16
OW	0	0	0	0	0	0	0
HN	15	7	5	5	4	3	3
BE	*335	*484	*361	*15	1	0	0
XF	79	98	40	0	0	0	0
XR	0	6	0	0	0	0	0
PG	*244064	124529	70934	47516	11179	198	23
WG	0	0	78	0	0	0	0
EW	588	1952	2350	1438	0	0	1
NW	*13609	428	139	*14287	14	0	0
LC	2	1	0	0	1	0	0
JI	*76945	19655	21914	20024	3075	452	1887
JH	1578	226	*5052	244	419	221	651
JE	27971	24836	17931	12126	6244	6189	10421
ZL	378	550	294	241	89	84	85
HD	11	15	5	11	12	10	15
SJ	47	25	9	17	6	2	14
RJ	0	0	0	0	0	0	0
EM	15	15	23	22	21	9	15
CG	54210	48911	37005	26286	13017	10317	20781
ZI	45	31	27	29	30	5	7
LQ	0	0	0	0	0	0	0
NE	0	1	0	0	0	0	0
YN	*50117	292	1755	*56383	409	3	2
YS	*25855	*23723	*21317	*20035	*23173	*5283	8
YE	457	791	914	416	214	141	270
ZH	1	0	0	3	2	1	2
BG	43	67	48	5	50	2	1
DB	74071	77847	84681	76462	20037	12495	50
QN	120	48	32	73	151	0	0
QS	2413	2082	1297	415	25	1	2
BB	0	4	4	2	0	0	0
EB	1	1	2	1	0	0	1
UO	0	0	0	0	2	0	22
ZM	11	2	2	0	0	0	0
EG	134	188	139	152	165	188	227
OJ	1	0	0	0	0	0	0
UE	0	0	0	1	0	0	0
UD	5	7	0	1	0	2	4
SU	46997	46767	46782	35647	19237	14672	16428
MY	44	49	50	18	24	20	27
DC	3	4	4	2	4	6	7
MN	396	448	402	239	83	83	92
WN	441754	422045	357073	227963	6692	310	161
AW	1	1	3	1	0	0	0
HL	0	1	0	0	0	0	1
GA	15606	15990	14507	9018	3061	1725	2651
T.	167874	163820	116347	67068	11410	541	530
TA	3	3	3	3	2	0	1
KQ	2	0	2	1	0	0	0
MA	131200	121633	93435	62635	25901	22484	31483
ZF	583	348	384	391	202	149	169
QB	0	0	0	0	0	0	0
PT	27935	20158	24233	13814	388	19	13
PN	0	0	1	1	1	3	2
GY	0	2	1	16	12	52	27
TB	0	0	0	0	0	0	1

Table 3. continued

	Species	Jul	Aug	Sep	Oct	Nov
	Number of sites visited	*741*	*823*	*1442*	*1548*	*1614*
SV	Shoveler	469	3304	10287	12396	12308
ZR	Hybrid Anas	0	0	0	0	0
IE	Ringed Teal	1	1	0	1	0
RQ	Red-crested Pochard	22	20	156	170	69
PO	Pochard	2692	5505	8652	14316	17925
NG	Ring-necked Duck	0	1	1	0	3
FD	Ferruginous Duck	1	0	0	0	0
TU	Tufted Duck	18494	32215	43793	49699	55067
SP	Scaup	9	24	61	689	554
AY	Lesser Scaup	0	0	0	0	1
ZD	Aythya hybrid	0	6	5	4	6
E.	Eider	12596	13143	13236	14857	15940
KE	King Eider	0	0	0	0	0
LN	Long-tailed Duck	0	1	2	1063	451
CX	Common Scoter	1998	1343	3479	8117	2587
DX	Black Scoter	0	0	0	0	0
FS	Surf Scoter	0	0	0	3	2
VS	Velvet Scoter	62	1	390	977	630
UX	Unidentified scoter	0	0	0	0	0
GN	Goldeneye	49	124	455	1886	4555
HO	Hooded Merganser	0	0	0	0	0
SY	Smew	0	0	0	5	20
RM	Red-breasted Merganser	1114	1131	1151	2235	1713
GD	Goosander	650	830	747	917	1058
RY	Ruddy Duck	679	1293	2235	2675	2646
OI	Argentine Bluebill	1	0	0	1	0
WQ	White-headed Duck	0	1	1	1	1
UM	Unidentified duck	0	21	0	0	0
RH	Red-throated Diver	35	42	253	297	265
BV	Black-throated Diver	5	0	0	20	5
ND	Great Northern Diver	1	1	3	7	70
UL	Unidentified diver	0	0	0	0	0
LG	Little Grebe	1386	2831	5199	5237	4569
GG	Great Crested Grebe	3471	5239	8459	8441	7538
RX	Red-necked Grebe	1	34	11	31	20
SZ	Slavonian Grebe	0	0	15	137	98
BN	Black-necked Grebe	32	40	55	65	39
CA	Cormorant	5982	10199	15890	16124	15462
SA	Shag	211	513	858	1422	1408
XU	Unidentified Cormorant/Shag	2	0	2	4	6
BI	Bittern	2	4	3	7	11
NT	Night Heron	0	0	0	0	0
EC	Cattle Egret	0	0	2	1	0
ET	Little Egret	1473	2391	3105	2985	2468
HW	Great White Egret	1	0	1	4	1
H.	Grey Heron	1878	2497	4002	3969	3214
UR	Purple Heron	0	1	0	0	0
OR	White Stork	1	2	3	2	1
IS	Sacred Ibis	0	1	1	0	1
NB	Spoonbill	6	4	1	18	9
WA	Water Rail	57	89	163	365	405
AK	Spotted Crake	1	2	3	1	0
JR	Sora	0	0	0	1	0
JC	Little Crake	0	0	1	0	0
CE	Corncrake	0	0	0	0	0
MH	Moorhen	4351	6880	12522	13648	12721
CO	Coot	34103	49380	91791	101931	98642
AO	American Coot	0	0	1	0	0
AN	Crane	0	0	2	0	0
KF	Kingfisher	135	239	484	485	419
	TOTAL WILDFOWL	**223751**	**318185**	**673894**	**1075873**	**1196670**

Table 3. continued

sites	Dec	Jan	Feb	Mar	Apr	May	Jun
	1635	*1683*	*1757*	*1561*	*879*	*760*	*742*
SV	11306	11754	12347	11478	3327	583	685
ZR	2	12	11	6	0	0	0
IE	1	0	1	0	0	1	0
RQ	164	270	193	128	37	19	19
PO	23857	23609	22707	15955	1110	649	692
NG	3	5	5	6	7	0	0
FD	1	1	2	0	0	0	2
TU	55148	54169	47103	40074	16397	7458	6759
SP	1829	1921	2392	671	178	64	2
AY	2	1	3	1	2	0	0
ZD	5	4	5	4	4	2	0
E.	16099	14436	30984	9221	12299	9222	10683
KE	0	0	1	0	0	0	0
LN	2371	11889	12094	401	142	34	1
CX	11615	12827	13055	8641	8420	3371	2978
DX	0	1	0	0	0	0	0
FS	2	0	1	0	0	1	0
VS	1204	1705	1869	387	315	246	37
UX	0	20	0	0	0	0	0
GN	9454	11503	12763	10696	3260	115	28
HO	1	0	0	0	0	0	0
SY	82	214	155	90	5	0	0
RM	3262	2784	3277	2615	1657	637	333
GD	3087	2678	2963	2124	487	211	289
RY	2991	2780	2166	1926	776	655	410
OI	0	1	0	0	0	0	1
WQ	0	0	0	0	0	0	0
UM	42	1	701	1	1	1	1
RH	372	264	472	178	389	182	44
BV	42	28	164	85	35	17	0
ND	71	51	334	68	41	17	0
UL	1	0	0	0	2	0	0
LG	4533	3800	3746	3185	1347	851	795
GG	8377	7212	6566	5756	2801	2595	2256
RX	62	17	26	12	5	3	1
SZ	152	166	224	84	57	0	0
BN	47	42	98	81	44	32	25
CA	14379	11374	11819	9890	4984	4198	4190
SA	1267	1232	1755	609	315	334	227
XU	4	2	0	0	0	0	0
BI	25	22	20	13	3	3	1
NT	0	0	0	1	0	0	0
EC	0	3	4	6	3	1	0
ET	1531	1148	1168	1008	866	547	703
HW	0	0	0	0	0	0	0
H.	3454	3382	3333	2607	1580	1541	1508
UR	0	0	0	0	0	2	0
OR	0	1	1	2	1	2	2
IS	0	0	1	0	0	0	0
NB	12	7	7	2	2	6	3
WA	540	483	516	360	78	33	40
AK	0	0	0	0	0	0	0
JR	0	0	0	0	0	0	0
JC	0	0	0	0	0	0	0
CE	0	0	0	0	1	1	3
MH	14530	12247	12636	11419	5306	3596	3383
CO	106790	96341	73971	54001	17171	11813	14378
AO	0	0	0	0	0	0	0
AN	0	0	0	0	0	0	0
KF	424	298	256	207	89	55	78
	1442959	**1412170**	**1206188**	**825175**	**220951**	**129502**	**145964**

Table 3. continued

Species		Jul	Aug	Sep	Oct	Nov
Number of sites visited		*741*	*823*	*1442*	*1548*	*1614*
OC	Oystercatcher	71842	175625	174560	197726	206275
IT	Black-winged Stilt	0	0	0	0	0
AV	Avocet	2591	2373	2828	5357	5450
TN	Stone-curlew	0	0	0	0	0
LP	Little Ringed Plover	189	51	27	0	0
RP	Ringed Plover	2020	15485	10303	9689	7478
KP	Kentish Plover	0	0	0	0	0
GP	Golden Plover	12990	40986	60155	139279	142793
GV	Grey Plover	3924	23837	28295	26477	40306
L.	Lapwing	26373	67967	103268	144930	195683
KN	Knot	58820	169530	186315	167078	269154
SS	Sanderling	8403	10626	7083	8780	8768
LX	Little Stint	12	14	56	22	13
TK	Temminck's Stint	0	0	1	0	0
WU	White-rumped Sandpiper	0	0	2	3	0
BP	Baird's Sandpiper	0	0	1	1	0
PP	Pectoral Sandpiper	0	0	3	0	0
CV	Curlew Sandpiper	127	168	164	17	4
PS	Purple Sandpiper	10	136	90	186	360
DN	Dunlin	86039	67651	73848	148721	188080
RU	Ruff	245	558	684	475	314
JS	Jack Snipe	0	0	5	80	135
SN	Snipe	260	1057	2986	5427	6429
LD	Long-billed Dowitcher	0	0	0	3	2
WK	Woodcock	0	0	0	13	18
BW	Black-tailed Godwit	13579	22278	35130	34468	27531
BA	Bar-tailed Godwit	5650	18589	24141	16316	18317
WM	Whimbrel	1205	638	168	32	34
CU	Curlew	47729	66655	74713	73996	60210
DR	Spotted Redshank	74	118	164	118	118
RK	Redshank	29209	53073	76738	85133	80345
GK	Greenshank	834	1360	1345	788	387
GE	Green Sandpiper	239	348	287	213	142
OD	Wood Sandpiper	2	18	1	1	1
CS	Common Sandpiper	643	462	496	91	72
TT	Turnstone	3580	8056	10150	14389	12277
NK	Red-necked Phalarope	0	0	3	0	0
PL	Grey Phalarope	0	0	0	0	13
U.	Unidentified wader	0	0	1	0	0
JW	Unidentified small wader	0	0	0	0	0
	TOTAL WADER	**376589**	**747659**	**874011**	**1079809**	**1270709**

Table 3. continued

sites	Dec	Jan	Feb	Mar	Apr	May	Jun
	1635	*1683*	*1757*	*1561*	*879*	*760*	*742*
OC	188091	160514	170779	118246	51059	33037	22041
IT	0	0	0	0	1	3	2
AV	5575	4187	4012	3604	1470	1233	1151
TN	0	0	0	0	2	1	4
LP	0	0	0	12	102	169	140
RP	7533	5501	5569	4009	3321	11034	1577
KP	0	0	0	0	0	1	0
GP	210530	227182	148164	44345	7550	239	27
GV	32827	28186	28634	24230	22818	16285	2878
L.	308643	475236	334405	57484	5540	4784	7045
KN	211517	209226	166331	124274	79805	28384	23128
SS	7727	6688	6206	5084	5821	8827	948
LX	7	5	4	7	1	7	9
TK	0	0	0	0	0	1	0
WU	0	0	0	0	0	0	0
BP	0	0	0	0	0	0	0
PP	0	0	0	0	0	1	0
CV	0	0	1	0	0	6	2
PS	840	1281	1019	595	228	17	0
DN	301392	286467	265400	152933	50136	88052	5676
RU	713	602	673	424	49	15	4
JS	107	128	127	135	26	3	10
SN	7372	6511	5584	5363	1039	100	84
LD	1	0	1	0	0	0	0
WK	65	36	41	33	7	1	2
BW	22241	18408	18957	20247	11968	1714	2155
BA	32255	28329	31393	21195	8947	1475	1449
WM	18	10	6	15	194	956	70
CU	53814	70866	60249	49955	19444	4005	4322
DR	60	64	65	47	35	10	3
RK	72347	64348	59025	48704	34274	3108	2636
GK	281	257	251	175	171	88	25
GE	117	68	86	83	47	3	23
OD	0	0	0	0	3	11	1
CS	65	34	36	36	139	318	196
TT	13701	12039	10644	8233	6936	2376	517
NK	0	0	0	0	0	0	4
PL	0	0	0	0	0	0	0
U.	0	290	0	0	0	0	0
JW	120	0	0	0	0	0	0
	1477959	**1606463**	**1317662**	**689468**	**311133**	**206264**	**76129**

Table 3. continued

	Species	Jul	Aug	Sep	Oct	Nov
	Number of sites visited	*623*	*689*	*1140*	*1212*	*1292*
MU	Mediterranean Gull	247	326	149	88	81
LF	Laughing Gull	0	0	0	0	4
FG	Franklin's Gull	0	0	0	0	1
LU	Little Gull	47	117	171	12	60
AB	Sabine's Gull	0	0	0	0	1
BH	Black-headed Gull	86477	114518	147447	122822	150842
IN	Ring-billed Gull	0	1	2	1	3
CM	Common Gull	4441	15433	16672	20385	33444
LB	Lesser Black-backed Gull	26976	20155	15853	14177	10701
YG	Yellow-legged Gull	15	178	92	55	25
YC	Caspian Gull	0	0	0	2	2
HG	Herring Gull	29923	36235	42757	37057	34774
IG	Iceland Gull	0	0	0	0	0
GZ	Glaucous Gull	0	0	0	1	1
GB	Great Black-backed Gull	2252	4228	7355	9740	8160
KI	Kittiwake	2404	2327	1551	553	623
ZU	Hybrid gull	0	0	0	0	0
UU	Unidentified gull	1987	2406	928	19	738
	TOTAL GULL	**154769**	**195924**	**232977**	**204912**	**239460**

	Species	Jul	Aug	Sep	Oct	Nov
	Number of sites visited	*624*	*693*	*1123*	*1162*	*1211*
AF	Little Tern	962	185	3	0	0
TG	Gull-billed Tern	0	0	0	0	0
BJ	Black Tern	0	27	4	2	1
WJ	White-winged Black Tern	0	0	1	0	0
TE	Sandwich Tern	8628	4650	2096	95	15
CN	Common Tern	5186	4031	545	32	8
RS	Roseate Tern	3	0	0	0	0
AE	Arctic Tern	2365	1229	55	6	20
UI	Common/Arctic Tern	7	0	0	0	0
UT	Unidentified tern	26	0	0	0	2
	TOTAL TERN	**17177**	**10122**	**2704**	**135**	**46**

Table 3. continued

sites	Dec	Jan	Feb	Mar	Apr	May	Jun
	1277	*1327*	*1371*	*1248*	*721*	*626*	*607*
MU	56	193	72	143	121	48	43
LF	4	2	2	1	0	0	0
FG	0	0	0	0	0	0	0
LU	1	6	0	50	340	36	44
AB	0	0	0	0	0	0	0
BH	158265	183960	153656	139391	48376	33660	39601
IN	2	2	2	2	1	0	0
CM	41616	40551	35919	34906	5529	2441	3067
LB	9277	12655	6763	10471	21092	23502	29065
YG	22	29	17	7	2	1	2
YC	3	4	1	1	0	0	0
HG	43085	38202	47346	38450	28688	27214	29495
IG	3	5	9	4	0	1	0
GZ	0	2	4	2	0	0	0
GB	6898	7495	4999	3207	2163	1846	1769
KI	379	186	63	229	974	1020	1462
ZU	0	0	0	1	0	0	0
UU	376	1287	40	0	203	195	0
	259987	**284579**	**248893**	**226865**	**107489**	**89964**	**104548**

sites	Dec	Jan	Feb	Mar	Apr	May	Jun
	1205	*1260*	*1286*	*1174*	*711*	*630*	*605*
AF	0	0	0	0	6	620	612
TG	0	0	0	0	0	1	0
BJ	0	0	0	0	0	4	0
WJ	0	0	0	0	0	0	0
TE	3	2	0	14	1340	4048	5540
CN	0	0	0	70	99	2687	3694
RS	0	0	0	0	0	0	0
AE	0	0	0	0	7	163	259
UI	0	0	0	0	0	41	1
UT	0	0	0	0	0	18	50
	3	**2**	**0**	**84**	**1452**	**7582**	**10156**

Table 4. Total numbers of waterbirds recorded by WeBS Core Counts in Northern Ireland in 2005/06. . Census totals are indicated by '*'.

	Species	Jul	Aug	Sep	Oct	Nov
	Number of sites visited	*1*	*3*	*14*	*19*	*18*
MS	Mute Swan	3	3	1304	1288	1269
AS	Black Swan	0	0	1	0	0
BS	Bewick's Swan	0	0	0	0	0
WS	Whooper Swan	0	0	10	385	1754
PG	Pink-footed Goose	0	0	0	4	0
NW	Greenland White-fronted Goose	0	0	0	28	0
JE	Re-established Greylag Goose	0	0	121	146	398
CG	Greater Canada Goose	0	0	172	1	190
YE	Naturalised Barnacle Goose	0	0	251	238	240
QN	Nearctic Light-bellied Brent Goose	0	0	12816	23047	7142
SU	Shelduck	3	11	512	1464	4185
DC	Wood Duck	0	0	0	0	0
WN	Wigeon	0	0	2057	10288	3671
GA	Gadwall	0	0	173	101	210
T.	Teal	0	4	3139	3880	3599
TA	Green-winged Teal	0	0	0	0	0
MA	Mallard	52	209	7785	6668	5125
PT	Pintail	0	0	35	275	91
SV	Shoveler	0	0	48	32	183
RQ	Red-crested Pochard	0	0	0	0	0
PO	Pochard	0	0	315	1289	4587
NG	Ring-necked Duck	0	0	0	0	0
TU	Tufted Duck	0	0	1985	7766	7818
SP	Scaup	0	0	122	116	3426
E.	Eider	11	0	2103	1581	790
LN	Long-tailed Duck	0	0	0	2	1
CX	Common Scoter	0	0	0	8	1
VG	Barrow`s Goldeneye	0	0	0	0	0
GN	Goldeneye	0	0	16	401	1943
SY	Smew	0	0	0	0	0
RM	Red-breasted Merganser	0	118	431	489	323
GD	Goosander	0	0	0	0	1
RY	Ruddy Duck	0	0	16	14	36
RH	Red-throated Diver	0	0	0	29	33
BV	Black-throated Diver	0	0	1	1	0
ND	Great Northern Diver	0	0	1	2	31
UL	Unidentified diver	0	0	0	0	0
LG	Little Grebe	0	0	435	448	535
GG	Great Crested Grebe	0	69	1689	2310	1158
RX	Red-necked Grebe	0	0	0	0	0
SZ	Slavonian Grebe	0	0	5	3	10
CA	Cormorant	26	238	2684	2321	1487
SA	Shag	0	0	301	492	394
XU	Unidentified Cormorant/Shag	0	0	292	147	128
ET	Little Egret	0	7	5	1	7
H.	Grey Heron	3	75	422	325	303
WA	Water Rail	0	0	0	4	3
MH	Moorhen	0	2	199	211	115
CO	Coot	0	0	3084	2491	2071
KF	Kingfisher	0	1	3	0	1
	TOTAL WILDFOWL	**98**	**737**	**42533**	**68296**	**53259**

Table 4. continued

	Dec	Jan	Feb	Mar	Apr	May	Jun
sites	*18*	*22*	*17*	*19*	*2*	*2*	*2*
MS	1538	1622	1488	909	5	34	5
AS	0	0	1	1	0	0	0
BS	18	15	18	0	0	0	0
WS	2169	2764	2574	1727	0	0	0
PG	0	0	0	2	0	0	0
NW	74	68	83	1	0	0	0
JE	493	904	1783	2235	0	0	0
CG	895	968	870	18	0	0	0
YE	240	243	149	68	0	0	0
QN	3573	2807	4396	3694	441	8	0
SU	5514	4347	3854	2233	158	117	42
DC	0	0	0	1	0	0	0
WN	5326	5676	5802	2809	0	0	43
GA	134	102	182	61	0	0	0
T.	6129	5958	4778	2004	28	0	0
TA	0	0	0	1	0	0	0
MA	5832	6028	4194	2084	33	15	26
PT	591	334	348	155	0	0	0
SV	128	181	93	90	0	0	0
RQ	0	0	1	0	0	0	0
PO	5629	8664	5793	1831	0	0	0
NG	1	0	0	0	0	0	0
TU	8985	10138	8592	4118	0	0	0
SP	3156	4713	6495	3695	0	0	0
E.	1357	1824	1471	1756	7	0	0
LN	15	10	4	22	0	0	0
CX	8	8	0	6	0	0	0
VG	1	1	2	0	0	0	0
GN	3450	4493	6532	3449	10	0	0
SY	0	1	3	0	0	0	0
RM	423	310	282	332	8	0	0
GD	0	1	1	0	0	0	0
RY	36	24	36	4	0	0	0
RH	115	38	19	54	0	0	0
BV	0	1	0	2	0	0	0
ND	63	16	5	13	0	0	0
UL	0	1	0	0	0	0	0
LG	488	435	474	156	3	0	0
GG	1600	1879	2503	2131	1	0	0
RX	0	0	0	1	0	0	0
SZ	42	0	0	0	0	0	0
CA	1728	1953	1886	1255	17	36	12
SA	249	526	150	267	0	0	0
XU	103	30	41	0	0	0	0
ET	8	7	10	7	1	0	0
H.	301	241	192	77	11	5	18
WA	3	2	4	0	0	0	0
MH	214	260	173	101	0	0	0
CO	4139	4048	3355	1169	4	0	0
KF	7	4	2	0	0	0	0
	64775	**71645**	**68639**	**38539**	**727**	**215**	**146**

Table 4. continued

	Species	Jul	Aug	Sep	Oct	Nov
	Number of sites visited	*1*	*3*	*14*	*19*	*18*
OC	Oystercatcher	54	2181	14413	14392	15138
RP	Ringed Plover	8	146	372	602	717
GP	Golden Plover	0	19	4916	9835	14726
GV	Grey Plover	0	0	80	48	112
L.	Lapwing	0	178	2311	6799	6955
KN	Knot	2	69	375	586	4050
SS	Sanderling	0	4	2	18	0
PP	Pectoral Sandpiper	0	1	0	0	1
CV	Curlew Sandpiper	0	22	4	2	0
PS	Purple Sandpiper	0	0	4	5	0
DN	Dunlin	10	52	1120	1712	5704
BQ	Buff-breasted Sandpiper	0	0	2	0	0
RU	Ruff	0	2	21	0	3
JS	Jack Snipe	0	0	0	0	11
SN	Snipe	0	1	25	34	274
WK	Woodcock	0	0	0	0	0
BW	Black-tailed Godwit	0	18	1158	928	422
BA	Bar-tailed Godwit	1	14	994	403	302
WM	Whimbrel	0	8	1	1	0
CU	Curlew	140	1042	4181	4944	3963
DR	Spotted Redshank	0	1	0	1	0
RK	Redshank	69	1557	7895	10668	7779
GK	Greenshank	0	46	139	186	132
CS	Common Sandpiper	0	12	3	0	0
TT	Turnstone	0	278	867	2118	889
WF	Wilson's Phalarope	0	1	0	0	0
	TOTAL WADER	**284**	**5652**	**38883**	**53282**	**61178**

	Species	Jul	Aug	Sep	Oct	Nov
	Number of sites visited	*1*	*2*	*10*	*15*	*14*
MU	Mediterranean Gull	0	0	0	1	1
LU	Little Gull	0	0	0	0	0
BH	Black-headed Gull	280	453	8025	9239	7486
IN	Ring-billed Gull	0	0	0	0	0
CM	Common Gull	21	117	5362	1497	4002
LB	Lesser Black-backed Gull	10	82	1839	581	224
YG	Yellow-legged Gull	0	0	0	0	1
HG	Herring Gull	11	310	3541	4263	3343
IG	Iceland Gull	0	0	0	0	0
GB	Great Black-backed Gull	7	10	670	1076	741
KI	Kittiwake	0	0	25	0	12
	TOTAL GULL	**329**	**972**	**19462**	**16657**	**15810**

	Species	Jul	Aug	Sep	Oct	Nov
	Number of sites visited	*1*	*2*	*10*	*11*	*9*
BJ	Black Tern	0	0	2	1	0
TE	Sandwich Tern	20	167	312	19	2
CN	Common Tern	0	0	36	2	1
	TOTAL TERN	**20**	**167**	**350**	**22**	**3**

Table 4. continued

	Dec	Jan	Feb	Mar	Apr	May	Jun
sites	*18*	*22*	*17*	*19*	*2*	*2*	*2*
OC	13936	12383	10974	8525	69	418	502
RP	656	642	182	205	0	48	13
GP	15578	24388	19856	9550	900	0	0
GV	113	157	203	76	0	0	0
L.	20918	21989	14996	1001	2	0	0
KN	6653	3960	5426	382	0	0	0
SS	18	19	0	6	268	30	0
PP	0	0	0	0	0	0	0
CV	0	0	0	0	0	0	0
PS	5	38	11	63	0	0	0
DN	6624	9967	9805	3533	60	333	206
BQ	0	0	0	0	0	0	0
RU	2	2	1	1	0	0	0
JS	0	1	1	0	0	0	0
SN	137	119	109	148	0	0	0
WK	0	1	3	0	0	0	0
BW	434	475	595	1148	19	0	0
BA	748	1471	1518	318	12	0	0
WM	0	1	1	0	0	0	0
CU	4711	5323	4984	3141	172	3	11
DR	0	0	0	1	1	0	0
RK	6425	5611	4799	5932	829	0	2
GK	77	67	78	109	7	0	0
CS	0	0	0	0	0	0	0
TT	1013	2012	1139	1799	19	0	0
WF	0	0	0	0	0	0	0
	78048	**88626**	**74681**	**35938**	**2358**	**832**	**734**

	Dec	Jan	Feb	Mar	Apr	May	Jun
sites	*13*	*17*	*12*	*14*	*2*	*2*	*2*
MU	1	0	0	3	0	0	0
LU	0	0	0	2	0	0	0
BH	6986	16573	9776	11574	154	50	82
IN	0	0	0	1	0	0	0
CM	1554	4685	4158	3432	98	13	11
LB	117	444	108	140	6	22	4
YG	0	0	0	0	0	0	0
HG	7890	10802	6449	3758	65	22	27
IG	0	3	1	3	0	0	0
GB	1136	1918	1212	570	21	26	27
KI	0	0	1	2	0	0	0
	17684	**34425**	**21705**	**19485**	**344**	**133**	**151**

	Dec	Jan	Feb	Mar	Apr	May	Jun
sites	*8*	*10*	*8*	*9*	*1*	*2*	*2*
BJ	0	0	0	0	0	0	0
TE	0	6	0	56	143	190	60
CN	0	0	16	0	4	136	0
	0	**6**	**16**	**56**	**147**	**326**	**60**

Table 5. Total numbers of waterbirds recorded by I-WeBS in the Republic of Ireland in 2005/06.

Species	Sep	Oct	Nov	Dec	Jan	Feb	Mar
Number of sites visited	*90*	*133*	*151*	*157*	*226*	*158*	*115*
Mute Swan	1456	1722	2443	1647	3988	1647	1102
Black Swan	2	3	0	0	0	0	0
Bewick's Swan	0	4	11	20	121	20	0
Whooper Swan	11	3351	2478	2090	5369	2090	1312
Pink-footed Goose	4	13	5	1	10	1	2
Greenland White-fronted Goose	0	5749	6642	7894	9061	7894	8124
Greylag Goose	978	1201	624	1377	1989	1377	2210
Greater Canada Goose	213	107	11	4	336	4	1
Barnacle Goose	0	153	510	1060	3124	1060	791
Dark-Bellied Brent Goose	0	0	1	3	4	3	0
Black Brant	0	0	1	1	1	1	0
Light-bellied Brent Goose	2456	6979	16534	17408	18513	17408	7233
Feral/hybrid Goose	96	236	0	90	194	90	60
Shelduck	138	1027	2878	6272	7284	6272	2768
Unidentified Duck	0	0	0	0	139	0	0
Wigeon	2623	21697	24191	15690	50142	15690	10145
American Wigeon	0	1	1	0	2	0	0
Gadwall	61	159	221	171	245	171	140
Teal	2943	6738	11027	10539	25017	10539	6190
Green-winged Teal	0	1	0	1	1	1	0
Mallard	6927	7764	7135	4283	10204	4283	2428
Pintail	4	350	175	756	967	756	277
Blue-winged Teal	0	0	1	0	0	0	0
Shoveler	58	698	920	1168	1834	1168	768
Red Crested Pochard	0	0	0	1	0	1	0
Pochard	19	135	10640	2428	7875	2428	372
Ring-necked Duck	0	0	0	2	2	2	1
Ferruginous Duck	0	0	0	0	1	0	0
Tufted Duck	1398	1000	4202	3357	12339	3357	1404
Scaup	12	72	68	82	605	82	71
Eider	12	7	169	6	19	6	2
Long-tailed Duck	0	4	18	24	26	24	26
Common Scoter	1103	1442	4286	1205	3545	1205	1265
Surf Scoter	0	0	2	0	0	0	0
Velvet Scoter	0	0	1	0	1	0	0
Goldeneye	3	118	412	888	1853	888	387
Smew	0	0	1	2	5	2	2
Red-breasted Merganser	144	494	748	433	1142	433	482
Goosander	0	0	1	6	0	6	0
Ruddy Duck	0	0	1	2	2	2	1
Feral/hybrid Mallard type	0	0	0	0	0	0	1
Red-throated Diver	7	36	173	69	107	69	94
Black-throated Diver	0	2	5	1	4	1	10
Great Northern Diver	6	314	327	152	316	152	195
Little Grebe	503	359	625	374	682	374	232
Great Crested Grebe	411	368	745	373	1360	373	429
Red-necked Grebe	0	0	0	1	0	1	1
Slavonian Grebe	0	2	10	2	44	2	4
Cormorant	3079	1888	2783	1565	3925	1565	801
Shag	59	80	534	91	449	91	92
Grey Heron	533	427	639	245	640	245	197
Little Egret	274	148	220	116	187	116	94
Coot	1644	4046	18450	1921	7892	1921	905
Moorhen	309	428	392	387	523	387	347
Water Rail	7	17	18	11	23	11	14
Kingfisher	10	16	11	4	13	4	4
TOTAL WILDFOWL	**27493**	**69340**	**121279**	**84219**	**182112**	**84219**	**50980**

Table 5. continued

Species	Sep	Oct	Nov	Dec	Jan	Feb	Mar
Oystercatcher	22011	20793	24615	19270	25937	19270	9887
Avocet	0	0	0	1	0	1	0
Ringed Plover	1873	3135	3340	1444	3975	1444	708
Dotterel	1	0	0	0	0	0	0
American Golden Plover	0	0	1	0	0	0	0
Golden Plover	6493	26964	71591	63847	111474	63847	13457
Grey Plover	440	152	1517	1423	1736	1423	650
Lapwing	2134	9282	35031	36604	120760	36604	1505
Knot	3336	4769	5522	11052	10713	11052	10517
Sanderling	1029	763	1707	3503	1399	3503	917
Little Stint	6	0	0	0	0	0	0
Curlew Sandpiper	48	0	2	0	0	0	0
Purple Sandpiper	3	0	25	28	93	28	47
Dunlin	3582	5312	12614	24726	38166	24726	8761
Buff-breasted Sandpiper	2	0	0	0	0	0	0
Ruff	27	15	0	3	6	3	1
Jack Snipe	1	28	14	22	20	22	15
Snipe	104	282	450	470	742	470	229
Long-billed Dowitcher	0	0	1	0	0	0	0
Woodcock	0	0	0	4	25	4	0
Black-tailed Godwit	7784	8143	11063	8816	12271	8816	4629
Bar-tailed Godwit	2301	1606	3325	8603	9285	8603	1643
Whimbrel	126	12	3	2	0	2	0
Curlew	9442	9492	13496	13561	24141	13561	3945
Spotted Redshank	1	3	24	2	4	2	0
Redshank	14225	14815	13926	9389	12617	9389	8034
Greenshank	542	370	443	377	411	377	220
Green Sandpiper	7	0	0	0	17	0	1
Wood Sandpiper	2	0	0	0	0	0	0
Common Sandpiper	5	10	4	0	10	0	1
Turnstone	1432	2307	3240	1774	2688	1774	1482
Grey Phalarope	0	0	1	0	0	0	0
TOTAL WADERS	**76957**	**108253**	**201955**	**204921**	**376490**	**204921**	**66649**

Species	Sep	Oct	Nov	Dec	Jan	Feb	Mar
Mediterranean Gull	34	34	32	30	21	30	3
Laughing Gull	0	0	1	0	0	0	1
Little Gull	2	0	0	1	1	1	3
Black-headed Gull	18334	13109	11755	12194	26907	12194	10296
Ring-billed Gull	0	3	3	2	7	2	6
Common Gull	7872	2970	5660	4465	10590	4465	4197
Lesser Black-backed Gull	1365	665	491	258	295	258	628
Herring Gull	1898	2560	1987	1992	3789	1992	1149
Yellow-legged Gull	0	0	0	1	0	1	0
Iceland Gull	0	0	0	2	3	2	2
Glaucous Gull	0	0	0	0	4	0	1
Great Black-backed Gull	1687	1916	1330	1001	1835	1001	695
TOTAL GULLS	**31192**	**21257**	**21259**	**19946**	**43452**	**19946**	**16981**

Species	Sep	Oct	Nov	Dec	Jan	Feb	Mar
Sandwich Tern	925	10	5	0	2	0	8
Roseate Tern	4	0	0	0	0	0	0
Common Tern	351	4	1	0	0	0	0
Arctic Tern	92	1	0	0	0	0	0
Forster's Tern	0	0	0	0	1	0	1
Black Tern	1	0	0	0	0	0	0
TOTAL TERNS	**1373**	**15**	**6**	**0**	**3**	**0**	**9**

SPECIES ACCOUNTS

Key to symbols commonly used in the species accounts.

As footnotes to thresholds:

? population size not accurately known

+ population too small for meaningful threshold

* where 1% of the national population is fewer than 50 birds, 50 is normally used as a minimum threshold for national importance

** a site regularly holding more than 20,000 waterbirds (excluding non-native species) qualifies as internationally important by virtue of absolute numbers

† denotes that a qualifying level different to the national threshold has been used for the purposes of presenting sites in this report

In tables of important sites:

- no data available

( ) incomplete count

† same meaning as used for thresholds

▲ site was of a higher importance status in the previous five-year period

▼ site was of a lower importance status in the previous five-year period

1,2 count obtained using different survey methodology from WeBS Core Counts (see table below)

Sources of additional information used in compiling tables of important sites are listed below. Non-WeBS counts are identified in the tables by the relevant number below given in superscript following the count.

1	RSPB/Talisman Energy studies, *e.g.* Stenning (1998)	30	WWT report to DTI. Aerial survey of Greater Wash strategic area
2	WWT studies, *e.g.* Rees *et al.* (2000)	31	All Wales Common Scoter Survey. WWT reports to CCW
3	Bean Goose Working Group, *e.g.* Smith *et al.* (1994)	32	All-Ireland Light-bellied Brent Goose Census
4	RSPB *pers comm.*	33	Cormorant Roost Survey 2003
5	Lancashire Goose Report, *e.g.* Forshaw (1998)	34	Scottish Bird Report records
6	SNH 'adopted' counts	35	Worden *et al* 2004
7	WWT data	36	RSPB data
8	Greenland White-fronted Goose Study, *e.g.* Fox and Francis (2004)	37	SNH data
9	SOTEAG reports, *e.g.* Heubeck (1998)	38	WWT UK-breeding Greylag Goose Survey
10	WeBS Low Tide Counts	39	Frank Mawby *in litt.*
11	Roost counts	40	Shetland co-ordinated swan count
12	Supplementary daytime counts	41	Supplementary daytime counts
13	WWT/JNCC National Grey Goose Census	42	Winter Gull Roost Survey
14	Firth of Clyde Eider counts, *e.g.* Waltho, C.M. (2004)	43	BTO/CCW Carmarthen Bay surveys
15	R. Godfrey (*in litt.*)	44	KOS Great Crested Grebe records
16	SNH Greenland Goose Census	45	B McMillan (*in litt.*)
17	R. MacDonald (*in litt.*)	46	C Langton (*in litt.*)
18	Little Egret Roost counts	47	B Yates (*in litt.*)
19	D Carrington (*in litt.*)	48	Tiree non-estuarine counts, per J Bowler
20	C Hartley (*in litt.*)	49	A Stevenson (*in litt.*)
21	WWT unpublished data	50	D Tate (*in litt.*)
22	Woolmer *et al.* 2001,Common scoter count in Carmart	51	Uist Greylag Goose Management Committee
23	Dorset Bird Report	52	Uists SPA wader survey (Ecology UK Ltd 2005)
24	Judith Smith, Gr. Manchester County recorder	53	P Wilson / Lancs Bird Report
25	BTO/Lucy Smith	54	W Aspin (*in litt.*)
26	Paul Daw, County recorder for Argyll	55	Winter Swan Census
27	Steve Percival`s counts of Lindisfarne – Svalbard Light-bellied Brent Geese	56	JNCC shore-based count
28	JNCC report of aerial surveys for seaducks, divers	57	RSPB Bean Goose counts
29	WWT report to DTI. Aerial survey of Thames strategic area	58	SNH Argyll goose counts
		59	WWT Dark-bellied Brent supplementary counts

Mute Swan
Cygnus olor

International threshold (British population): 320
International threshold (Irish population): 100
Great Britain threshold: 375
All-Ireland threshold: 100

GB max: 21,307 Oct
NI max: 1,622 Jan

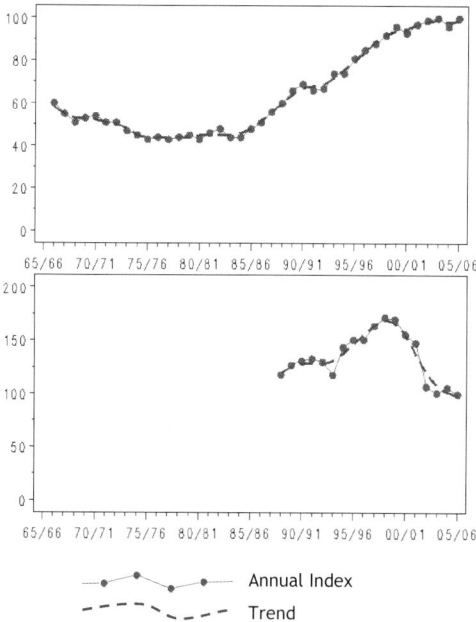

Figure 3.a, Annual indices & trend for Mute Swan for GB (above) & NI (below).

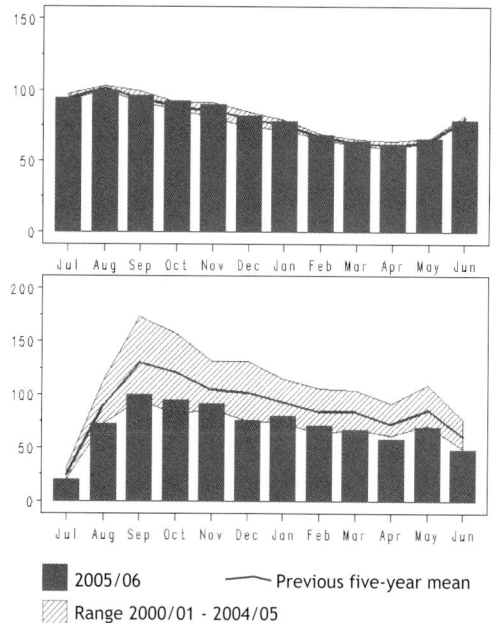

Figure 3.b, Monthly indices for Mute Swan for GB (above) & NI (below).

In Britain, the number of Mute Swans has shown considerable increase over the past twenty years leading to an estimated spring population of 31,700 in 2002 (Ward *et al.* 2007). The wintering population was estimated at 37,500 in the late 1990s and as movements of birds from the continent are rare, the differences between these estimates are likely to be due in part to late winter mortality, and in part to differences in estimation techniques. Since the 2002 national census was carried out, WeBS indices have remained roughly level, although the counted maximum in 2005/06 was the highest ever recorded and 7% higher than during 2001/02. It is widely believed that a reduction in the occurrences of lead poisoning, following the ban on the use of lead in fishing and restrictions for shooting, as well as milder winters, have contributed to the population increase seen in Britain over the past couple of decades (Rowell & Spray 2004).

Conversely, numbers in Northern Ireland remain low after falling from an all-time high in 1997/98; the Northern Ireland index fell again to its lowest level to date. However, numbers in the province remain relatively less well recorded due to the relatively small proportion of sites counted for WeBS, whilst the nature and extent of movement between sites in Northern Ireland and the Republic are also not fully understood.

The key sites for Mute Swan remain fairly consistent between years, although a low count at Loch Bee saw that site no longer support a mean peak in excess of the threshold for international importance. A very low total for the Ouse Washes was the lowest in a long, continuing decline, the peak now less than half that in 2001/02. The peak at the Loch of Harray was also very low. In Northern Ireland, a slight revival was noted at both Strangford Lough and Loughs Neagh & Beg.

	01/02	02/03	03/04	04/05	05/06	Mon	Mean
Sites of international importance in the UK							
Fleet and Wey	1,228	1,368	1,092	1,118	1,147	Sep	1,191
Loughs Neagh and Beg	1,423	1,510	920	949	1,024	Sep	1,165
Somerset Levels	1,084	(1,098)	(883)	1,076	1,024	Jan	1,071
Ouse Washes	1,110 [12]	782 [12]	606	693	469	Jan	732
Rutland Water	590	594	542	593	510	Aug	566
Tweed Estuary	464 [12]	414	582	614	460	Jul	507
Loch of Harray	597	672	522	467	251	Feb	502
Hornsea Mere	217	486	527 [12]	455	462	Aug	429
Loch Leven	506	550	526	202	319	Oct	421
Tring Reservoirs	282	447	322	404	346	Dec	360
Severn Estuary	339	284	(318)	390	390	Dec	351
Upper Lough Erne	306	323	272	449	300	Jan	330
Abberton Reservoir	187	387	379	318	373	Aug	329
Lower Lough Erne		199	(286)	(300)	309	Feb	274
Strangford Lough	183	180	193	94	133	Oct	157
Upper Quoile River	117	(71)	108	108	134	Jan	117
Sites no longer meeting table qualifying levels in WeBS-Year 2005/2006							
Loch Bee (South Uist)	200	297	407	394	267	Feb	313
Other sites surpassing table qualifying levels in WeBS-Year 2005/2006 in Great Britain							
Severn Estuary	339	284	(318)	390	390	Dec	351

Black Swan

Cygnus atratus

Escape[†]

Native Range: Australia

GB max:	43	Sep
NI max:	1	Sep

The British maximum of 43 was similar to that of previous years. The peak count at Arnot Park Lake reached nine in July and remained between eight and six throughout the rest of the year. In total the species was recorded at 79 sites, six less than during the previous year and of which 74 were in England, two in Wales and two in Scotland. The only record of this species from Northern Ireland was of a single bird at Clea Lakes, Strangford Lough during September, February and March.

Sites with three or more birds in 2005/06[†]

Arnot Park Lake	9	Jul
Fleet and Wey	8	Jul, Dec
River Kennet - Ramsbury to Chilton Foliat	5	Jul-Sep
Exe Estuary	4	Sep
Kingsmill Reservoir	4	Apr
Pagham Harbour	4	May
Abberton Reservoir	3	Oct, Nov, Jan, Feb
Aqualate Mere	3	Jul
Fairburn Ings	3	Jun
Thames Estuary	3	Nov

[†] as no British or All-Ireland thresholds have been set a qualifying level of three has been chosen to select sites for presentation in this report

Bewick's Swan (Paul Doherty)

Bewick's Swan
Cygnus columbianus

GB max: 7,663 Jan
NI max: 18 Dec

International threshold (*bewickii*):   200
Great Britain threshold: 81
All-Ireland threshold: 25*

50 is normally used as a minimum threshold

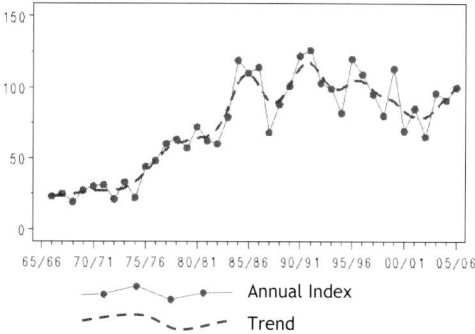

- Annual Index
- Trend

Figure 4.a, Annual indices & trend for Bewick's Swan for GB.

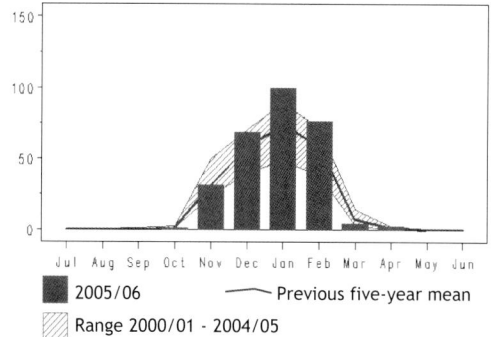

- 2005/06
- Previous five-year mean
- Range 2000/01 - 2004/05

Figure 4.b, Monthly indices for Bewick's Swan for GB.

Following two consecutive years of record counts, the peak on the Ouse Washes this winter was lower again. However, as the peak count on the Nene Washes was higher than usual this year, this appears to have been due to local redistribution around the East Anglian fens, as opposed to any overall decrease. Indeed, the national maximum count in mid-January was the highest recorded for many years and the annual index suggests a rising trend. In the west, numbers remained low on the Severn Estuary, Martin Mere / Ribble Estuary and the Dee Estuary, whilst a second year of low counts at the Somerset Levels means that the site no longer supports nationally important numbers. Even further west in Northern Ireland, the previous key site of Loughs Neagh & Beg saw an almost complete collapse in numbers, with the peak count only four birds in December, lower than the flock of 18 at Lough Foyle in February.

During the 2005/06 winter, the overall proportion of young in British flocks was assessed at 10.9%, below the mean value of the preceding five years (14.7%). However, mean brood size of successful pairs was slightly higher at 2.2 cygnets. Interestingly, there was a far higher proportion of young at WWT Slimbridge (17.7%) than elsewhere in the UK, although the total number of birds involved was low.

	01/02	02/03	03/04	04/05	05/06	Mon	Mean
Sites of international importance in the UK							
Ouse Washes	5,735 [11]	5,177 [11]	6,330 [11]	7,491 [11]	5,449 [11]	Dec	6,036
Nene Washes	347 [11]	1,068 [11]	790 [11]	262 [11]	1,649 [11]	Feb	823
Hickling Broad				282 [55]			282
Severn Estuary	310 [7]	345 [7]	230	223 [7]	225	Jan	267
Martin Mere and Ribble Estuary	296 [7]	315	221	175	(132)	Feb	252
St Benet`s Levels	147	287	280				238
Breydon Wtr & Berney Marshes	85	240	220	237	231	Feb	203 ▲
Sites of national importance in Great Britain							
Old Romney			184 [12]				184
Walland Marsh	180	220	148	140	135	Feb	165
Dee Estuary (England & Wales)	(78)	(70)	(92)	(101)	63	Feb	81
Sites no longer meeting table qualifying levels in WeBS-Year 2005/2006							
Somerset Levels	108	(69)	(112)	21	22	Jan	66

Whooper Swan
Cygnus cygnus

International threshold: 210
Great Britain threshold: 57
All-Ireland threshold: 100

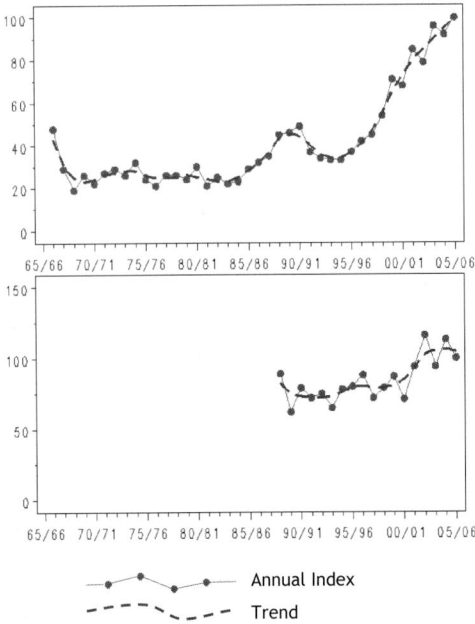

GB max: 8,051 Dec
NI max: 2,764 Jan

Annual Index
Trend

2005/06 — Previous five-year mean
Range 2000/01 - 2004/05

Figure 5.a, Annual indices & trend for Whooper Swan for GB (above) & NI (below).

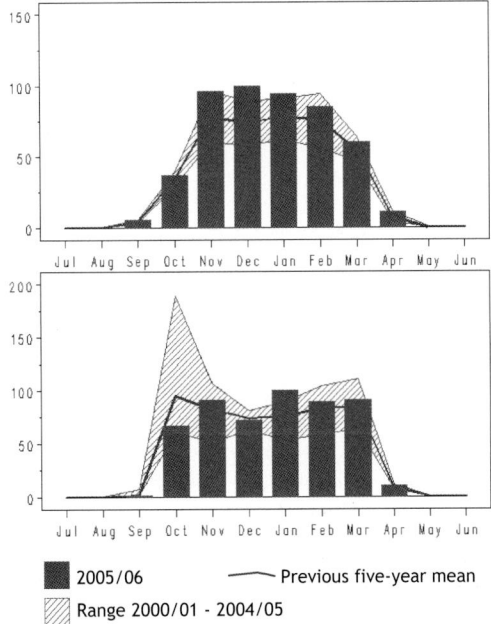

Figure 5.b, Monthly indices for Whooper Swan for GB (above) & NI (below).

Overall, counted numbers of Whooper Swans in the UK in 2005/06 were similar to the previous year, but in Britain, the annual index reached a new record high. The peak roost count on the Ouse Washes during the 2005/06 winter was lower than the two previous record peaks, but is still well above all counts prior to 2003/04. Elsewhere, relatively high counts were recorded at a number of sites, including Loch Eye & Cromarty Firth, River Clyde (Carstairs-Thankerton), River Earn (Lawhill Oxbows), Loch Heilen, Montrose Basin and the Humber Estuary. Conversely, peak counts were lower than usual at Loch of Wester and Vasa Loch (Shapinsay).

Small numbers of Whooper Swans in the summer months were almost all recorded from sites in Scotland, suggesting that most were of wild origin, probably largely birds which were injured or otherwise unfit to migrate to the breeding grounds. The main arrival started from October, with peak numbers from November to February. Many were still present in March but a rapid decline took place by the time of the April count.

Assessment of flocks around the UK found 12.1% young overall, well below the five year average of 15.9%, with a mean brood size of 2.5 cygnets per successful pair. Particularly low numbers of juveniles were found at the Ouse Washes (8.8%), continuing the decline noted there in recent winters in the breeding success of birds wintering there.

	01/02	02/03	03/04	04/05	05/06	Mon	Mean
Sites of international importance in the UK							
Ouse Washes	2,894 [11]	2,745 [11]	3,624 [11]	4,397 [11]	3,547 [11]	Dec	3,441
Martin Mere and Ribble Estuary	1,762 [7]	1,770 [7]	1,597	2,081 [55]	1,666	Jan	1,775
Loughs Neagh and Beg	(1,532)	1,514	(867)	1,543	1,268	Mar	1,464
Lough Foyle	548	3,284	680	950 [55]	1,030	Jan	1,298
Upper Lough Erne	1,228 [12]	658	855	1,123	822	Feb	937
Loch of Strathbeg	(223)	(67)	794	355	680	Nov	610
Solway Estuary	(309)	340 [7]	(250)	508 [55]	(150)	Jan	424
Loch Eye and Cromarty Firth	230	141	322	275	518	Oct	297
Loans of Tullich				253 [55]			253
Bridge of Crathies				(220) [55]			(220)
Sites of national importance in Great Britain							
Wigtown Bay	156	(135)	255	205	(165)	Jan	205
Norham West Mains				184 [55]	190	Mar	187
Loch of Wester	341			128	56	Jan	175 ▼
R Clyde: Carstairs to Thankerton	242	(101)	91	110	220	Nov	166
Black Cart Water	238 [2]	176 [2]	151 [2]	112	112	Dec	158
Lawers Pond					152	Oct	152 ▲
Loch Heilen			24	60	360	Oct	148 ▲
R. Nith - Keltonbank to Nunholm	125	(108)	165	(104)			145
Loch a` Phuill (Tiree)	83	168	118	194	132	Nov	139
Tarbat Ness		0	44	306	202	Feb	138
Nene Washes	110 [11]	143 [11]	111 [11]	104 [11]	215 [11]	Feb	137
Killimster Loch	135						135
Leven Cut				125 [55]			125
East Fenton Farm Reservoir				89	156	Jan	123
Inner Moray and Inverness Firth	173	60	165	27	166	Jan	118
Strathearn South Kinkell				111 [55]			111
Loch Insh and Spey Marshes	92	91	110	124	82	Jan	100
R. Eden: Grinsdale- Sandsfield					98	Mar	98 ▲
River Tweed: Kelso-Coldstream	60	116	109	75	132	Feb	98
Caistron Quarry	71	67	164	96	66	Feb	93
Lindisfarne	15	(90)	(139)	71	119 [10]	Jan	87
Loch of Lintrathen	10	166	93	69			85
Kinnordy Loch	116	82	35	96	58	Mar	77
Lower Derwent Ings	61	91	52	102	74	Dec	76
Morecambe Bay	125	6	(20)	63	(100)	Dec	74 ▲
Vasa Loch Shapinsay		68	96	119	12	Dec	74
Loch of Spiggie	47	86	89	69	77	Oct	74
Farmland near Monymusk				65 [55]			65
River Earn - Lawhill Oxbows	12	0	0	113	193	Nov	64 ▲
Loch Tuamister (Lewis)				63 [55]			63
Merryton Haughs					62	Mar	62 ▲
Drem Pools				8	115	Dec	62 ▲
Broubster Leans				75 [55]	49	Feb	62
Milldam and Balfour Mains Pools	98	41	86	0	84	Nov	62
Farmland near Whitekirk				61 [55]			61
Lower Teviot Valley	(50)	(29)		(58)			(58)
St Benet`s Levels	6	58	108				57
Sites of all-Ireland importance in Northern Ireland							
Strangford Lough	212	191	150	244	242	Nov	208
Lough McNean Lower	76 [12]			124	103	Dec	101 ▲
Sites no longer meeting table qualifying levels in WeBS-Year 2005/2006							
Dornoch Firth	53 [10]	23	94	18	84	Jan	54
Ravenstruther		0	48	56	56	Jan	40
Forth Estuary	(20)	(24)	62 [10]	19	(7)	Apr	41
Other sites surpassing table qualifying levels in WeBS-Year 2005/2006 in Great Britain							
Montrose Basin	1	10	24	28	181	Feb	49
Humber Estuary	9	(21)	14 [10]	8	(115)	Nov	33
Dornoch Firth	53 [10]	23	94	18	84	Jan	54
St Johns Loch				10	80	Nov	45
Linne Mhuirich		47	36	10	65	Oct	40
Uyea Sound	15	23	37	47	64	Nov	37
River Earn: Millands Msh & Flds	84	0	(76)	15	63	Nov	48
The Wash	9	21	18	15	57	Nov	24

Coscoroba Swan
Coscoroba coscoroba

Escape

Native Range: S America

GB max: 1 Nov
NI max: 0

A single Coscoroba Swan was on Burton Marsh on the Dee Estuary during November. This is the second ever recorded during WeBS, the first being on the River Usk, which stayed throughout winter 2000/01.

Chinese Goose
Anser cygnoides

Escape

Native Range: E Asia

GB max: 15 Dec
NI max: 0

Chinese Goose (the domesticated form of the Swan Goose) was recorded at ten sites across the UK, seven in England, two in Scotland and one in Wales. The national maximum was down on the previous year, being more in line with figures from 2003/04. The highest single site total was of six at Diss Mere in December.

Bean Goose
Anser fabalis

International threshold: 800
Great Britain threshold: 4*
   +

GB max: 484 Jan
NI max: 0

*50 is normally used as a minimum threshold

Following the relatively large arrival of Tundra Bean Geese (*Anser fabalis rossicus*) during the 2004/05 winter, the situation was more typical in 2005/06. In fact, away from the two main wintering sites of Taiga Bean Geese (*Anser fabalis fabalis*), the species was recorded from only ten other sites, with only the Ouse Washes and North Warren holding more than three birds. At a small number of sites, presumed escaped birds were recorded during the summer months.

In the Slamannan area, counts increased once again to a new peak of 300 birds, the highest count of this species in Britain since 1994 and the first site outside Norfolk ever to reach 300. The first few geese returned in September but numbers were fairly constantly high between October and February with virtually all birds gone by March. Although the peak at the Middle Yare Marshes represented a slight increase on the previous winter, peak counts there have remained at less than 200 for four years in succession. Numbers were lower than average until January, when a notable arrival occurred, followed by a similarly rapid disappearance by February.

At Slamannan, about 14.4% of geese were first-year birds, with an average of 1.25 young per successful pair recorded. These assessments were made in November and after this time it becomes increasingly difficult to assess the age of Bean Geese; for this reason, it is not possible to assess the breeding success of the Yare Valley flock.

	01/02	02/03	03/04	04/05	05/06	Mon	Mean
Sites of national importance in Great Britain							
Slamannan Area	192 [3]	231 [3]	235 [3]	262 [3]	300 [3]	Jan	244
Middle Yare Marshes	272 [4]	183 [4]	140	156 [36]	169 [57]	Jan	184
Ouse Washes	4 [12]	8 [12]	4	87 [11]	9	Jan	22
Walland Marsh	0	0	0	86	0		17
Nth Warren & Thorpeness Mere	0	0	3 [12]	38	10 [12]	Feb	10
Fleet and Wey	(26)	6	0	0	0		6
Abberton Reservoir	22	0	0	0	(0)		6
Somerset Levels	0	(0)	0	14	(0)		5
Balnakeil Bay			5 [12]				5
Whitemoor Haye	3	0	0	17	0		4

Pink-footed Goose
Anser brachyrhynchus

International threshold: 2,700
Great Britain threshold: 2,400
+

GB max: 258,247 Nov
NI max: 4 Oct

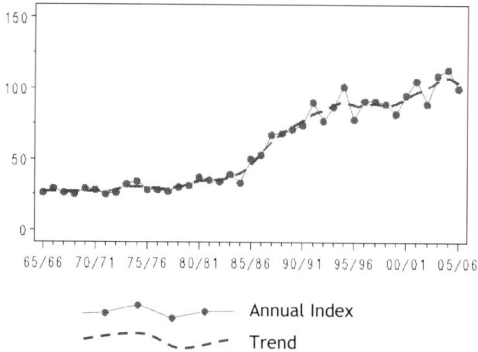

Figure 6.a, Annual indices & trend for Pink-footed Goose for GB.

Given that this species is not well covered by standard WeBS counts (as most birds use waterbodies only for roosting overnight spending the day on surrounding farmland), Pink-footed Geese are predominantly surveyed by dedicated coordinated roost counts at the key sites as well as the Goose and Swan Monitoring Program (www.wwt.org.uk/research/monitoring/goose_and_swan.asp). The national index is also based on these roost counts, and in autumn 2005 fell somewhat from the record level of the previous two years. Where possible, the British counts have been synchronised with counts from Iceland and the Faeroe Islands, to try to get as full a picture of the population as possible, although in autumn 2005 the vast majority of birds were present in Britain from October onwards, when the first coordinated count take place. The British peak total count in autumn 2005 was of 258,247 birds, although consideration of sites not counted during that

month, plus the addition of a small number on the Faeroes, increased the estimate of the population in that month to 268,650. This represented a decline of 8% on the peak estimate for autumn 2004.

As usual, a major arrival was noted at Loch of Strathbeg in September, with the huge count of 68,000 here being both the highest single site count of the year, and an all-time record for this site. A large proportion of birds was present in Scotland in October but by November, many had moved south to Norfolk and by December over half of the total population was present in this county. In Norfolk, a record roost total of 66,000 was noted at Holkham/Wells area, although this count could not be assigned to either single site in the table below, whilst the peak at Scolt Head was slightly lower than in recent years, albeit still extremely high. The January roost count at Snettisham was the highest there to date. In southwest Lancashire, In southwest Lancashire, the main arrival of geese was earlier than usual and October numbers were high; many appeared to have left by November but numbers were high again in December (Forshaw 2006). Higher than average peak counts were also noted from Loch Leven and Loch Spynie, whilst the Findhorn Bay roost was relatively low this autumn. Breeding success was estimated at 18.1% young, which was lower than for the preceding three years but still comparable to the mean of most recent 5 yrs. The average of 1.7 young per successful pair was also lower than average.

	01/02	02/03	03/04	04/05	05/06	Mon	Mean
Sites of international importance in the UK							
Scolt Head	33,900 [13]	62,500 [13]	80,000 [11]	66,000 [12]	55,000 [13]	Dec	59,480
Loch of Strathbeg	46,898 [13]	39,900	66,000 [13]	65,000 [13]	68,000 [13]	Sep	57,160
Holkham Bay	45,000 [13]	33,800 [13]	47,750 [13]	58,000 [12]	50,000 [13]	Jan	46,910
Snettisham	35,000 [13]	37,050 [13]	27,350 [13]	35,360 [12]	49,610 [13]	Jan	36,874
Southwest Lancashire	33,180 [13]	31,645 [13]	27,025 [13]	43,950 [5]	31,860 [13]	Oct	33,532
West Water Reservoir	23,270 [13]	(40,000) [13]	34,210 [13]		28,240 [13]	Oct	31,430
Montrose Basin	38,669 [13]	11,500 [13]	10,149 [13]	31,896 [13]	30,181 [13]	Nov	24,479
Morecambe Bay	14,100 [13]	14,600 [5]	17,050 [5]	26,910 [5]	20,980 [13]	Jan	18,728
Ythan Estuary and Slains Lochs	13,900 [13]	19,600 [13]	19,200 [13]	16,200	(1,800)	Jan	17,225
Loch Leven	16,200 [13]	(12,874) [13]	15,120 [13]	14,750	22,175 [13]	Oct	17,061
Aberlady Bay	13,740 [13]	22,200 [13]	15,040 [13]	18,430 [13]	14,250 [13]	Nov	16,732
Findhorn Bay	14,000 [13]		25,000 [13]	18,000 [13]	9,400 [13]	Oct	16,600

	01/02	02/03	03/04	04/05	05/06	Mon	Mean
Loch Spynie	9,100[13]	11,700[13]	11,100[13]	27,000[13]	23,000[13]	Nov	16,380
Loch of Skene	13,175[13]	(8,420)[13]	(8,500)[13]	12,000[13]	17,730[13]	Dec	14,302
Carsebreck and Rhynd Lochs	14,500[13]	10,320[13]	11,450[13]	8,770[13]	11,130[13]	Oct	11,234
Breydon Wtr & Berney Marshes	4,380	7,100	17,100	12,784	11,213	Dec	10,515
Dupplin Lochs	17,500[13]	9,500[13]	14,100[13]	2[13]			10,276
Easterton - Fort George					10,000[13]	Nov	10,000 ▲
Loch of Lintrathen	5,920[13]	(6,440)[13]	11,100[13]	8,921[13]	9,790[13]	Oct	8,933
Hule Moss	8,600[13]	5,850[13]	14,200[12]	7,950[13]	6,000	Oct	8,520
Loch Tullybelton				6,500[13]			6,500
Horsey Mere	(5,000)[13]	4,000[13]	8,200[13]	7,231[12]	6,240[13]	Dec	6,418
Cameron Reservoir	15,823	3,000	8,900[13]	2,692[13]	521	Nov	6,187
R Clyde: Carstairs to Thankerton	11,000	3,350	5,300	(3,050)	4,500	Oct	6,038
Solway Estuary	(5,550)	(4,075)	(10,243)	2,612[13]	(6,862)	Feb	5,868
Wigtown Bay	5,316[13]	(4,747)	8,662[13]	(7,219)	802	Feb	5,500
Humber Estuary	4,300[13]	4,620[13]	6,562	(5,638)	3,909	Nov	5,006
Holme and Thornham					5,000[13]	Jan	5,000 ▲
Lindisfarne	6,450[13]	(3,679)	1,496	5,300[13]	5,800[13]	Nov	4,762
Norton Marsh					4,500[13]	Oct	4,500 ▲
Tay Estuary	11,385[13]	2,700[13]	2,425[13]	4,560[13]	0		4,214
Fala Flow	7,500[13]	2,790[13]	5,450[13]	741[13]			4,120
Strathearn (West)		4,100[13]					4,100
Heigham Holmes	2,500[13]				5,670[13]	Jan	4,085 ▲
Loch Eye and Cromarty Firth	367[13]	14,050[13]	546	900	3,226	Dec	3,818
River Tay - Haughs of Kercock				4,000[13]	3,500[13]	Nov	3,750
River Nith - Keltonbank/Nunholm	(1,200)	(470)	(3,710)	(950)			(3,710)
Tay and Isla Valley	2,133[13]	2,497[13]	4,134[13]	4,000	3,500	Nov	3,253
Holburn Moss	20	4,250[13]	6,500[13]	2,300[13]	2,950[13]	Oct	3,204
Skinflats	3,800[13]	1,900[13]	3,250[13]	2,530[13]	3,980[13]	Oct	3,092
Rossie Bog	50[13]			6,290[13]	2,250[13]	Nov	2,863 ▲
Lake of Menteith	3	4,515[13]	4,026[13]	5,357[13]	11	Mar	2,782
Sites no longer meeting table qualifying levels in WeBS-Year 2005/2006							
Thornham				950[13]			950
Other sites surpassing table qualifying levels in WeBS-Year 2005/2006 in Great Britain							
Folly Loch and Fairnington Fields	1,800[13]	32	5,500	4	4,563[13]	Nov	2,380
Lochhill				0	3,525[12]	Sep	1,763

European White-fronted Goose
Anser albifrons albifrons

International threshold: 10,000
Great Britain threshold: 58
All-Ireland threshold: +

GB max: 2,350 Feb
NI max: 0

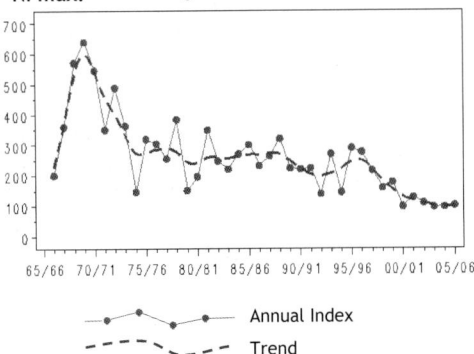

Figure 7.a, Annual indices & trend for European White-fronted Goose for GB.

- Annual Index
- Trend

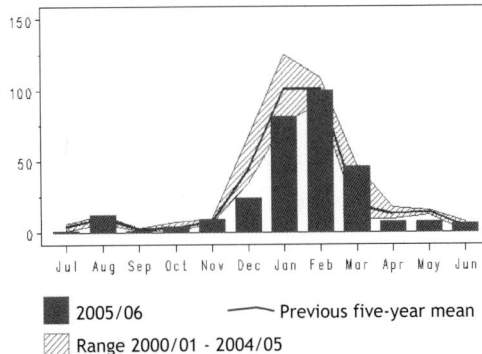

- 2005/06
- Range 2000/01 - 2004/05
- Previous five-year mean

Figure 7.b, Monthly indices for European White-fronted Goose for GB.

With the notable exception of WWT Slimbridge on the Severn Estuary, the key sites for European White-fronted Goose are extremely concentrated in the southeast of Britain, from Norfolk to Kent and seldom more than 20 km inland. Away from the tabulated key sites, European White-fronted Geese were recorded from a further 32 sites, virtually all in England. Although large numbers of European White-fronted Geese continue to migrate from their breeding grounds in arctic Russia to continental northwest Europe, the annual British index shows that numbers here remain at a very low level compared to previous decades, although are roughly stable since the turn of the century. A look at the monthly indices reveals that this is a mid-winter bird here now, with the peak clearly occurring in January and February and numbers outside these two months very much lower. However, the peak of 2350 in February was the highest recorded by WeBS since the 2001/02 winter suggesting that the recent downturn in numbers has at least halted.

At the main sites, peak numbers were below their recent averages on the Severn Estuary, the Middle Yare Marshes and the Suffolk

coastal strip of Minsmere, North Warren and the Alde Estuary. A flock of 86 birds on the Thames Estuary was sufficient to cause the five-year peak mean for the site to exceed the national importance threshold. The peak count at the Ouse Washes was the highest there since 1996, whilst the Stodmarsh flock was the highest count ever recorded at this site.

The overall proportion of young, assessed during January & February 2006 at six sites, was 34.3%, with a mean brood size of 3.01 young per successful pair. Prior to 2004/05, productivity estimates for European White-fronted Geese were only carried out at WWT Slimbridge; the proportion of young at this site in 2005/06 (31.6%) was the highest recorded in the past 13 years.

	01/02	02/03	03/04	04/05	05/06	Mon	Mean
Sites of national importance in Great Britain							
Severn Estuary	1,250 [12]	990 [10]	780 [12]	745 [7]	750	Jan	903
Swale Estuary	360	655	327	(398)	430	Feb	443
North Norfolk Coast	380	347	540	340	404	Jan	402
Dungeness Gravel Pits	355	460	205 [12]	110	250	Feb	276
Walland Marsh	450	300	140	137	310	Mar	267
Breydon Wtr & Berney Marshes	110	181	455	267	290	Mar	261
Nth Warren & Thorpeness Mere	250	310 [12]	190 [12]	302	200	Mar	250
Alde Complex	5 [10]	385	54	25	12	Jan	96
Middle Yare Marshes	74	89	120	109	76	Jan	94
Thames Estuary	(41)	89	42	(16)	86	Feb	72 ▲
Minsmere	120	1	175	9	2	Mar	61
Other sites surpassing table qualifying levels in WeBS-Year 2005/2006 in Great Britain							
Stodmarsh NNR & Collards Lagn	5	0	0	0	122	Feb	25
Ouse Washes	1	1	15	19	79	Feb	23

Greenland White-fronted Goose
Anser albifrons flavirostris

International threshold:	270
Great Britain threshold:	209
All-Ireland threshold:	140

GB max: 14,287   Mar
NI max: 83   Feb

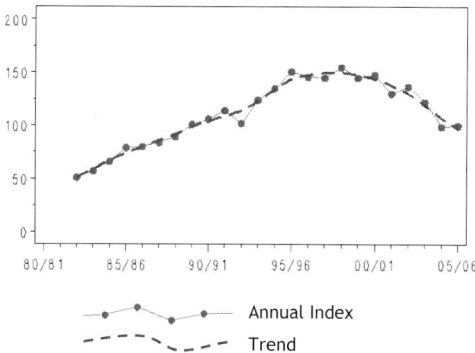

Figure 8.a, Annual indices & trend for Greenland White-fronted Goose for GB.

The annual censuses by Greenland White-fronted Goose Study were carried out in December 2005 and March 2006. In Britain, these located 13,609 and 14,287 birds

respectively. In March 2006, 10,608 were also counted in Ireland (75% of these at Wexford Slobs), leading to a world population of 24,895, the lowest spring total since 1988. The decision to ban hunting of Greenland White-fronted Geese on migration in Iceland took effect from autumn 2006, and so will not have affected the numbers reported here. The first indications of any effect of this ban may become apparent in next year's *Waterbirds in the UK* report.

In Scotland, the most important site remains Islay but peak numbers fell once again to their lowest levels for many years. Other sites where the count fell below the recent average included Rhunahaorine, Loch Lomond and Colonsay, latterly, where numbers returned to normal after the high count of 2004/05. Mean peak numbers at Clachan & Whitehouse and

Loch of Mey were no longer sufficient to exceed the threshold for national importance, although counts for Eriska/Benderloch did reach this level. The isolated flock at the Dyfi Estuary in Wales declined again slightly in numbers, from 92 in 2004/05 to a peak of 84 this winter.

Although breeding success increased slightly compared to 2004, the value of 8.6% young remained well below the long-term average. The mean brood size of 3.08 young per successful pair (with the largest broods on Islay with a value of 3.58 young per successful pair) was higher than the long-term average. The fact that fewer pairs are nesting successfully, but those that do so are still productive, could imply that it is the available area for nesting that is declining, and not its quality. This would fit in with observations that Lesser Canada Geese are increasingly occupying some of the breeding range of the Greenland White-fronted Geese and exhibit behavioural dominance over the latter.

	01/02	02/03	03/04	04/05	05/06	Mon	Mean
Sites of international importance in the UK							
Island of Islay	12,261 [16]	12,254 [6]	11,272 [6]	8,350 [8]	7,456 [8]	Dec	10,319
Machrihanish	1,448 [8]	1,501 [8]	1,377 [6]	1,407 [8]	1,433 [8]	Mar	1,433
Rhunahaorine	1,594 [8]	1,450 [8]	1,156 [6]	894 [8]	955 [8]	Mar	1,210
Tiree	1,076 [8]	1,093 [8]	1,093 [16]	1,133 [36]	1,112 [8]	Mar	1,101
Isle of Coll	705 [8]	611 [8]	495 [8]	814 [8]	778	Oct	681
Isle of Colonsay	104 [8]	87 [8]	79 [6]	1,718 [6]	111 [58]	Mar	420
Keills Peninsula & Isle of Danna	403 [8]	411 [8]	377 [6]	338 [8]	344 [8]	Mar	375
Stranraer Lochs	500 [8]	365 [8]	281 [8]	257 [8]			351
Isle of Lismore	295 [8]	310 [8]	290 [8]	310 [8]	320 [8]	Mar	305
Loch Lomond	294 [8]	450 [8]	260 [8]	240 [8]	210 [8]	Dec	291
Sites of national importance in Great Britain							
Loch Ken	326 [8]	275 [8]	300 [8]	215 [8]	220 [8]	Mar	267 ▼
Sites no longer meeting table qualifying levels in WeBS-Year 2005/2006							
Clachan and Whitehouse	100 [8]	250 [8]	215 [6]	209 [8]	193 [8]	Dec	193
Loch of Mey	260 [8]	208 [8]	196 [8]	193 [8]	184 [8]	Feb	208
Other sites surpassing table qualifying levels in WeBS-Year 2005/2006 in Great Britain							
Eriska/Benderloch	159 [8]	133 [8]	154 [8]	145 [8]	239 [8]	Mar	166

Lesser White-fronted Goose

Anser erythropus

Vagrant and escape
Native Range: SE Europe, Asia

GB max: 2 Dec
NI max: 0

Lesser White-fronted Geese were present at four sites during 2005/06; all records were of single birds and are likely to refer to escapes. These were at Testbourne Estate in August, River Avon - Ringwood to Christchurch in September and December, East Park Hull in December and Llyn Alaw in January.

Greylag Goose
Anser anser

Icelandic Population

International threshold: 870
Great Britain threshold: 819
All-Ireland threshold: 40*

GB max: 76,945 Dec
NI max: 0

50 is normally used as a minimum threshold

Greylag Geese breeding in Iceland winter almost exclusively in northern Britain, apart from small numbers in Ireland, Norway, the Faeroes and Iceland. The peak number recorded by the Icelandic-breeding Goose Census (IGC) during 2005/06 was 94,359 over the weekend of 5th-6th November. This total was adjusted (by taking into account missing sites and potential overlap with birds assumed not to be of Icelandic origin) to give a population estimate of 95,938 birds, a 10.5% decline on the equivalent estimate for the autumn of 2004. Despite this decline, however, this was still the second-highest total to date. The total number of birds assumed to refer to the Icelandic population actually counted at British sites (without adjustment for non-counted sites) was 76,490 birds in November 2005.

Relatively few birds had arrived by the time of an IGC census in early October, with most birds making an appearance over the course of the following month. By the November census, about two-thirds of birds were found in northern Scotland (especially Orkney). For the first time, the IGC trialled a third coordinated count, this time in early December when 76,945 were counted in Britain. The purpose of this extra count, which will be further trialled over two subsequent winters, is to ascertain whether birds remaining later into the autumn in Iceland are having a major impact on the census and whether December is now a better month in which to conduct a coordinated census.

Counts in Orkney remained high, with West Mainland the key area, supporting about 40% of the total for the islands. Other sites with relatively high counts included Loch Eye/Cromarty Firth, Loch Fleet, Loch Ussie

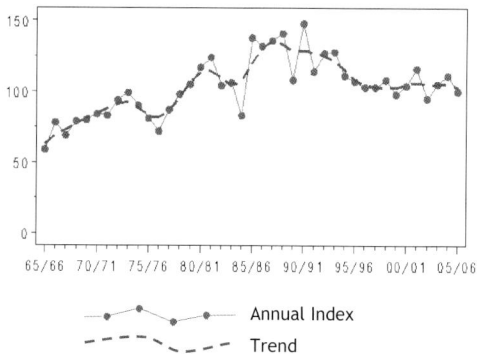

Figure 9.a, Annual indices & trend for Icelandic Population for GB.

and the Forth Estuary, whilst counts at Loch Ken and the Dornoch Firth were lower than usual. Assessment of breeding success revealed an overall proportion of 22.7% young, higher than average although the mean brood size, at 2.3 young per successful pair, was a little lower than the recent five-year mean.

	01/02	02/03	03/04	04/05	05/06	Mon	Mean
Sites of international importance in the UK							
Orkney	22,665 [13]	26,505 [13]	43,097 [13]	42,697 [13]	40,403 [13]	Dec	35,073
Loch Eye and Cromarty Firth	5,680 [13]	(7,028) [13]	6,523 [13]	8,313 [13]	13,269	Dec	8,446
Caithness Lochs	(7,854)	2,792 [13]	2,971 [13]	11,755 [13]	8,727 [13]	Nov	6,820
Loch of Skene	2,100 [13]	(1,021) [13]	(2,600) [13]	4,500 [13]	4,700 [13]	Nov	3,767
Easterton - Fort George					3,500 [13]	Dec	3,500 ▲
Loch Spynie	5,300 [13]	3,200 [13]	2,200 [13]	1,000 [13]	2,600 [13]	Nov	2,860
Dornoch Firth	2,386 [13]	2,916	2,259	1,720	1,632 [13]	Nov	2,183
Tay and Isla Valley	2,092 [13]	(1,700)	2,425 [13]	1,930	2,155	Dec	2,151
Loch Fleet Complex	4,210 [13]	817 [13]	905 [13]	990 [13]	3,000	Oct	1,984
Bute	2,300 [13]	1,380 [13]	2,000 [13]	1,780 [13]	2,110 [13]	Dec	1,914
Lochs Davan and Kinord	5,277 [13]	2,700 [13]	920 [13]	135	105	Apr	1,827
Loch Garten	2,800 [13]	1,000 [13]	1,000 [13]	2,100 [13]	1,700 [13]	Oct	1,720
Munlochy Bay	3,500 [13]	3,130 [13]	110 [13]	20 [13]	1,000 [13]	Dec	1,552
Kilconquhar Loch	1,380 [13]	1,552	1,620	1,200 [13]	1,500 [13]	Nov	1,450
Forth Estuary	826	1,564	792	802	2,105	Mar	1,218 ▲
East Mains Flood				1,131 [13]			1,131
Inner Firth of Tay	1,900 [13]		754 [13]	842 [13]	850 [13]	Dec	1,087
Strathearn (West)		1,050 [13]	1,050 [13]				1,050
Loch Ken	(1,368)	(1,106)	(1,280)	1,023	380 [13]	Oct	1,031
Beauly Firth	840 [13]	2,010 [13]	280 [13]	600 [13]	1,380 [13]	Nov	1,022
Lower Teviot Valley	(598)	(1,800)	525	(833)	1,250	Nov	896
Sites of national importance in Great Britain							
Findhorn Bay	1,950 [13]		190 [13]	1,100 [13]	200 [13]	Nov	860 ▼
Cochrage Loch	850 [13]						850
Haddo House Lakes	980	975	1,100 [13]	603	520	Nov	836 ▼
Loch of Strathbeg	1,744 [13]	415 [13]	295 [13]	801 [13]	(853) [13]	Dec	822
Sites no longer meeting table qualifying levels in WeBS-Year 2005/2006							
Gadloch	685	994	650	650	1,020 [13]	Oct	800
Caistron Quarry	1,100	1,000	800	513	660	Feb	815
Whitrig Bog		1,000 [13]		850	450	Mar	767
Upper Tay	1,022 [13]	943 [13]	1,197 [13]	181 [13]	385 [13]	Nov	746
Marlee Loch				880 [13]	160 [13]	Oct	520
Other sites surpassing table qualifying levels in WeBS-Year 2005/2006 in Great Britain							
Loch Ussie	520 [13]	2	261	0	3,280 [13]	Nov	813
River Devon - Tullibody Bridge	160	175	658	223	1,270	Feb	497

	01/02	02/03	03/04	04/05	05/06	Mon	Mean
Shetland Isles		259 [13]		586 [13]	1,126 [13]	Oct	657
Island of Rousay				165 [13]	1,024 [13]	Nov	595
Gadloch	685	994	650	650	1,020 [13]	Oct	800
Isle of Cumbrae				500 [13]	1,000 [13]	Nov	750
Tyninghame Estuary	160 [13]	830	80	210	870	Mar	430
Monk Myre	0	0	0	200	850 [13]	Nov	210

Northwest Scottish Population

International threshold: 100
Great Britain threshold: 90

GB max: 9,530 Aug
NI max: 0

Greylag Geese breeding in the Hebrides and adjacent coastal areas of northwest Scotland are a remnant of the native population of this species, which once bred across much of the UK. Although once extirpated throughout the remainder of the range, the species has become re-established in many areas and is now spreading northwards towards the range of the native breeders. As a result, it is becoming increasingly difficult to separate the two groups with confidence. A full survey of Greylag Geese in Scotland is planned for summer 2008, which should help to address this issue.

The most important areas for the presumed native breeders are censused at the end of the breeding season each August, with counts repeated in many of these areas each February.

The August total of 9,530 birds this year was a considerable increase on that of August 2004 and, moreover, surveyors considered that the Uists total of 4,642 birds was probably an undercount due to poor weather conditions at the time; this opinion is backed up by the fact that a higher total of 4,689 birds were counted on the Uists in February 2006, despite overwinter mortality. The February 2006 Uists total was 17% higher than the equivalent count in February 2005. On Tiree, counts were made throughout the winter and revealed numbers to be similar to those of the last few years. Breeding success on Tiree appeared to be very low in summer 2005, with the proportion of 18.8% young well below that of the previous year and the mean brood size of 2.0 the lowest on record.

	01/02	02/03	03/04	04/05	05/06	Mon	Mean	
Sites of international importance in the UK								
Tiree	3,674 [36]	3,516 [36]	3,563 [36]	4,005 [36]	3,892 [36]	Jan	3,730	
North Uist	2,076 [17]	2,261 [17]	2,642 [17]	2,970 [51]	2,671 [51]	Aug	2,524	
South Uist	2,303 [17]	2,095 [17]	2,102 [17]	2,111 [51]	2,119 [51]	Feb	2,146	
Isle of Coll		675 [16]	740	960	980 [58]	Mar	839	
Benbecula	376 [17]	488 [17]	319 [17]	414 [51]	473 [51]	Feb	414	
Machrihanish					272 [58]	Dec	272	▲
Moine Mhor and Add Estuary					254 [58]	Mar	254	▲
Loch Gorm				4	420	Oct	212	▲
Melbost / Tong / Broad Bay	197		(4)	(86)	137	Jul	167	
Tayinloan					141 [58]	Dec	141	▲
Loch Gruinart Floods		0	3	21	509	Aug	133	▲
Kentra Moss & Lower Loch Shiel	90	93	102	136	107	Sep	106	
Branahuie Saltings	101						101	
Sites no longer meeting table qualifying levels in WeBS-Year 2005/2006								
Loch Broom					7	Feb	7	
Isle of Colonsay			116 [6]		0 [58]		58	
Loch Urrahag	(27)			31			31	

Re-established Population

GB max: 27,971   Dec
NI max: 2,235   Mar

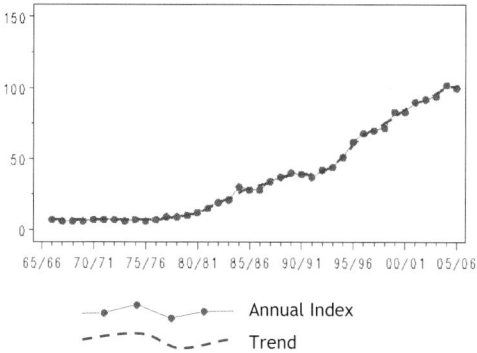

Figure 10.a, Annual indices & trend for Re-established Population for GB.

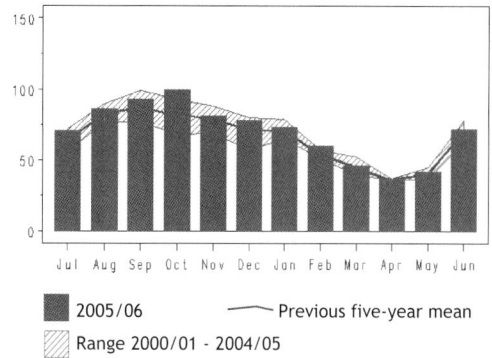

2005/06   Previous five-year mean
Range 2000/01 - 2004/05

Figure 10.b, Monthly indices for Re-established Population for GB.

Although the British index fell slightly for the first time 1991/92, the maximum count in December was the highest on record. The Northern Ireland peak count was also higher than for 2004/05 though remained around average. Thirty-two sites in Britain have mean peaks in excess of 500 birds, two more than in 2004/05, whilst a further eleven sites surpassed this level during the current WeBS year. The biggest increases were at the Swale Estuary and Hornsea Mere which both held four figure counts for the first time, whilst counts at Tophill Low Reservoirs and Lower Derwent Ings were also the highest recorded there during WeBS counts. Conversely, the peak counts at both Breydon Water and Nosterfield Gravel Pits were at their lowest since 2001/02.

In Northern Ireland, the peak count at Lough Foyle was slightly lower than that of 2004/05 although still well above the average, whilst counts from other sites in the province were about average for the five-year period. The overlap between re-established and Icelandic Greylag Geese in both Northern Ireland and Scotland remains difficult to assess, with forthcoming surveys planned to help resolve some of these issues.

	01/02	02/03	03/04	04/05	05/06	Mon	Mean
Sites with mean peak counts of 500 or more birds in Great Britain[†]							
Nosterfield Gravel Pits	1,084	1,746	(1,338)	2,215	1,663	Sep	1,677
North Norfolk Coast	1,850	1,657	(1,767)	1,371	(1,435)	Sep	1,661
Lower Derwent Ings	807	1,219	1,047	927	1,401	Jan	1,080
Tophill Low Reservoirs	1,183	828	683	867	1,400	Oct	992
The Wash	967	895	1,011	1,038	1,005	Nov	983
Eccup Reservoir	760	1,000	1,084	750	546	Dec	828
Baston and Langtoft Gravel Pits	(330)	(600)	(803)				(803)
Sutton and Lound Gravel Pits	1,057	1,176	407	950	424	Apr	803
Swale Estuary	830	760	718	625	1,062	Oct	799
Bolton-on-Swale Gravel Pits	699	1,060	710	729	774	Sep	794
Kirkby-on-Bain Gravel Pits	635	900	1,072	925	387	Oct	784
Morecambe Bay	867	741	629	786	881	Nov	781
Ouse Washes	958	691 [12]	883 [12]	782	574	Oct	778
Humber Estuary	648	1,053	(769)	821	525	Aug	763
Hornsea Mere	745	465	642	785	1,145	Jun	756
Livermere	490	806	280	1,176	879	Feb	726
Abberton Reservoir	(2,500)	80	77	278	665	Oct	720
Tattershall Pits	400	730	1,015	445	950	Jan	708
Breydon Wtr & Berney Marshes	340	723	720	1,148	491	Nov	684
Hickling Broad	412	28	1,106	831	909	Sep	657

	01/02	02/03	03/04	04/05	05/06	Mon	Mean
Lavan Sands	609	1,037	623	406	560	Sep	647
Orwell Estuary	604 [10]	587 [10]	677 [10]	(543)	(618)	Dec	623
Llyn Traffwll	700	769	891	341	395	Sep	619
Dungeness Gravel Pits	554	(502)	667	529	676	Aug	607
Martham Broad		925	750	215	500	Dec	598
Thames Estuary	(329)	(378)	(456)	593	(400)	Nov	593
Medway Estuary	(311)	(135)	(146)	589 [10]	(122)	Oct	589
Little Paxton Gravel Pits	467	746	652	518	511	Sep	579
WWT Martin Mere	438	580	600	620	530	Oct	554
Besthorpe & Girton GPs & Fleet	(236)	(4)	(331)	(539)	(388)	Jan	(539)
Alton Water	490	577	571	419	612	Feb	534
Hardley Flood		515	487	515			506
Sites with mean peak counts of 50 or more birds in Northern Ireland[†]							
Loughs Neagh and Beg	915	1,179	1,270	1,005	(630)	Feb	1,092
Lough Foyle	786	1,207	518	1,291	1,129	Mar	986
Strangford Lough	405	577	373	307	355	Mar	403
Belfast Lough	188	144	132	125	137	Nov	145
Lower Lough Erne		(71)	(54)	137	(140)	Jan	139
Ballysaggart Lough		70	66				68
Other sites surpassing table qualifying levels in WeBS-Year 2005/2006 in Great Britain[†]							
Hagnaby Fen Restoration		17	160	5	(1,004)	Sep	297
North Cave Wetlands	369	102	232	244	850	Oct	359
River Cam - Kingfishers Bridge	300	223	(193)	(248)	800	Dec	441
Llyn Alaw	275	538	557	273	765	Sep	482
Seaton Gravel Pits and River	240	185	365	110	683	Dec	317
Lackford GPs	362	151	903	249	583	Jan	450
Dee Flood Meadows	480	540	480	350	577	Nov	485
Grimsthorpe Lake	425	350	272	405	535	Sep	397
Point of Ayre Gravel Pit	140	185	402	550	530	Aug	361
Nunnery Lakes	170	208	390	329	515	Nov	322
Scaling Dam Reservoir	280	315	377	405	503	Jun	376
Other sites surpassing table qualifying levels in WeBS-Year 2005/2006 in Northern Ireland[†]							
Upper Quoile River	36	0	17	0	74	Mar	25
Upper Lough Erne	41	18	(15)	52	62	Feb	43

[†] as no British or All-Ireland thresholds have been set qualifying levels of 500 and 50 have been chosen to select sites, in Great Britain and Northern Ireland respectively, for presentation in this report

Bar-headed Goose
Anser indicus

Escape
Native Range: S Asia

GB max: 18 Sep
NI max: 0

The national maximum was slightly down on the previous year despite birds being reported from 11 more sites; 44 in total. The highest counts were of five at both Chichester Harbour in September and Spade Oak Gravel Pit (Little Marlow) in June. Kilmardinny Loch was the only site at which the species was recorded year-round. As usual, no birds were recorded in Northern Ireland.

Snow Goose
Anser caerulescens

Vagrant and escape
Native Range: N America

GB max: 58 Sep
NI max: 0

Snow Geese were recorded in every month during 2005/06. The maximum count was higher than for 2004/05 though the number of sites at which Snow Geese were recorded decreased slightly to 25. Away from the resident flock on the Isle of Coll, the highest site total again came from the Lower Windrush Valley, where 14 were recorded, whilst at Blenheim Park Lake, the maximum count decreased slightly on 2004/05 to 12 in June. Whilst most of the records doubtless refer to birds escaped from captivity, one at Loch of Strathbeg in September accompanying the large flock of Pink-footed Geese was probably a wild bird.

Ross's Goose
Anser rossii

Escape and possible vagrant
Native Range: N America

GB max: 2 Aug
NI max: 0

Up to four Ross's Geese were reported between July and September, with singles (perhaps the same bird) at Norton Marsh and Stiffkey Fen on the North Norfolk Coast in July and September respectively, plus two at Tyninghame Estuary in August. Both of these sites have hosted this species previously. There are several records from the North Norfolk Coast, the last being in June 2004, while the only previous record at Tyninghame Estuary was in August 2004.

Emperor Goose
Anser canagicus

Escape
Native Range: Alaska, NE Siberia

GB max: 23 Feb
NI max: 0

Emperor Goose was recorded from just two sites. The usual flock at South Walney Island in Morecambe Bay peaked at 23 in February. The lowest count at this site was of nine in May. The only other report of this species was of a single bird at Lackford Lakes in October and January.

Greater Canada Goose
Branta canadensis

Naturalised introduction[†]
Native Range: N America

GB max: 54,210 Dec
NI max: 968 Jan

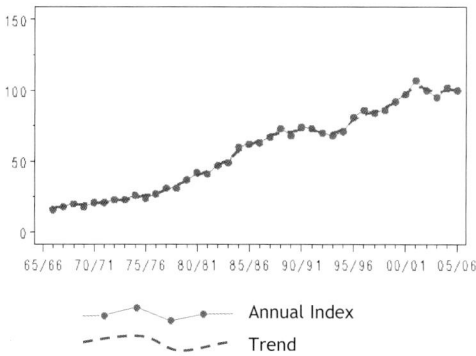

Figure 11.a, Annual indices & trend for Greater Canada Goose for GB.

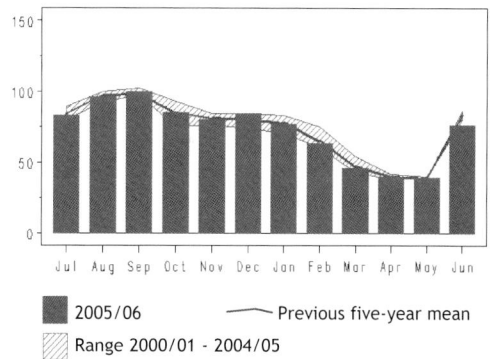

Figure 11.b, Monthly indices for Greater Canada Goose for GB.

During the past five years the British index has shown signs of stabilising following a long-term increase since the late 1960s. Throughout the year, the monthly indices reveal that numbers were extremely similar to average, as would be expected for a resident species, which is stable in numbers. However, the maximum counted total for Northern Ireland sites increased by about 50% over the previous year, with particularly significant increases at Upper and Lower Lough Erne, although the species is still absent from Loughs Neagh & Beg.

Despite the national picture of stability in Britain, there is considerable between-year variation in peak counts at individual sites. The highest single-site total in 2005/06 was at the Dyfi Estuary, this being the second highest ever recorded at a single site for WeBS. Numbers at the Dyfi Estuary have peaked at over 2,000 every year during the past five years and it is now the principal site for this species. Numbers at the Mersey Estuary have increased over the past five years, and are now at an all-time high. The September count at Rostherne Mere was exceptionally high and presumably represented a local concentration of birds more usually dispersed over several sites in the area. Conversely, there were some particularly low counts from Tring Reservoirs, Fairburn Ings and Walthamstow Reservoirs.

	01/02	02/03	03/04	04/05	05/06	Mon	Mean
Sites with mean peak counts of 600 or more birds in Great Britain[†]							
Dyfi Estuary	2,156 [10]	3,029	2,437	2,380	2,947	Dec	2,590
Dee Estuary (England & Wales)	(2,268)	(2,568)	1,529	2,316	1,987	Jan	2,134
Mersey Estuary	737	1,437	1,177	2,088	2,188	Jul	1,525
Colliford Reservoir	894	1,884	1,284	1,477	841	Jun	1,276
Arun Valley	1,550	(1,754)	860	1,235	742	Nov	1,228
Rutland Water	1,120	1,276	1,369	1,244	1,070	Oct	1,216
Taw-Torridge Estuary	(888)	1,179	526	(912)	(1,109)	Dec	923
Ellesmere Lakes	906	751	812	1,348	668	Sep	897
Alde Complex	510 [10]	(514)	(896)	1,246	780	Feb	858
Tring Reservoirs	893	962	560	1,550	308	Sep	855
Bewl Water	500	885	960	986	900	Aug	846
Abberton Reservoir	(2,000)	270	639	(616)	607	Oct	826
Southampton Water	(1,084)	609	777	(548)	(674)	Oct	823
Stour Estuary	713	983	1,135	601	622	Jan	811
Doxey Marshes SSSI	638	(637)	881	893	(601)	Dec	804
Chew Valley Lake	810	830	785	810	650	Jul	777
Harewood Lake	700	700	686	870	888	Jan	769
Cleddau Estuary	1,000	765	655	622	585 [10]	Jan	725
College Lake Reserve	476	444	773	919	973	Oct	717
Pitsford Reservoir	722	967	727	441	682	Aug	708
Somerset Levels	399	1,378	555	432	778	Dec	708
Fairburn Ings	709	823	893		381	Jan	702
Walthamstow Reservoirs	662	945	837	784	278	Dec	701
Lee Valley Gravel Pits	955	678	699	577	564	Sep	695
King`s Bromley Gravel Pits	669	712	776	721	586	Jul	693
Thames Estuary	598	(706)	(329)	786	672	Sep	691
Arlington Reservoir	823	523	703	750	642	Sep	688
Medway Estuary	(229)	(150)	(234)	365 [10]	935 [10]	Nov	650
Llangorse Lake	600	700	415	936	537	Dec	638
Humber Estuary	425	(456)	525	868	729	Sep	637
Watermead Country Park South	574	610	632	723	608	Jul	629
Carsington Water	511	848	680	500	490	Sep	606
Roadford Reservoir	507	501	611	763	650	Sep	606
Exe Estuary	445	510	617	772	680	Nov	605
Sites with mean peak counts of 50 or more birds in Northern Ireland[†]							
Lower Lough Erne		(110)	(343)	217	532	Jan	375
Upper Lough Erne	347	293	263	384	484	Feb	354
Strangford Lough	238	323	307	229	260 [10]	Dec	271
Lough McNean Lower				40	147	Jan	94
Other sites surpassing table qualifying levels in WeBS-Year 2005/2006 in Great Britain[†]							
Rostherne Mere	31	71	227	81	2,328	Sep	548
Eccup Reservoir	220	300	478	90	969	Jan	411
Dolydd Hafren	(500)	308	(500)	(500)	(800)	Jan	522
Royd Moor Reservoir				23	750	Feb	387
Hallington Reservoir	468	683	287	663	748	Sep	570
Lower Derwent Ings	564	642	594	429	712	Dec	588
Stanford Reservoir	433	357	336	454	705	Jul	457
Tamar Lakes	743		400	45	675	Sep	466
Teifi Estuary	193	380	300	294	671	Dec	368
River Trent: Stoke Lk/Radcliff viaduct	62	26	255	84	652	Sep	216
Ingbirchworth Reservoir				0	650	Mar	325
Hornsea Mere	89	285	184	510	640	Jun	342
Withington Hall Pool	67	520	620	540	635	Nov	476
Ribble Estuary	324	443	(493)	552	626	Jan	488
Bar Mere	13	480	400	473	600	Jan	393

[†] *as no British or All-Ireland thresholds have been set qualifying levels of 600 and 50 have been chosen to select sites, in Great Britain and Northern Ireland respectively, for presentation in this report*

Lesser Canada Goose
Branta hutchinsii

Vagrant and escape
Native Range: N America

GB max: 1 Oct
NI max: 0

This is only the second year that Lesser Canada Goose has featured as a full species following the recent taxonomic review in 2005 (Sangster *et al.* 2005). The only record of Lesser Canada Goose was of a single bird at King`s Bromley Gravel Pits in October and was presumably an escape.

Hawaiian Goose
Branta sandvicensis

Escape
Native Range: Hawaii

GB max: 1 Jan
NI max: 0

A single Hawaiian Goose was at Leisure Lakes, Lancashire in January. This was a new WeBS species for this site, the only other records have been of singles at Harewood Lake in 2003/04 and Lower Derwent Ings in 1996/97.

Barnacle Goose
Branta leucopsis

Greenland Population

International threshold: 560
Great Britain threshold: 450
All-Ireland threshold: 75

GB max: 56,383 Mar
NI max: 0

Barnacle Geese breeding along the east coast of Greenland winter exclusively in northwest Scotland and Ireland. Due to the dispersed nature of the winter population, a full census is carried out every five years (the next due in spring 2008). However, many of the main resorts are surveyed every year, largely by SNH and the Uists Greylag Goose Management Committee. Such counts showed that numbers in Argyll and on the Uists continue to increase, whilst those on Orkney (at South Walls) were more stable. The key site remains Islay, with the spring count of 47,303 by far the highest ever recorded for the island. There were also notably high counts on Coll and Tiree. On North Uist, coordinated counts in February revealed a total of 4,659 birds (with 1,327 on Berneray alone), over 60% higher than equivalent counts the previous year. At the Dyfi Estuary in west Wales, the regular wintering flock of birds assumed to be from this population peaked at 165 birds, a slight rise on the low number present in 2004/05. The breeding success of Greenland Barnacle Geese was assessed at Islay and Tiree in November 2005, with an average of 6.6% young and a mean brood size of 1.75 young per successful pair. This represented a poor breeding season, perhaps linked to low rodent abundance (as shown in www.arcticbirds.ru) in its Arctic breeding areas leading to high levels of predation.

	01/02	02/03	03/04	04/05	05/06	Mon	Mean
Sites of international importance in the UK							
Island of Islay	35,213 [16]	36,478 [37]	40,018 [6]	44,186 [6]	47,303 [58]	Mar	40,640
North Uist	3,326 [17]	2,732 [17]		2,836 [51]	4,649 [51]	Feb	3,386
Tiree	2,132 [36]	2,786 [36]	2,796 [36]	3,273 [36]	3,474 [36]	Jan	2,892
South Walls (Hoy)	2,600 [37]	1,800 [37]			2,901 [37]	Mar	2,434
Isle of Coll	933 [36]	1,010 [36]	792 [6]	1,297	2,240 [58]	Mar	1,254
Balnakeil Bay			826 [12]		970	Fcb	898
Colonsay/Oronsay		510 [36]	793 [6]	1,000 [6]			768
Sound of Harris (NW) (Harris)		706 [35]					706
North Sutherland		669 [35]					669
Sites of national importance in Great Britain							
Keills Peninsula & Isle of Danna	420 [6]	400 [6]	640 [6]	708 [6]	468 [58]	Dec	527
Sites no longer meeting table qualifying levels in WeBS-Year 2005/2006							
Rispond Bay				550	0		275

Svalbard Population

GB max: 28,920 Nov
NI max: 0

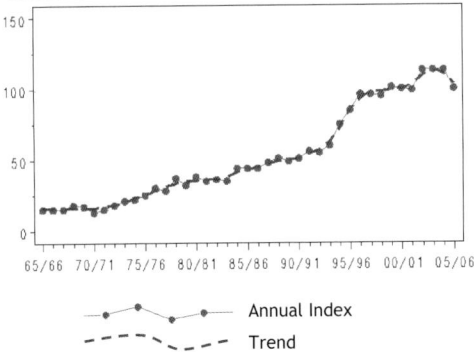

Figure 12.a, Annual indices & trend for Svalbard Population for GB.

The peak combined total for the Solway Firth, the wintering grounds for the vast majority of Svalbard-breeding Barnacle Geese, reached a new record high (albeit by only three birds!) However, this count was carried out over two consecutive dates and it is feared that some double-counting may have occurred. The "adopted population estimate" for the Solway was therefore calculated as 23,900 birds (taken as the average of the next highest count and all those within 10% of it). This represented a decrease of 11% (about 3000 birds) compared to the adopted population estimate of the previous year.

The first flock arrived on the Solway on 5th October 2006, but the main arrival was late, with many grounded in Norway, Shetland and at Loch of Strathbeg on their way. Most birds had arrived on the Solway by early November, however, and most remained until late April, with some into early May. WeBS Core Counts suggested that Rockcliffe Marsh was the most important area of the Solway throughout the winter, with other large flocks noted at Calvo Marsh, Herdhill Scar to Cardurnock and Anthorn-Newtown, as well as at Auchencairn Bay. Within the Solway flocks, the proportion of young birds was estimated as 7.9%, with 2.5 young per successful pair on average. The brood size estimate was the highest since 1980 but the percentage of young, whilst at its highest value since 2002, was still lower than the ten-year mean.

As usual, the only other sites supporting substantial numbers of Barnacle Geese assumed to be of this population were Loch of Strathbeg and Lindisfarne. At the former, a large flock was present for the October WeBS count and several hundred birds remained there throughout the winter. Conversely, at Lindisfarne there were flocks of 100 in September and 300 in December, but only very small numbers in other winter months.

	01/02	02/03	03/04	04/05	05/06	Mon	Mean
Sites of international importance in the UK							
Solway Firth	23,524 [7]	28,447 [7]	27,510 [7]	28,270 [7]	28,450 [7]	Nov	27,240
Loch of Strathbeg	10,390 [36]	276	190	1,100 [36]	4,336	Oct	3,258
Lindisfarne	(800)	(280)	1,572	320	600	Dec	831

Barnacle Geese (Rob Robinson)

Naturalised Population

GB max: 958   Sep
NI max: 251   Sep

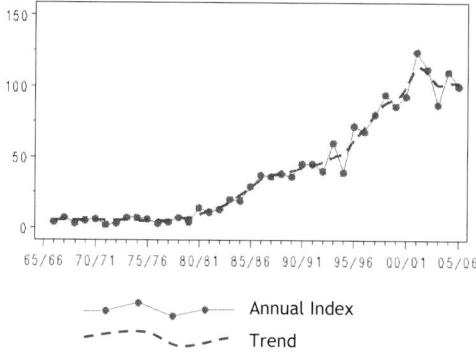

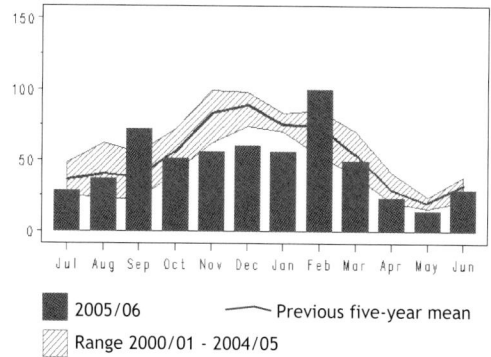

------•------ Annual Index

- - - - - - Trend

■ 2005/06   ⌒ Previous five-year mean

▨ Range 2000/01 - 2004/05

Figure 13.a, Annual indices & trend for Naturalised Population for GB.

Figure 13.b, Monthly indices for Naturalised Population for GB.

The British index for naturalised Barnacle Goose has fluctuated in recent years and saw a slight drop from 2004/05, although numbers remain high. The peak count was the second highest maximum count recorded by WeBS, exceeded only by the record peak of 2004/05. The number of WeBS sites at which presumed naturalised Barnacle Geese were recorded rose to nearly 250 with the peak counts at Lound Waterworks, Derwent Water, Frampton Pools, Minsmere and Ullswater being the highest on record for these sites. Conversely, two sites formerly supporting high numbers - Hornsea

Mere and Eversley Cross & Yateley Gravel Pits - have both seen a steady decline in numbers over the past five years. The peak count for Northern Ireland, coming in September, was comprised entirely of birds at Strangford Lough.

Counts of Barnacle Geese are assigned as naturalised birds purely on geographical location, and as a result, some extralimital birds from the Svalbard and Greenland populations may have been incorrectly assigned.

	01/02	02/03	03/04	04/05	05/06	Mon	Mean
Sites with mean peak counts of 50 or more birds in Great Britain[†]							
Lound Waterworks					393	Jan	393
Willington	115	84	298				166
Roxton Gravel Pits	105	107	262	120	195	Aug	158
Eversley Cross & Yateley GPs	236	219	158	107	62	Dec	156
Duddon Estuary	(150)	(1)	(65)	(0)	(88)	Oct	(150)
Humber Estuary	(53)	(74)	80	(200)	88	Feb	123
Benacre Broad	42	120	250	130	52	Sep	119
Hornsea Mere	202	132	96	73	71	Sep	115
Middle Yare Marshes	141	104	72	82	74	Dec	95
Severn Estuary	73	96	(94)	101	111	Jan	95
Derwent Water	61	90	82	98	105	Jun	87
Frampton Pools	(75)	79	98	52	113	Mar	86
Minsmere	29	62	73	4	249	Sep	83
Ullswater	0	2	135	110	143	Feb	78
The Hen Reedbeds	(0)		(0)	(0)	(68)	Feb	(68)
Barcombe Mills Reservoir	76	64	73	52	47	Aug	62
Sites with mean peak counts of 50 or more birds in Northern Ireland[†]							
Strangford Lough	214	223	232	248	251	Sep	234
Other sites surpassing table qualifying levels in WeBS-Year 2005/2006 in Great Britain[†]							
Dungeness Gravel Pits	13	22	15	31	136	Feb	43
Osmaston & Shirley Park Lakes	2	2	1	23	53	Sep	16

[†] *as no British or All-Ireland thresholds have been set a qualifying level of 50 has been chosen to select sites for presentation in this report*

Dark-bellied Brent Goose
Branta bernicla bernicla

International threshold: 2,000
Great Britain threshold: 981
All-Ireland threshold: +

GB max: 84,681 Feb
NI max: 0

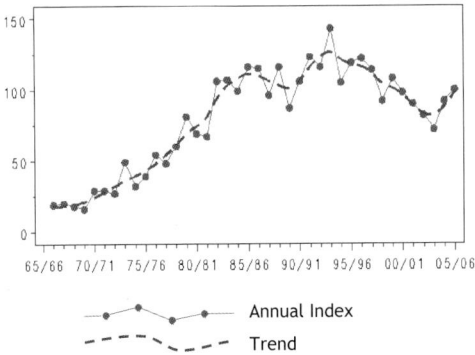

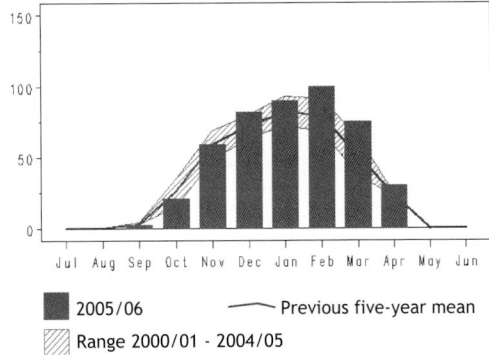

- Annual Index
- Trend

■ 2005/06
⌢ Previous five-year mean
▨ Range 2000/01 - 2004/05

Figure 14.a, Annual indices & trend for Dark-bellied Brent Goose for GB.

Figure 14.b, Monthly indices for Dark-bellied Brent Goose for GB.

The national maximum count of Dark-bellied Brent Geese was only slightly lower than during the previous year and typically for this species peak numbers were recorded during the latter half of the winter. During the ten years to 2003/04 the national index for Dark-bellied Brent Geese fell steadily by a third and dipped to a level akin to that of 1980/81. However, numbers have shown a revival over the past two years and in 2005/06 the index rose by around ten percent, following the excellent breeding season in 2005. Initial numbers, during the early part of the winter, were similar to average, but rose in December

and January and from February onwards were the highest of the past five years.

In the five-years to 2005/06 thirteen sites held internationally important numbers and a further thirteen sites held nationally important numbers. The majority of the key sites for this species are on the south and east coasts of England with only Burry Inlet elsewhere. The count at The Wash was the highest since December 1999, whilst numbers at Chichester and Langstone Harbours, and at Newtown Estuary on the Isle of Wight, were also higher than in recent years. The count at Lindisfarne was the highest ever at this site being over twice that of the previous peak.

	01/02	02/03	03/04	04/05	05/06	Mon	Mean
Sites of international importance in the UK							
The Wash	17,924	20,314	18,734	21,969	24,490	Feb	20,686
Thames Estuary	12,157	(8,908)	(6,741)	9,455	12,567	Oct	11,393
Chichester Harbour	7,470	7,358	8,290	7,436	9,018	Mar	7,914
North Norfolk Coast	8,033	9,180	5,722	6,607	8,831	Feb	7,675
Blackwater Estuary	(7,195)	6,100	4,892	7,178	5,946	Dec	6,262
Hamford Water	4,331	3,567	3,336	5,890	5,952	Jan	4,615
Langstone Harbour	4,813	4,686	1,765	5,069	5,496	Jan	4,366
Crouch-Roach Estuary	3,471	3,083	2,914	4,635 [10]	3,520	Mar	3,525
Portsmouth Harbour	(1,682)	(2,185)	(2,293)	(1,725)	2,925	Feb	2,925
Colne Estuary	2,572	(409)	(1,959)	(2,538)	(2,123)	Dec	2,572
Pagham Harbour	3,178	2,252	1,210	2,654	2,819	Jan	2,423
Humber Estuary	1,432	(2,351)	2,118 [10]	(2,667)	(2,636)	Feb	2,241
North West Solent	2,350	1,500	1,790	(2,208)	2,377	Feb	2,045 ▲
Sites of national importance in Great Britain							
Newtown Estuary	1,660	1,779	(1,235)	(1,444)	2,033	Dec	1,824
Dengie Flats	1,798	1,160	1,507	(1,538)	2,445	Dec	1,728
Stour Estuary	1,412	1,753	1,914	1,782	1,617	Apr	1,696
Deben Estuary	2,218	1,251	2,234	984	(1,449)	Feb	1,672
Swale Estuary	1,690 [10]	1,278	1,210	2,111	1,861	Jan	1,630

	01/02	02/03	03/04	04/05	05/06	Mon	Mean
Beaulieu Estuary	2,015	1,512	835	1,498	2,173	Feb	1,607
Fleet and Wey	2,188	398	1,337	2,625	1,436	Dec	1,597
Exe Estuary	1,183	1,714	1,368	1,645	1,531	Feb	1,488
Medway Estuary	(1,725)	(1,179)	836	1,834 [10]	(1,515)	Feb	1,478
Southampton Water	1,455	1,326	1,274	1,386	(783)	Mar	1,360
Orwell Estuary	1,215 [10]	1,525 [10]	1,396 [10]	976	1,477	Jan	1,318
Poole Harbour	(599)	(740)	(868)	(491)	1,160	Feb	1,160
Burry Inlet	1,174	917	(1,255)	811	1,121	Mar	1,056

Light-bellied Brent Goose
Branta bernicla hrota

East Canadian High Arctic Population

International threshold: 260
Great Britain threshold: +[†]
All-Ireland threshold: 200

GB max: 151 Apr
NI max: 23,047 Oct

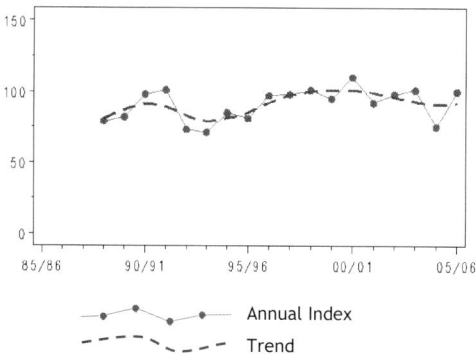

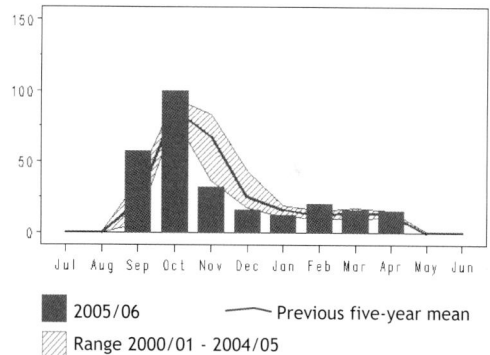

Figure 15.a, Annual indices & trend for East Canadian High Arctic Population for NI.

Figure 15.b, Monthly indices for East Canadian High Arctic Population for NI.

The all-Ireland Light-bellied Brent Goose Census was carried out at all known sites in mid-October 2005, including aerial coverage of staging areas in western Iceland, and located a total of 31,442 geese, about 5% lower than in 2004. Only about 1,500 were still in Iceland at this time, with the largest numbers by far being at Strangford Lough as usual. The peak count there was down somewhat on that of the previous year but still the second highest ever at the site. October counts at Lough Foyle revealed higher numbers than usual however. The monthly indices reveal larger than usual numbers in Northern Ireland in September and October, followed by a very rapid dispersal and lower numbers than usual from November to January. Interestingly, all of the other key Northern Ireland sites saw peak numbers in the late winter, with a record count at Dundrum Bay in March. The overall percentage of young in October 2005 was assessed as 12.9%, a relatively low value and presumably

contributing to the small decline in the overall population size. There was an average of 2.95 young per successful pair.

In Britain, following the discussion in *Waterbirds in the UK 2004/05* about a September arrival in the Hebrides, the same was seen again in autumn 2005. Following a peak WeBS count of 76 in September at Loch Gruinart on Islay, up to 1,000 were recorded on the island in late September and early October. By the WeBS count on 17th October 2005 only 28 remained and most had presumably moved on to Strangford by this date. Small numbers were also noted in September and October on Skye, Tiree and along the Ayrshire coast. Peak counts from sites around the Irish Sea, as well as the Channel Islands, were later in the winter following the main dispersal from Strangford Lough. The peaks at the Dee Estuary, Loch Ryan and the Guernsey Shore were the highest on record for each of these sites.

	01/02	02/03	03/04	04/05	05/06	Mon	Mean
Sites of international importance in the UK							
Strangford Lough	19,583 [32]	17,520 [32]	21,500 [32]	26,250 [32]	21,885 [32]	Oct	21,348
Lough Foyle	1,841	1,563 [32]	3,277 [32]	1,603 [32]	3,968	Oct	2,450
Outer Ards Shoreline	210 [32]	700	642	762	618	Mar	586
Carlingford Lough	(259)	319	(570)	538	508	Mar	484
Killough Harbour	489 [10]	472	383	434	516	Mar	459
Dundrum Bay	320	242	188	302	640	Mar	338
Sites of all-Ireland importance in Northern Ireland							
Larne Lough	235	139	235	254	218	Feb	216
Sites with mean peak counts of 25 or more birds in Great Britain[†]							
Traeth Melynog			117	146			132
Jersey Shore	127						127
Loch Gruinart		2	0	284	76	Sep	91
Inland Sea, Beddmanarch & Alaw Est	80	76	95				84
Dee Estuary (England & Wales)	32 [10]	25	66	121	138	Mar	76
Foryd Bay	43	9	96	115	54	Apr	63
Loch Ryan	28 [12]	25 [12]	45	67	89	Dec	51
Derbyhaven Bay				39			39
Morecambe Bay	7	62 [10]	53	31	22	Dec	35
Guernsey Shore	(0)	19	(0)	(0)	35	Mar	27

[†] as no British threshold has been set a qualifying level of 25 has been chosen to select sites for presentation in this report

Svalbard Population

International threshold: 70
Great Britain threshold: 30*

GB max: 3,756 Nov
NI max: 0

*50 is normally used as a minimum threshold

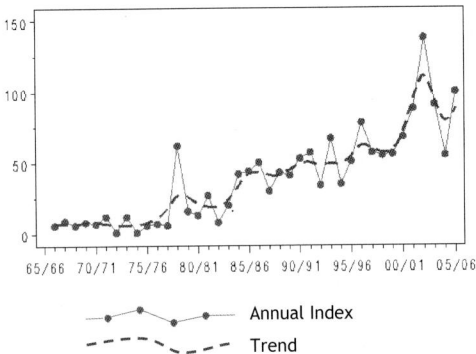

Figure 16.a, Annual indices & trend for Svalbard Population for GB.

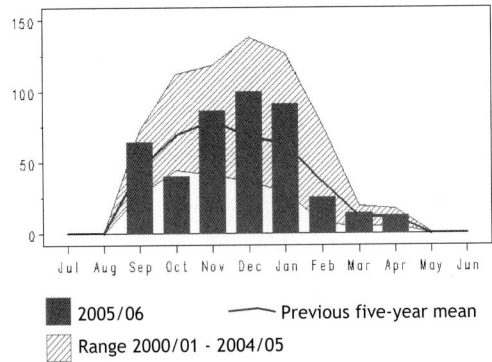

2005/06
Range 2000/01 - 2004/05
Previous five-year mean

Figure 16.b, Monthly indices for Svalbard Population for GB.

The maximum count of 3,756 Svalbard-breeding Light-bellied Brent Geese in 2005/06 was almost twice that seen during the previous winter but approximately the same as in 2003/04. The general rise in the population size recorded over several decades, therefore, appears to be maintained. Monthly indices suggest that the principal arrival was in September, but numbers rose to peak around New Year. However most birds departed Britain between January and February, presumably to their spring staging grounds in Denmark, with only a few hundred remaining into the spring. As ever, the vast majority of this population occurring in Britain winters at Lindisfarne. Geese at this site were aged on three occasions over the course of the 2005/06 winter. The proportion of young in the flocks was again low at 6.5%, with a mean brood size of 2.12 young per successful pair.

Away from Lindisfarne, Light-bellied Brent Geese found along the east coast are generally assumed to relate to this population, whereas west coast birds tend to be assigned to the East Canadian High Arctic population, although the division is somewhat artificial. During the

2005/06 winter, at sites away from Lindisfarne numbers peaked in February, presumably as part of the dispersal from the key site. The peak on the Moray Firth was the highest for three years, and five-year peak mean numbers at Seahouses to Budle Point exceeded the threshold for national importance. The flock of 30 birds at St Andrews Bay in February was the first WeBS record at this site since January 1987.

	01/02	02/03	03/04	04/05	05/06	Mon	Mean
Sites of international importance in the UK							
Lindisfarne	(4,845)	(3,150)	3,716	2,505 [10]	3,688	Nov	3,689
Sites of national importance in Great Britain							
Inner Moray and Inverness Firth	41	100	55	18	81	Feb	59
Seahouses to Budle Point	70	0	(0)	23	48	Feb	35 ▲
Other sites surpassing table qualifying levels in WeBS-Year 2005/2006 in Great Britain							
St Andrews Bay	0	0	0	0	30	Feb	6

Black Brant
Branta bernicla nigricans

Vagrant
Native Range: N America and E Asia

GB max: 4 Jan
NI max: 0

Most records of Black Brant were from the Solent area, with records from Portsmouth, Langstone, Chichester and Pagham Harbours between November and March. Away from the south coast, single birds were at Holkham and Burnham Overy Fresh Marshes (North Norfolk Coast) in January and Capel Fleet (Swale Estuary) in February, with two at Titchwell (North Norfolk Coast) in February.

Red-breasted Goose
Branta ruficollis

Vagrant and escape
Native Range: SE Europe, Asia

GB max: 2 Jul
NI max: 0

Red-breasted Geese were recorded at just three sites, all of these records likely to refer to escaped birds. A maximum of two were at Diss Mere between July and December. Singles were also recorded at Clifford Hill Gravel Pits in November and The Hen Reedbeds in February.

Egyptian Goose
Alopochen aegyptiaca

Naturalised introduction
Native Range: Africa

GB max: 384 Sep
NI max: 0

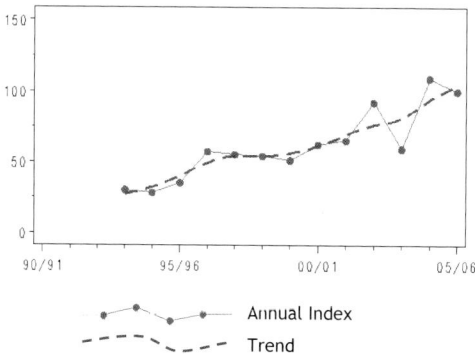

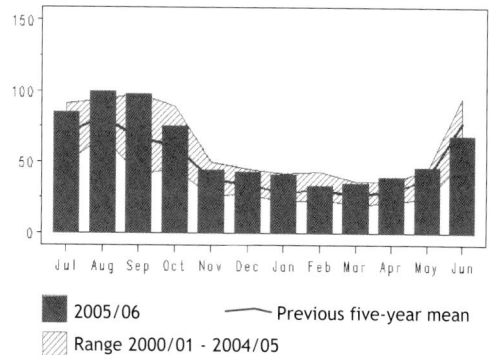

Figure 17.a, Annual indices & trend for Egyptian Goose for GB.

Figure 17.b, Monthly indices for Egyptian Goose for GB.

The British peak was somewhat lower than that of the previous year and this was reflected in the national index. Despite this slight fall, the underlying trend is one of increase, which has been the case since at least 1993/94. Egyptian Goose numbers were high throughout the year, particularly during autumn when numbers surpassed those of the previous five-year period.

East Anglia remains the most important area for this species with 14 out of the 17 key sites being in this region. Numbers at the principal site, the North Norfolk Coast, were the same as the low figure of two years ago; however, these figures have been influenced by incomplete coverage of the key location – Holkham Lake - during the past three years. Numbers at Breydon Water have been increasing over recent years and in 2005/06 the WeBS total was the highest ever recorded at the site. Similarly, record site totals were also received from Eversley Cross and Yateley Gravel Pits, East Wretham Meres, Tundry Pond, Tattershall Pits and Thorpe Water Park.

	01/02	02/03	03/04	04/05	05/06	Mon	Mean
Sites with mean peak counts of 10 or more birds in Great Britain[†]							
North Norfolk Coast	318	233	(126)	(144)	(126)	Oct	276
Sennowe Park Lake Guist		98	85				92
Breydon Wtr & Berney Marshes	48	63	65	82	85	Aug	69
Rutland Water	60	58	70	46	53	Sep	57
St Benet's Levels	52	88	23				54
Middle Yare Marshes	45	72	24	(47)	(26)	Oct	47
Nunnery Lakes	19	21	51	36	31	Oct	32
Spade Oak GP (Little Marlow)	23	33	6	37	49	Sep	30
Weybread Pits	18	31	30	41			30
Cranwich Gravel Pits	19	34					27
Whitlingham Country Park	10	7	18	59	27	Jul	24
Trinity Broads	18	20	10	22			18
Hickling Broad	0	0	(0)	21	42	Oct	16
Lound Waterworks					16	Sep	16
Barton Broad	15	18	14	13	16	Aug	15
Ampton Water		2	25				14
Stanford Training Area	13						13
Other sites surpassing table qualifying levels in WeBS-Year 2005/2006 in Great Britain[†]							
Eversley Cross & Yateley GPs	0	2	9	6	24	Mar	8
East Wretham Meres	2	3	2	2	14	Aug	5
Tundry Pond	0	0	0	4	13	Sep	3
Tattershall Pits	3	4	7	10	11	Aug	7
Thorpe Water Park	(4)	(4)	4	4	11	Jan	6

[†] as no British or All-Ireland thresholds have been set a qualifying level of 10 has been chosen to select sites for presentation in this report

Orinoco Goose

Escape

Neochen jubata

Native Range: S America

GB max: 1 Oct
NI max: 0

A single Orinoco Goose was recorded at Roath Park Lake in October and December. This was the first time that this escaped species has been recorded by WeBS.

Cape Shelduck

Escape

Tadorna cana

Native Range: S Africa

GB max: 3 Oct
NI max: 0

Cape Shelduck were reported at just two sites in 2005/06. Three were at Cley Marsh, North Norfolk Coast in October and a single bird was at Crosslanes, Shropshire the following March. This species had not previously been recorded during WeBS at either of these sites.

Ruddy Shelduck
Tadorna ferruginea

Escape and possible vagrant
Native Range: Asia, N Africa, S Europe

GB max: 22   Oct
NI max: 0

Ruddy Shelduck was recorded in ten months and from 13 sites during 2005/06. The British maximum of 22 was over twice that of the previous year. Half of the reports were of single birds and a further nine were of two birds. All reports of more than two birds were from the North Norfolk Coast. These were of three at Holkham and Burnham Overy Fresh Marshes in January, and a record WeBS count of 20 at Cley Marsh in October. It is most likely that the majority of Ruddy Shelduck records originate from naturalised populations in mainland Europe. For example, numbers have increased dramatically in Switzerland over the last decade, with over 300 recorded during waterbird counts there in November 2005, mostly in the northern half of the country (V. Keller *pers. comm.*)

Shelduck
Tadorna tadorna

International threshold: 3,000
Great Britain threshold: 782
All-Ireland threshold: 70

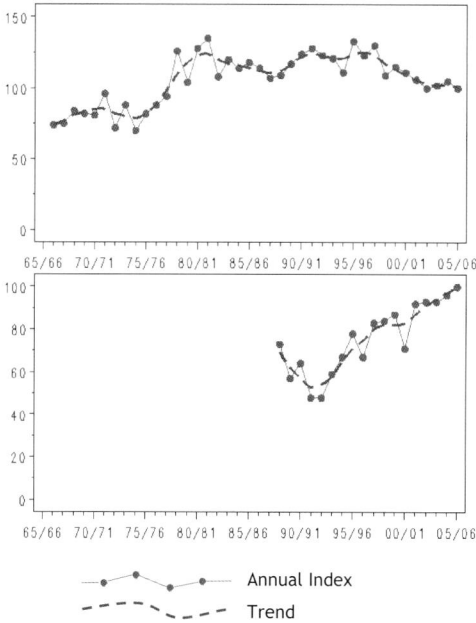

GB max: 46,997   Dec
NI max: 5,514   Dec

Figure 18.a, Annual indices & trend for Shelduck for GB (above) & NI (below).

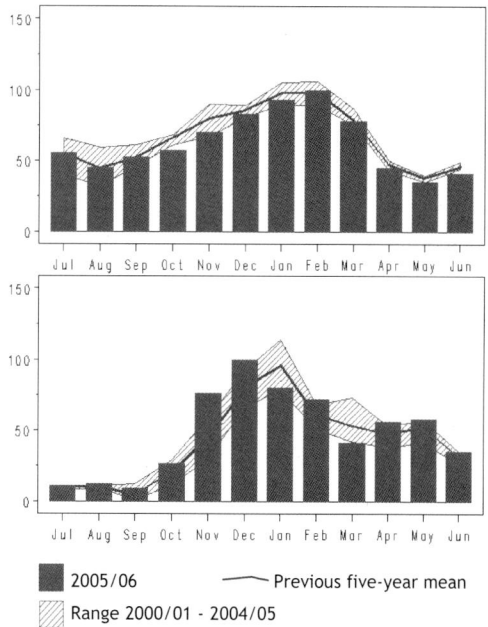

Figure 18.b, Monthly indices for Shelduck for GB (above) & NI (below).

The British Index for Shelduck dropped just slightly in 2005/06, whilst the maximum count was at its lowest for 30 years. The Northern Ireland index, however, continues to rise as it has done steadily since the low in the early 1990s. The British monthly index shows a peak in February following a gradual increase through the winter as many birds return from their moulting grounds in northwest Germany.

The Mersey Estuary remained the key site for the species, with the peak count again coming during the moult period in July, whilst counts on the adjacent Dee Estuary peak in the autumn, and there is presumably a certain degree of redistribution between these two sites. Counts from the other sites supporting internationally important numbers were about in line with their respective five-year means, with the exception of the Wash where numbers continue to fall annually. A large proportion of the birds recorded in Northern Ireland come from Strangford Lough where the low tide peak count was slightly above average for the period.

	01/02	02/03	03/04	04/05	05/06	Mon	Mean
Sites of international importance in the UK							
Mersey Estuary	5,740	19,810	17,823	13,420	15,605	Jul	14,480
Dee Estuary (England & Wales)	10,200	10,533	12,630	13,334	(8,872)	Oct	11,674
The Wash	11,783	7,834	7,341	7,451	6,904	Feb	8,263
Morecambe Bay	6,137	7,164	8,228	7,728	(6,609)	Nov	7,314
Humber Estuary	3,655	(4,819)	6,426 [10]	(4,188)	5,223	Aug	5,101
Solway Estuary	(2,213)	(4,324)	3,131	5,359	(858)	Oct	4,271
Strangford Lough	4,162	4,199 [10]	4,475	3,801	4,451 [10]	Nov	4,218
Severn Estuary	3,776	3,495 [10]	2,579	3,460	4,482	Dec	3,558
Ribble Estuary	3,190	3,063	3,829	3,850	2,935	Sep	3,373
Forth Estuary	(2,920)	3,531	3,452	3,164 [11]	3,063	Jul	3,303
Sites of national importance in Great Britain							
Thames Estuary	2,940	3,285	1,584	(2,318)	(1,295)	Jan	2,603
Blackwater Estuary	(1,808)	2,572	1,904	2,073	(1,828)	Mar	2,183
Medway Estuary	(2,045)	(1,257)	(2,177)	2,360 [10]	1,949	Dec	2,162
Swale Estuary	2,342	2,290	1,818	2,207	2,138	Feb	2,159
Poole Harbour	(2,221)	2,385	(2,072)	1,547	(1,857)	Jan	2,056
Hamford Water	1,737 [10]	1,903	1,657	1,951	1,493	Jan	1,748
Crouch-Roach Estuary	(478)	(385)	(342)	1,661 [10]	(397)	Feb	1,661
Lindisfarne	1,546 [10]	1,826	1,323 [10]	1,773 [10]	1,180 [10]	Nov	1,530
North Norfolk Coast	2,012 [10]	1,182	1,112	1,110	1,283	Dec	1,340
WWT Martin Mere	950	1,435	1,150	1,510 [11]	965	Jan	1,202
Stour Estuary	1,258	805	957	1,296	(1,421)	Feb	1,147
Montrose Basin	776	1,191	(1,240)	690	1,239 [10]	Jan	1,027
Alde Complex	881	945	1,124	1,025	925	Feb	980
Colne Estuary	920	(263)	(804)	(701)	(471)	Dec	920
Chichester Harbour	1,014 [10]	1,019	810	825	793	Feb	892
Deben Estuary	676	864	802	883	707	Jan	786
Sites of all-Ireland importance in Northern Ireland							
Larne Lough	776	637	633	808	880	Feb	747
Carlingford Lough	(365)	493	423	452	560	Jan	482
Belfast Lough	437	199 [10]	494 [10]	489 [10]	347 [10]	Feb	393
Lough Foyle	536	232	(315)	250	392	Nov	353
Loughs Neagh and Beg	102	146	205	260	98	Feb	162
Dundrum Bay	93	99	138	330	81	Dec	148
Bann Estuary	138	87	104	92	86	Apr	101
Sites no longer meeting table qualifying levels in WeBS-Year 2005/2006							
Burry Inlet	963	570	(847)	804	637	Dec	764

Muscovy Duck

Cairina moschata

Escape[†]

Native Range: S America

GB max: 65 Nov
NI max: 0

Muscovy Ducks were recorded at 31 sites in 2005/06, all but four of which were in England, the remainder being in Scotland. The British maximum was just five below that of 2004/05. The highest single-site total of 25 at Fort Henry Ponds and Exton Park Lakes and was the highest ever at this site. The summed site maximum of 97 was slightly below that of the previous year.

	01/02	02/03	03/04	04/05	05/06	Mon	Mean
Sites with mean peak counts of 5 or more birds in Great Britain[†]							
Fort Henry Ponds & Exton Pk Lks	14	0	0	14	25	Nov	11
Wilderness Pond	10	12	7				10
Par Sands Pools & St Andrews Rd	20	6	8	6	4	Sep	9
Derwent Water	10	6	6	11	7	Oct	8
High Batts Recording Area				8	8		8
Other sites surpassing table qualifying levels in WeBS-Year 2005/2006 in Great Britain[†]							
Ewell Court Pond		0	1 [12]	3	6	Nov	3
Cotswold Water Park (West)	0	0	0	0	5	Sep	1

[†] *as no British or All-Ireland thresholds have been set a qualifying level of 5 has been chosen to select sites for presentation in this report*

Wood Duck
Aix sponsa

Escape
Native Range: N America

GB max: 7   Jun
NI max: 1   Mar

Wood Ducks were recorded at ten sites in total, seven in England and one each in Scotland, Northern Ireland and the Channel Islands. The highest single-site total was six at Stanton Lake in June. Peak counts of two were noted at Mere Sands Wood Nature Reserve in October and River Avon – Salisbury in November, January and February. All other records were of single birds. The record of a single bird in Belfast Lough in March was the first time that this species was recorded by WeBS in Northern Ireland.

Mandarin
Aix galericulata

Naturalised introduction[†]
Native Range: E Asia

GB max: 448   Jan
NI max: 0

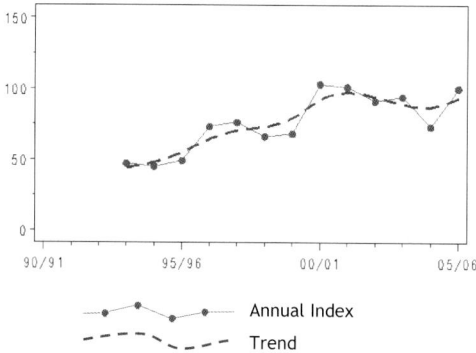

Figure 19.a, Annual indices & trend for Mandarin for GB.

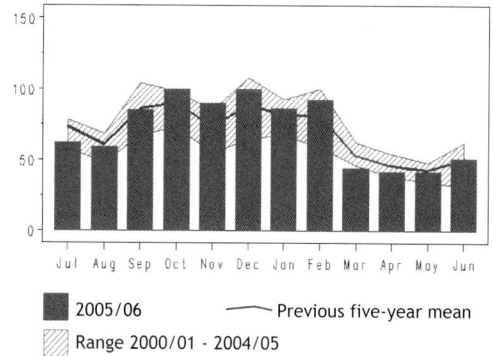

Figure 19.b, Monthly indices for Mandarin for GB.

Following a decline over the past five years, the British index showed a sharp rise in 2005/06 to a peak similar to the record peak in 2001/02. The British maximum count was also high and the monthly indices were above average for most of the winter months. There were two three-figure counts, from Bradley Ponds and Headley Mill Pond, but the supplementary counts from the combined Forest of Dean Ponds were much lower than usual (although there was only a single survey in December). Typically, peak counts fluctuate between years, but other sites recording notably high peak counts were Darwell Reservoir, Arun Valley, Allestree Park Lakes, Osmaston and Shirley Park Lakes, Kedleston Park Lake and Blackbrook Reservoir. Again, none were recorded in Northern Ireland.

	01/02	02/03	03/04	04/05	05/06	Mon	Mean
Sites with mean peak counts of 10 or more birds in Great Britain[†]							
Forest of Dean Ponds	159 [12]	120 [12]	160 [12]	221 [15]	66 [15]	Dec	145
Bradley Pools	85	55	188	65	144	Jan	107
Headley Mill Pond	70	76	76	23	132	Feb	75
Cuttmill Ponds	98	51	59	61	66	Oct	67
Wraysbury Pond	78	63 [11]	61	51	48	Dec	60
Busbridge Lakes	54	47	72				58
Stockgrove Country Park	54	70	43				56
Bough Beech Reservoir	77 [46]	33 [46]	56 [46]	60 [46]	45 [46]	Oct	54
Dee Flood Meadows	79	49	32	42	36	Sep	48
Darwell Reservoir	43	25	56	13	58	Oct	39
Passfield Pond	14	67	73	16	15	Feb	37
Arun Valley	28	41	32	27	47	Dec	35
Connaught Wtr (Epping Forest)	26	31	44	32	27	Nov	32

	01/02	02/03	03/04	04/05	05/06	Mon	Mean
River Thames at Staines Bridge		31					31
Harewood Lake	11	53	35	31	15	Sep	29
Sutton Place	44	21	32	20	4	Sep	24
Fonthill Lake	17	18	20	38	22	Feb	23
Lost/Golding & Baldwins Hill Pnds	10	78	5	12	1	Aug	21
Paultons Bird Park	21	20					21
Severn Estuary	65	28	3	4	5	Sep	21
Osterley Park Lakes	13	20	19	31	18	Jul	20
Strawberry Hill Ponds	7	30	23	15	11	Nov	17
Linacre Reservoirs	5	14	17	23	23	Nov	16
Panshanger Estate	16	24	12	11	16	Oct	16
Woburn Park Lakes	12	24	8	6	18	Jun	14
Allestree Park Lakes		4	5	5	37	Dec	13
Swanbourne Lake	4	13	5	18	18	Aug	12
Eversley Cross & Yateley GPs	9	6	9	25	4	Jan	11
Other sites surpassing table qualifying levels in WeBS-Year 2005/2006 in Great Britain[†]							
Osmaston & Shirley Park Lakes	0	2	0	0	28	Sep	6
Kedleston Park Lake	0	1	0	14	24	Jan	8
Blackbrook Reservoir	1	0	2	2	17	Nov	4
River Test - Broadlands Estate	6	9	4	9	15	Mar	9
Hampstead and Highgate Ponds	4	5	9	8	13	Oct	8
Longueville Marsh	0	0	6	25	13	Dec	9
Aldenham Reservoir	0		3	5	11	Jan	5
Buckhurst Hill Ponds	0	11	11	2	10	Nov	7
River Test: Fullerton/Stockbridge	8	2	0	2	10	Jan	4

[†] as no British or All-Ireland thresholds have been set a qualifying level of 10 has been chosen to select sites for presentation in this report

Wigeon
Anas penelope

International threshold: 15,000
Great Britain threshold: 4,060
All-Ireland threshold: 1,250

GB max: 441,754 Dec
NI max: 10,288 Oct

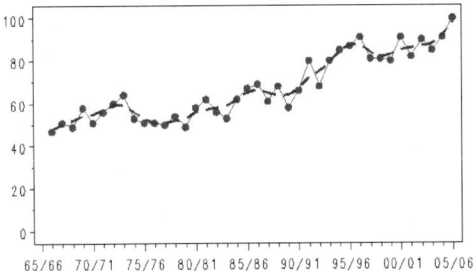

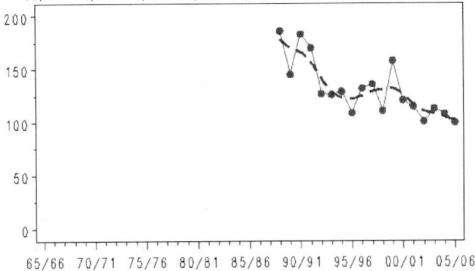

Figure 20.a, Annual indices & trend for Wigeon for GB (above) & NI (below).

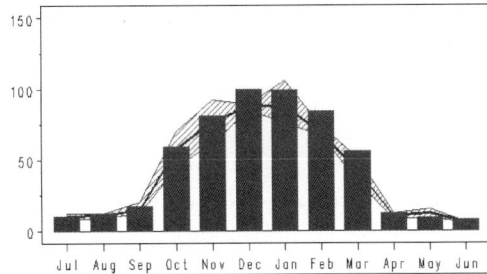

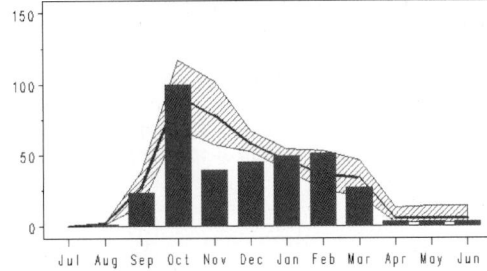

Figure 20.b, Monthly indices for Wigeon for GB (above) & NI (below).

The appearance of the British annual index has changed somewhat to those published in previous volumes of *Waterbirds in the UK*. This is due to the fact that the index used to be calculated using only January counts, but now all months from September to March inclusive are used for the calculation. This now means that the trend reflects numbers throughout the winter rather than just counts during January, which may have overlooked peaks in earlier or later months. The index has revealed a steady increase in Wigeon in Britain since the beginning of WeBS and it reaches a new high point in 2005/06. Additionally, the counted British maximum reached its second highest level to date. In contrast, numbers in Northern Ireland have undergone a clear decline with numbers currently at their lowest ever level. Much of this decline is due to unusually low numbers during November and December, with numbers only returning to average in January and February.

Counts at key sites were generally similar to average. However, the February peak at the Ouse Washes was by far the highest ever recorded at the site and indeed the only count at a site that even approaches those made on the Ribble Estuary, by far the pre-eminent site over many years. One site seeing a sustained rise in numbers in recent times is the Cleddau Estuary, where the peak count was about three times that seen in 2001/02, with a high count also on the nearby Burry Inlet. A spectacular count at Abberton Reservoir in December, which does not (on average) support nationally important numbers, was the eighth highest peak count in the UK in 2005/06. Few sites show sustained declines, although numbers have dropped significantly on the Mersey Estuary over the last five years.

	01/02	02/03	03/04	04/05	05/06	Mon	Mean
Sites of international importance in the UK							
Ribble Estuary	68,661	75,617	82,627	86,157	79,659	Dec	78,544
Ouse Washes	26,623	26,753 [12]	33,773 [12]	30,794	55,816	Feb	34,752
Somerset Levels	28,779	(39,546)	29,397	15,346	18,142	Jan	26,242
Breydon Wtr & Berney Marshes	21,700	15,999 [10]	16,811	19,019	22,134	Dec	19,133
North Norfolk Coast	19,078	16,056	20,694	17,444	18,426	Jan	18,340
Swale Estuary	15,303	22,827	20,772	13,832	16,651	Jan	17,877
Lindisfarne	(12,435)	(20,016)	(12,321)	15,960	13,614	Oct	16,530
Dornoch Firth	17,967	16,979	12,485	14,746	13,749	Oct	15,185
Sites of national importance in Great Britain							
Cromarty Firth	11,987	(6,041)	12,877	13,487	12,652	Oct	12,751
Lower Derwent Ings	8,970	11,217	13,171	10,215	14,320	Dec	11,579
Blackwater Estuary	5,789	10,976	7,057	7,385	6,708	Dec	7,583
Severn Estuary	5,579	7,019	9,110	8,058	6,249	Jan	7,203
Alde Complex	6,647 [10]	7,387	(4,956)	7,274	7,182	Jan	7,123
Nene Washes	5,053	11,866	8,190	4,998	5,380	Feb	7,097
Morecambe Bay	5,861	5,634	7,151	8,095	8,631	Jan	7,074
Inner Moray and Inverness Firth	7,070	7,820	7,587	5,595	6,078	Jan	6,830
Thames Estuary	5,808	9,798	5,565	4,343	6,449	Oct	6,393
Cleddau Estuary	3,192	3,720	6,045	8,468	9,441	Nov	6,173
Middle Yare Marshes	5,668	5,508	4,998	7,846	6,291	Jan	6,062
Fleet and Wey	5,337	5,360	(5,105)	4,469	6,122	Oct	5,322
Montrose Basin	4,381	4,752	5,488	(4,147)	5,065	Dec	4,922
Dee Estuary (England & Wales)	4,941 [10]	3,979	5,658	2,464	6,695	Nov	4,747
Arun Valley	4,010	(6,237)	5,073	2,956	3,375	Dec	4,330
Solway Estuary	(3,085)	(5,497)	(3,671)	(2,841)	3,341	Nov	4,170 ▲
The Wash	2,287	5,630	3,476	3,424	5,812	Jan	4,126 ▲
Loch of Harray	4,255	(2,682)	4,823	4,265	3,095	Feb	4,110
Sites of all-Ireland importance in Northern Ireland							
Lough Foyle	5,696	2,609	3,978	4,589	6,559	Oct	4,686
Strangford Lough	2,414	3,400	4,299	3,281	2,636	Oct	3,206
Loughs Neagh and Beg	2,707	1,908	3,060	3,611	2,701	Feb	2,797
Sites no longer meeting table qualifying levels in WeBS-Year 2005/2006							
Humber Estuary	2,514	(5,513)	4,734 [10]	(3,570)	3,233	Jan	3,913
Mersey Estuary	9,150	4,280	2,044	2,085	1,554	Dec	3,823
River Avon: Ringwood/Christch	(1,450)	6,394	(1,783)	(999)	1,038	Feb	3,716
Other sites surpassing table qualifying levels in WeBS-Year 2005/2006 in Great Britain							
Abberton Reservoir	(1,380)	945	436	573	13,954	Dec	3,977
Crouch-Roach Estuary	2,571	3,442	2,247	4,115	4,213	Dec	3,318
Burry Inlet	1,710	(2,106)	3,137	1,492	4,066	Feb	2,601

American Wigeon

Anas americana

Vagrant
Native Range: N & C America

GB max: 3 Feb
NI max: 0

American Wigeon were noted at five sites in four different months. The first bird of the winter was at Ouse Fen & Pits in December, followed by one at Roadford Reservoir in January then singles in February at Angle Bay (Cleddau Estuary), Loch of Strathbeg and Belvide Reservoir, the latter remaining into March.

Chiloe Wigeon

Anas sibilatrix

Escape
Native Range: S America

GB max: 2 Jul
NI max: 0

A maximum of two Chiloe Wigeon was recorded at Blagdon Lake in July. Two were also at Mansands Ley in September, with one there in June. In January a single was recorded at the Camel Estuary.

Gadwall

Anas strepera

International threshold:	600
Great Britain threshold:	171
All-Ireland threshold:	+[†]

GB max: 15,990 Jan
NI max: 210 Nov

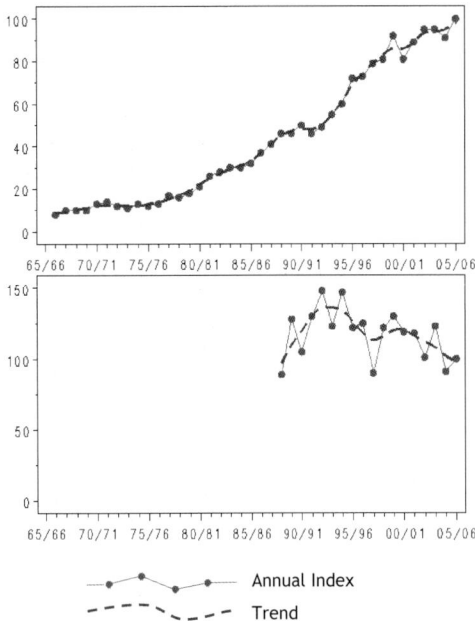

Figure 21.a, Annual indices & trend for Gadwall for GB (above) & NI (below).

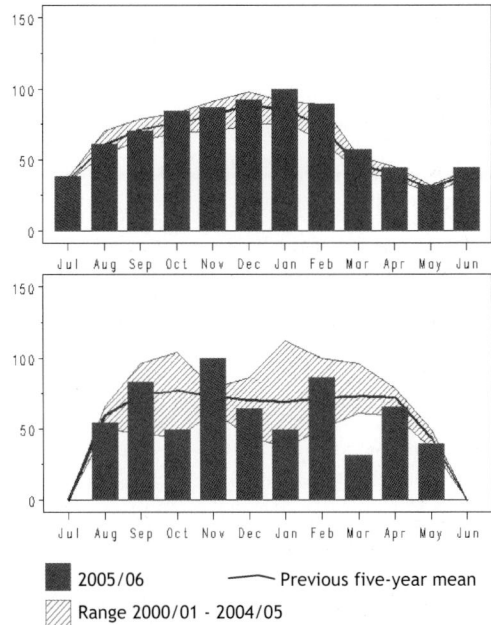

Figure 21.b, Monthly indices for Gadwall for GB (above) & NI (below).

The increase in Gadwall that has been apparent since the mid-1970s shows little sign of stopping. The British index illustrates the continual rise yet the Northern Ireland index shows a steady decrease since the early 1990s.

The monthly indices for Great Britain also show that counts were higher or equal to the five-year mean during 2005/06, with a peak in January following a steady rise since August.

Despite the increase nationally, only five sites supported internationally important numbers, with Thames Estuary and Wraysbury Gravel pits no longer qualifying as such. Several sites supported record peak numbers, with the peak counts at both the Ouse Washes and Abberton Reservoir quite exceptional; the former exceeded all other WeBS site totals on record by a considerable margin. Conversely, peak counts at Wraysbury Gravel Pits and Hoveton Great Broad were incredibly low, the former potentially due to a major increase in disturbance at the site (V Chambers *pers.*

comm.) Other sites with peak counts well below expected levels included Chichester Gravel Pits, Woolston Eyes, Alde Complex and Staines Reservoirs, with two sites - Brent Reservoir and Buckden & Stirtloe Pits - no longer supporting nationally important numbers. However, Dinton Pastures and Blunham Gravel Pit newly qualify as supporting nationally important numbers. Loughs Neagh and Beg remained the key site in Northern Ireland and the low tide counts at Strangford Lough resulted in the highest count there for over a decade.

	01/02	02/03	03/04	04/05	05/06	Mon	Mean	
Sites of international importance in the UK								
Ouse Washes	433 [12]	782	889 [12]	1,242	1,916	Feb	1,052	
Rutland Water	747	867	1,096	491	670	Oct	774	
Somerset Levels	754	(1,077)	430	729	704	Nov	739	
Lee Valley Gravel Pits	717	808	560	622	878	Dec	717	
River Avon: Fordingbridg/Ringwd	525	824	701	684	653	Dec	677	
Sites of national importance in Great Britain								
Abberton Reservoir	173	730	519	425	1,024	Jan	574	
Thames Estuary	(512)	815	(554)	471	377	Dec	554	▼
Loch Leven	320	840	635	360	392	Oct	509	
Wraysbury Gravel Pits	552	745	516	706	21	Jan	508	▼
Pitsford Reservoir	581	164	898	124	482	Aug	450	
Cotswold Water Park (West)	267	403	375	327	427	Jan	360	
Fen Drayton Gravel Pits	362	336	219	400	378	Jan	339	
Orwell Estuary	160 [10]	465 [10]	446	234 [10]	347 [10]	Dec	330	
Hoveton Great Broad	310	278	667		49	Oct	326	
Alton Water	268	270	360	182	495	Jan	315	
Chichester Gravel Pits	569	349	319	176	149	Jan	312	
Minsmere	212	394	239	309	398	Nov	310	
Chew Valley Lake	230	360	410	315	200	Sep	303	
Little Paxton Gravel Pits	287	275	339	225	315	Dec	288	
Hickling Broad	162	407	(200)	216	340	Jan	281	
Eversley Cross & Yateley GPs	292	305	230	256	315	Jan	280	
Stodmarsh NNR & Collards Lagn	259	360	264	217	252	Nov	270	
Tees Estuary	(107)	208	231	(289)	(332)	Oct	265	
Burghfield Gravel Pits		312	325	255	156	Dec	262	
Hornsea Mere	240	285	219	235	315	Dec	259	
Severn Estuary	250	253	292	194	297	Jan	257	
Woolston Eyes	124	182	297	470	196	Oct	254	
Sutton and Lound Gravel Pits	370	58	198	307	(304)	Jan	247	
Meadow Lane Gravel Pits St Ives	211	321	190	153	354	Dec	246	
North Norfolk Coast	(221)	215	(262)	231	262	Jan	243	
Alde Complex	277	163	(244)	352	172	Dec	242	
Blagdon Lake	257	17	335	204	287	Jul	220	
Thrapston Gravel Pits	(98)	(218)	207				213	
Lackford GPs	68	432	(225)	118	206	Sep	210	
Lower Derwent Ings	208	319	215	147	108	Jan	199	
Fairburn Ings	150	154	367		122	Jun	198	
Whitlingham Country Park	177	222	358	72	149	Jan	196	
Staines Reservoirs	101	162	126	455	126	Sep	194	
Dinton Pastures	291	144	97	138	260	Dec	186	▲
Nth Warren & Thorpeness Mere	84 [12]	229 [12]	113	353	148	Feb	185	
Ravensthorpe Reservoir	372	288	98	53	69	Dec	176	
Colne Valley Gravel Pits	(211)	149	238	126	144	Jan	174	
Lakenheath Fen		179	263	139	113	Oct	174	
Blunham Gravel Pit	148	214	152				171	▲
Sites no longer meeting table qualifying levels in WeBS-Year 2005/2006								
Brent Reservoir	295	109	68	93	102	Sep	133	

	01/02	02/03	03/04	04/05	05/06	Mon	Mean
Buckden and Stirtloe Pits	118	208					163
Sites with mean peak counts of 10 or more birds in Northern Ireland[†]							
Loughs Neagh and Beg	178	149	173	130	172	Nov	160
Strangford Lough	58	57	73	48	113 [10]	Jan	70
Hillsborough Main Lake	27	12	41	9	6	Jan	19
The Wash	117	172	67	87	271	Mar	143
Swillington Ings	155	113	52	89	265	Sep	135
Potteric Carr	(62)	140	114	36	216	Sep	127
Bainton Pits	76	118	168	200	206	Nov	154
Welbeck Estate	66	135	103	(98)	186	Nov	123
Holme Pierrepont Gravel Pits			166	123	(179)	Oct	156
Ditchford Gravel Pits	118		192	180	178	Feb	167
Dogmersfield Lake	39	81	130	170	177	Jan	119
Other sites surpassing table qualifying levels in WeBS-Year 2005/2006 in Northern Ireland							
Upper Lough Erne	0	4	4	10	20	Feb	8

[†] *as no All-Ireland threshold has been set a qualifying level of 10 has been chosen to select sites for presentation in this report*

Teal

Anas crecca

International threshold: 5,000
Great Britain threshold: 1,920
All-Ireland threshold: 650

GB max: 167,874 Dec
NI max: 6,129 Dec

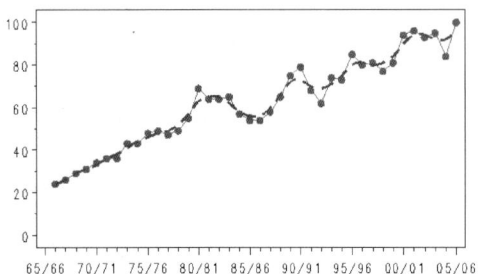

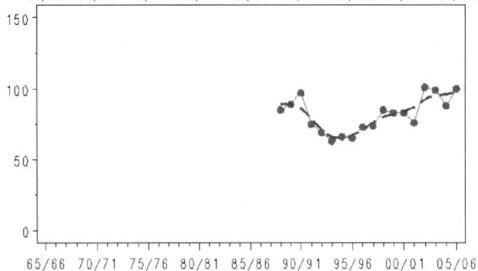

Annual Index
Trend

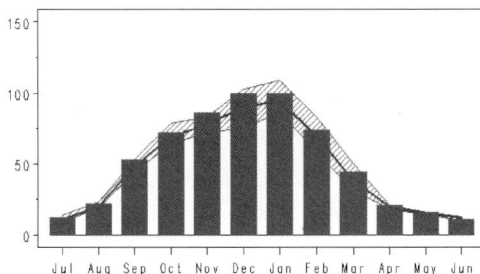

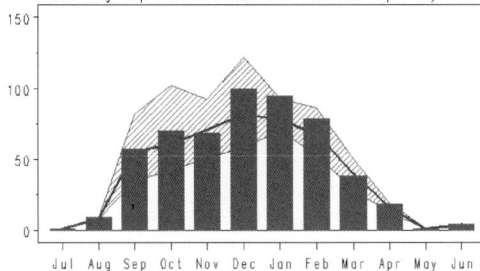

2005/06 — Previous five-year mean
Range 2000/01 - 2004/05

Figure 22.a, Annual indices & trend for Teal for GB (above) & NI (below).

Figure 22.b, Monthly indices for Teal for GB (above) & NI (below).

The long-term trend for Teal in Great Britain continues to rise and the index reached its highest ever level in 2005/06; the maximum counted total was also about 10% higher than the 2004/05 maximum. As usual, numbers peaked in December and January. The Northern Ireland index also continues to display a rising trend, following a decline to a low in the mid 1990s.

Peak counts at individual sites do tend to be relatively erratic for this species compared with many others. Numbers at the key British site of the Somerset Levels remained low, although counts at other internationally important sites such as Mersey Estuary, Ribble Estuary and Ouse Washes, were higher than in recent years (the latter site seeing high numbers of many duck species this winter).

Due to a re-evaluation upwards of the international threshold, WWT Martin Mere, Hamford Water, North Norfolk Coast, Lower Derwent Ings, Breydon Water and the Dee Estuary no longer qualify as sites of international importance. The Great Britain threshold remains the same with Abberton Reservoir and Stodmarsh NNR & Collards Lagoon now qualifying as sites of national importance as a consequence of notably high peak counts in 2005/06. In Northern Ireland, Strangford Lough, Lough Foyle and Carlingford Lough recorded their highest peaks for over five years, whereas counts at Loughs Neagh and Beg, and Upper Lough Erne were at their lowest for the same period.

	01/02	02/03	03/04	04/05	05/06	Mon	Mean
Sites of international importance in the UK							
Somerset Levels	29,586	(33,390)	17,225	7,161	8,719	Jan	19,216
Mersey Estuary	17,660	7,855	8,364	6,023	9,200 [10]	Nov	9,820
Ribble Estuary	5,316	4,671	7,421	8,688	9,571	Nov	7,133
Thames Estuary	6,994	9,838	6,691	5,433	5,137	Dec	6,819
Ouse Washes	5,757	4,433 [12]	5,102	6,731	9,772	Feb	6,359 ▲
Loch Leven	4,100	6,562	4,847	6,060	4,840	Sep	5,282 ▲
Swale Estuary	4,297	5,752	5,428	4,187	(5,783)	Nov	5,089 ▲
Sites of national importance in Great Britain							
Hamford Water	9,055 [10]	3,628	6,579	2,164	3,276	Dec	4,940
WWT Martin Mere	4,460	2,750	5,100	8,300	3,800	Oct	4,882 ▼
North Norfolk Coast	5,718	5,281	3,436	3,730	4,994	Nov	4,632
Dee Estuary (England & Wales)	6,887 [10]	4,361	5,459	2,752	2,854	Dec	4,463 ▼
Lower Derwent Ings	3,419	4,797	4,061	3,476	4,479	Jan	4,046
Severn Estuary	4,449	3,748	3,006	3,466	5,293	Jan	3,992
Breydon Wtr & Berney Marshes	6,487	3,124	1,982	4,733	2,372	Feb	3,740
Otmoor	(856)	3,633		(2,138)	(369)	Jan	3,633
Abberton Reservoir	(1,871)	736	3,863	1,224	7,741	Dec	3,391 ▲
Alde Complex	3,690	2,609	(2,530)	3,028	3,913	Jan	3,310
Mersehead RSPB Reserve	4,390	3,100	2,850	2,900			3,310
Humber Estuary	1,284	2,681	(5,111)	2,349	3,739	Nov	3,033
Inner Moray and Inverness Firth	2,289	2,948 [10]	3,439	3,397	2,995	Jan	3,014
The Wash	2,217	1,918	4,223	2,578	4,097	Dec	3,007
Blackwater Estuary	2,517	3,721	(2,873)	(2,064)	2,751	Dec	2,996
Hickling Broad	3,005	2,879	1,814	2,400	4,550	Nov	2,930
Morecambe Bay	2,519	2,261	2,808	(3,699)	2,538	Nov	2,765
Solway Estuary	(750)	(2,813)	(1,286)	1,941	3,152	Nov	2,635
Holburn Moss	1,700	2,250 [12]	3,500	3,000			2,613
Minsmere	2,227	2,189	4,381	1,984	1,796	Oct	2,515
Dornoch Firth	2,797	2,502	2,619	2,451	2,044	Dec	2,483
Arun Valley	2,194	(3,934)	1,912	1,229	2,390	Nov	2,332
Loch Gruinart Floods		2,095	2,476	2,549	2,058	Nov	2,295
Forth Estuary	(1,585)	1,984	(2,511)	1,880	2,165	Jan	2,135
Cleddau Estuary	1,621	2,095	2,129	2,269	2,435	Dec	2,110
Middle Yare Marshes	(2,097)	(1,300)	(1,372)	(1,067)	(1,034)	Jan	(2,097) ▲
Woolston Eyes	3,675	1,320	2,072	2,170	1,207	Jan	2,089
River Avon: Ringwood/Christch	(654)	4,841	695	(309)	410	Nov	1,982
Stodmarsh NNR & Collards Lagn	1,126	1,170	1,183	2,500	3,633	Jan	1,922 ▲
Sites of all-Ireland importance in Northern Ireland							
Strangford Lough	2,121	2,177	2,232	2,015	2,573	Dec	2,224
Loughs Neagh and Beg	1,633	1,887	2,732	2,019	1,427	Feb	1,940
Lough Foyle	684	2,275	582	1,038	1,405	Dec	1,197
Upper Lough Erne	333	1,635	407	723	174	Dec	654
Sites no longer meeting table qualifying levels in WeBS-Year 2005/2006							
Poole Harbour	1,667	2,235	(2,357)	1,806	1,402	Mar	1,893
Nene Washes	940	4,046	2,730	726	584	Jan	1,805
Other sites surpassing table qualifying levels in WeBS-Year 2005/2006 in Great Britain							
Chew Valley Lake	840	445	1,520	845	3,050	Sep	1,340
Loch of Strathbeg	1,435	(634)	1,708	998	2,457	Oct	1,650
Pagham Harbour	1,197	1,236	1,455	2,321	2,122	Dec	1,666
Crouch-Roach Estuary	1,340	1,249	1,129	2,981 [10]	1,926	Jan	1,725
Other sites surpassing table qualifying levels in WeBS-Year 2005/2006 in Northern Ireland							
Carlingford Lough	450	352	498	647	710	Jan	531

Green-winged Teal
Anas carolinensis

Vagrant
Native Range: N America

GB max: 3   Dec
NI max: 1   Mar

Green-winged Teal were recorded in eight months from October to June and from twelve sites. All records were of single birds and there were only three long-stayers. These were at Dee Estuary (England & Wales) from October to June, Langford Lowfields Gravel Pits in November and December and Loch Gruinart Floods from December to March. The only record from Northern Ireland was a single at Belfast Lough in March.

Speckled Teal
Anas flavirostris

Escape
Native Range: S America

GB max: 2   Nov
NI max: 0

Two Speckled Teal were present at Bramshill Park Lake between November and February; in March only one was reported. Birds have been recorded annually at this site since the first in early 1998.

Mallard
Anas platyrhynchos

International threshold: 20,000**
Great Britain threshold: 3,520[†]
All-Ireland threshold: 500

GB max: 131,200   Dec
NI max: 7,785   Sep

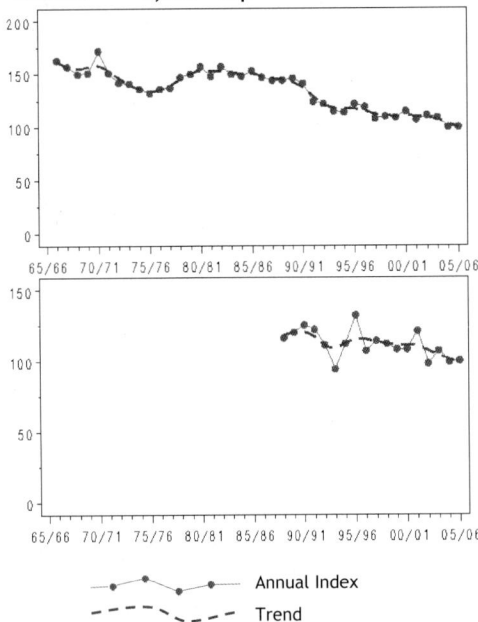

Annual Index
Trend

2005/06   Previous five-year mean
Range 2000/01 - 2004/05

Figure 23.a, Annual indices & trend for Mallard for GB (above) & NI (below).

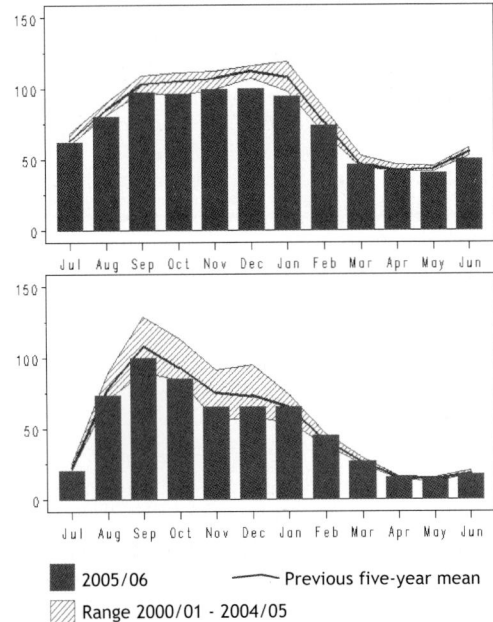

Figure 23.b, Monthly indices for Mallard for GB (above) & NI (below).

The long-term decline continued for this familiar species although totals were much the same as last year. This decline is reflected in the maximum count, which was lower than the maximum for 2004/05 for the same month. In the main winter period between September and January, monthly index values were below the five-year mean, though from February through to June the values were largely the same as their recent averages. This could imply one or more of a number of reasons: a reduction in winter immigrants; poor breeding success leading to a lower 'surplus' of young; reduced numbers of birds released for shooting purposes; or a mild winter meaning that many birds remained on smaller, uncounted, waterbodies rather than congregating on larger sites, however, further research would be needed. The Northern Ireland index showed a slight increase and the maximum count there was increased since the previous year, although the longer-term trend also continues to be one of decline.

The Ouse Washes still remains the only site in Britain supporting nationally important numbers of Mallards. Despite the record numbers of many other species of ducks at this site during the 2005/06 winter, the Mallard peak was one of the lowest on record and at the current rate it is unlikely that average numbers will remain above the threshold for national importance for much longer. Only seven other sites in Great Britain had means in excess of 2,000 birds and, interestingly, four of these sites showed an increase in peak numbers from 2004/05. Peak counts for Northern Ireland were much the same as last year although at Loughs Neagh and Beg, and Lough Foyle peak counts were higher than for 2004/05.

	01/02	02/03	03/04	04/05	05/06	Mon	Mean
Sites of national importance in Great Britain							
Ouse Washes	4,457	3,580 [12]	3,988 [12]	3,505	2,454	Feb	3,597
Sites of all-Ireland importance in Northern Ireland							
Loughs Neagh and Beg	4,243	4,763	4,774	4,027	4,612	Sep	4,484
Strangford Lough	2,227	1,851	1,568	1,621	1,586	Dec	1,771
Lough Foyle	1,181	705	791	1,025	1,133	Sep	967
Lower Lough Erne		533	(494)	754	556	Jan	614
Sites with mean peak counts of 2,000 or more birds in Great Britain[†]							
WWT Martin Mere	3,800	3,280	3,350	2,930	3,150	Oct	3,302
Ampton Water		2,535	3,735				3,135
Severn Estuary	2,761	2,936	2,701	3,353	3,884	Sep	3,127
The Wash	1,781	2,384	2,639	2,437	2,529	Dec	2,354
Humber Estuary	1,524	2,957	(2,347)	2,455	(1,621)	Nov	2,321
Tring Reservoirs	1,834	2,800	2,000	1,557	2,500	Jul	2,138
Morecambe Bay	(1,683)	2,455	2,208	1,891	1,740	Dec	2,074
Other sites surpassing table qualifying levels in WeBS-Year 2005/2006 in Great Britain[†]							
Swale Estuary	1,635	1,452	1,800	2,010	2,247	Dec	1,829

[†] as few sites exceed the British threshold a qualifying level of 2,000 has been chosen to select sites for presentation in this report

Chestnut Teal
Anas castanea

Escape

Native Range: S Australia

GB max: 2 Aug
NI max: 0

Two Chestnut Teal were recorded at Liden Lagoon, Wiltshire in August. Previously, two birds were noted at this site in October 2003.

Pintail
Anas acuta

International threshold: 600
Great Britain threshold: 279
All-Ireland threshold: 60

GB max: 27,935 Dec
NI max: 591 Dec

The Great Britain annual index reached the highest level since the mid 1980s following a steady increase since a low point in 2000/01. This was reflected with a notably high maximum GB total count. The monthly indices were above the five year mean in all months and showed a peak in February, where as in previous years this peak has been earlier in the winter. Eighteen sites hold internationally important numbers of Pintail, though the River Avon (Ringwood-Christchurch) and Arun Valley no longer qualify. Following a high peak, Pagham Harbour now qualifies for this status. The peak count on the Dee Estuary was the highest there since 1992 whilst relatively high numbers were also recorded from Burry Inlet, Ribble Estuary, Duddon Estuary, Lindisfarne, North West Solent, Foryd Bay, Chichester Harbour, Rutland Water and Lavan Sands. Conversely, peak counts from the Wash, Swale Estuary, Lower Derwent Ings, and in particular River Avon where the peak was of a single bird in February, were notably lower than usual, although at the last of these sites, numbers have always varied hugely depending upon the amount of winter flooding.

In Northern Ireland, the index rose sharply to a new high following two years of small declines, largely due to the increased numbers at Strangford Lough and Lough Foyle. The Northern Ireland maximum in December was also much higher than 2004/05, helped considerably by record numbers at Strangford Lough.

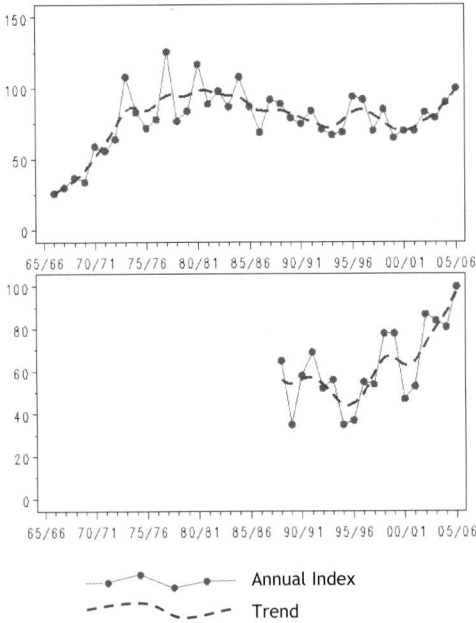

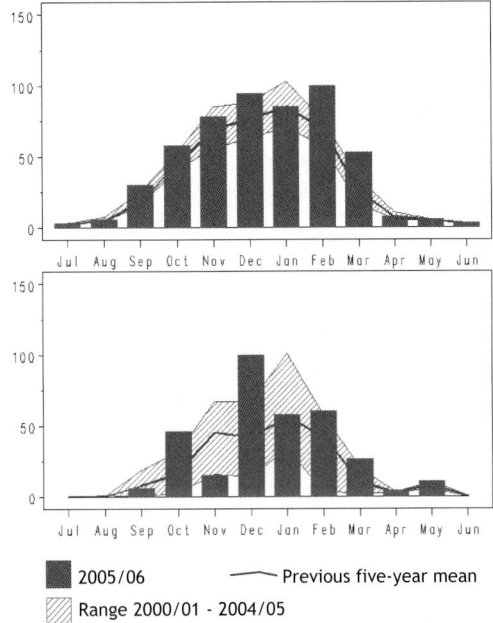

Annual Index

Trend

2005/06   Previous five-year mean

Range 2000/01 - 2004/05

Figure 24.a, Annual indices & trend for Pintail for GB (above) & NI (below).

Figure 24.b, Monthly indices for Pintail for GB (above) & NI (below).

	01/02	02/03	03/04	04/05	05/06	Mon	Mean	
Sites of international importance in the UK								
Dee Estuary (England & Wales)	6,023	6,000 [12]	6,317	4,312	6,330	Nov	5,796	
Solway Estuary	(8,070)	(3,357)	4,183	(4,352)	(1,575)	Nov	5,535	
Burry Inlet	1,305	4,410	5,772	2,745	4,837	Feb	3,814	
Morecambe Bay	3,471	3,628	(3,942)	3,620	3,045	Oct	3,541	
Ouse Washes	2,606 [12]	2,844 [12]	2,277 [12]	3,330	2,848	Feb	2,781	
Ribble Estuary	619	1,405	(2,562)	(3,058)	3,579	Dec	2,245	
Nene Washes	1,250	3,478	1,779	327	281	Mar	1,423	
Duddon Estuary	391	(415)	(1,299)	1,626 [10]	2,210 [10]	Feb	1,409	
Severn Hams	2,000 [12]	(80)	(250)	70	(101)	Dec	1,035	
Medway Estuary	(1,118)	(333)	(95)	812 [10]	(809)	Feb	965	
Severn Estuary	(780)	(891)	(354)	(784)	905	Jan	905	
North Norfolk Coast	1,296	(475)	(768)	712	657	Feb	888	
The Wash	516	1,253	1,086	915	567	Jan	867	
Swale Estuary	998	946	962	672	579	Dec	831	
Mersehead RSPB Reserve	410	1,140	480	970			750	
Somerset Levels	1,084	(1,315)	494	261	333	Dec	697	
Dee Flood Meadows	1,050	(628)	580	250	(329)	Mar	627	
Pagham Harbour	587	304	477	834	893	Jan	619	▲
Sites of national importance in Great Britain								
WWT Martin Mere	635	487	463	710 [11]	(535)	Mar	574	
River Avon: Ringwood/Christch	280	2,013	25	46	1	Feb	473	▼
Blackwater Estuary	352	498	461	555	(387)	Dec	467	
Wigtown Bay	(195)	(320)	(359)	(654)	349	Nov	454	
Arun Valley	(413)	(775)	403	293	290	Dec	435	▼
Alde Complex	705	403	(330)	313	307	Jan	432	
Lower Derwent Ings	411	660	573	296	167	Jan	421	
Blyth Estuary	368						368	
Lindisfarne	272 [10]	330	384	301	536	Oct	365	
North West Solent	233	96	391	412	670	Mar	360	▲
Stour Estuary	629	193	111	(263)	473	Dec	352	

	01/02	02/03	03/04	04/05	05/06	Mon	Mean
Malltraeth Cob and Pools			207	421	397	Dec	342
Inner Moray and Inverness Firth	313	· 310	258	518	281	Jan	336
Orwell Estuary	473 [10]	372 [10]	325 [10]	165 [10]	308 [10]	Nov	329
Breydon Wtr & Berney Marshes	329	571	271 [10]	248	202	Dec	324
Otmoor	160 [12]	481 [12]		(156)	(46)	Jan	321
Poole Harbour	424	191	316	338	(208)	Feb	317
Crouch-Roach Estuary	192	385	267	(281)	380	Dec	306
Fleet and Wey	281	149	(281)	420	360	Feb	303 ▲
Sites of all-Ireland importance in Northern Ireland							
Strangford Lough	348	378	582	349	643 [10]	Dec	460
Sites no longer meeting table qualifying levels in WeBS-Year 2005/2006							
Thames Estuary	(223)	355	(126)	(100)	162	Oct	259
Other sites surpassing table qualifying levels in WeBS-Year 2005/2006 in Great Britain							
Foryd Bay	132	248	80	136	449	Feb	209
Chichester Harbour	190	69	233	297	375	Feb	233
Rutland Water	190	120	145	61	327	Oct	169
Lavan Sands	(80)	101	236	(144)	283	Nov	207
Other sites surpassing table qualifying levels in WeBS-Year 2005/2006 in Northern Ireland							
Lough Foyle	21	21	22	52	94	Mar	42

Bahama Pintail

Escape

Anas bahamensis

Native Range: S America

GB max: 3   May
NI max: 0

Bahama Pintails were reported from just two sites in 2005/06. Two were present at Stanton Lake in August with singles present in May and June. Single birds were recorded at Doddington Pool between February and June. This species has been recorded by WeBS in almost every year since 1993.

Garganey

Anas querquedula

International threshold: 20,000**
Great Britain threshold: +[†]
All-Ireland threshold: +[†]

GB max: 52   May
NI max: 0

As a summer visitor, Garganey are considered here for the calendar year 2005. The peak count was recorded during May and was one less than the previous British maximum. None were recorded in Northern Ireland. The highest single site total was again recorded at Wraysbury Gravel Pits where 12 were present in September. The only other double-figure count was at Ouse Washes in April. In Britain Garganey are rarely found in any concentrations and birds are most visible during spring and autumn. Garganey were recorded at a total of 61 sites during 2005, with most records between April and October, although there was an exceptional record of a single bird at Northward Hill RSPB Reserve (Thames Estuary) in January. The latest record was of one at Mere Sands Wood Nature Reserve in November.

	2001	2002	2003	2004	2005	Mon	Mean
Sites with mean peak counts of 4 or more birds in Great Britain[†]							
Wraysbury Gravel Pits	0	15	12	14	12	Sep	11
Stodmarsh NNR & Collards Lagn	30	5	7	0	2	Apr	9
Rye Harbour and Pett Level	7	9	2	3	8	Apr	6
Breydon Wtr & Berney Marshes	4	4	4	8	4	Jul	5
Dungeness Gravel Pits	12	(0)	3	4	2	Jul	5
North Norfolk Coast	(4)	(4)	(3)	4	5	Apr	5
Ouse Washes	3	7	2	4	10	Apr	5
Thames Estuary	(2)	(5)	(2)	(3)	(2)	Apr	(5)
Other sites surpassing table qualifying levels in Summer 2005 in Great Britain[†]							
Humber Estuary	0	2	2	1	6	Sep	2
Loch of Strathbeg	0	0	(0)	1	6	Aug	2
Chew Valley Lake	2	1	3	1	4	Aug	2
Nene Washes	2	0	8	0	4	Apr	3
Pitsford Reservoir	2	0	1	0	4	Sep	1

[†] *as no British or All-Ireland thresholds have been set a qualifying level of four has been chosen to select sites for presentation in this report*

Blue-winged Teal

Anas discors

Vagrant and escape
Native Range: Americas

GB max: 1 Jun
NI max: 0

The only Blue-winged Teal recorded during 2005/06 was a single at the Avon Estuary in June, the first time that this species has been noted at this site. Although many other vagrant ducks are most frequent in winter, summer Blue-winged Teal records are not especially unusual.

Shoveler

Anas clypeata

International threshold: 400
Great Britain threshold: 148
All-Ireland threshold: 65

GB max: 12,396 Oct
NI max: 183 Nov

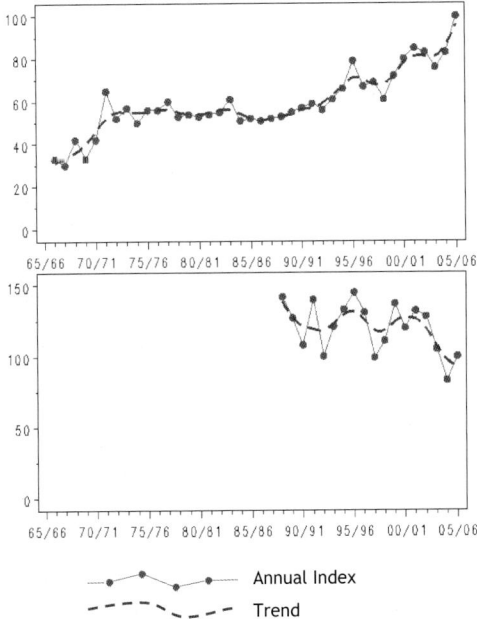

Figure 25.a, Annual indices & trend for Shoveler for GB (above) & NI (below).

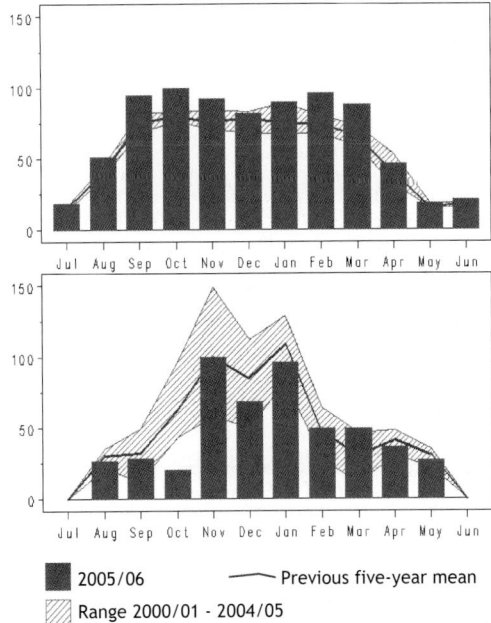

Figure 25.b, Monthly indices for Shoveler for GB (above) & NI (below).

The British index jumped to its highest ever level following a steady rise seen since the early 1990s; the maximum counted total was also the highest on record. The Northern Ireland index and the maximum count showed a slight rise too, though the overall trend there remains one of decline.

The Ouse Washes has now replaced the Somerset Levels as the most important site in the UK for Shoveler, following a fourth consecutive four-figure count at the former and a third successive below-average count at the latter. Two sites, Swale Estuary and Loch Leven, no longer qualify as sites supporting internationally important numbers of Shoveler following a succession of decreasing counts, although following a record high count, Dungeness Gravel Pits now does qualify for this status. The peak count on the Severn Estuary was also a record WeBS total, whilst counts at Abberton Reservoir, Burry Inlet, Staines Reservoirs, Stodmarsh NNR & Collards Lagoon, Grafham Water and Woolston Eyes were also high. There was a particularly low peak at Wraysbury Gravel Pits, as also noted for Gadwall at this site and was potentially due to an increase in disturbance (V Chambers *pers. comm.*)

Numbers at Northern Ireland's key site, remained well below the five-year mean. Strangford Lough, rose from 2004/05 but

	01/02	02/03	03/04	04/05	05/06	Mon	Mean
Sites of international importance in the UK							
Ouse Washes	968 [12]	1,125	1,104 [12]	2,414	1,279	Mar	1,378
Somerset Levels	1,170	(2,190)	784	(902)	845	Feb	1,247
Chew Valley Lake	805	535	565	395	660	Dec	592
Rutland Water	608	504	475	663	680	Sep	586
Thames Estuary	(605)	697	415	402	357	Nov	495
Abberton Reservoir	440	422	488	355	(674)	Nov	476
Breydon Wtr & Berney Marshes	679	415	322	468	333	Mar	443
Dungeness Gravel Pits	504	320	378	340	574	Sep	423 ▲
Sites of national importance in Great Britain							
Severn Estuary	366	368 [10]	325	266	603	Jan	386
Swale Estuary	587	440	330	292	199	Mar	370 ▼
Loch Leven	400	550	295	386	204	Sep	367 ▼
Staines Reservoirs	356	377	261	308	469	Sep	354
Burry Inlet	215	397	327	344	437	Dec	344
Lee Valley Gravel Pits	321	308	246	275	282	Oct	286
Lower Derwent Ings	241	442	319	314	107	Jan	285
Alde Complex	(407)	229	(106)	175	253	Mar	266
Stodmarsh NNR & Collards Lagn	206	244	202	272	384	Oct	262
Nene Washes	374	262	200	177	213	Mar	245
Chichester Gravel Pits	317	238	321	173	165	Dec	243
North Norfolk Coast	289	182	212	234	278	Jan	239
Llynnau Y Fali	176	337	233	232	210	Mar	238
Morecambe Bay	380	(82)	184	167	159	Nov	223
Ribble Estuary	179	197	231	219	286	Nov	222
Pitsford Reservoir	153	91	378	70	347	Oct	208
Minsmere	207	233	180	227	183	Nov	206
Fairburn Ings	153	159	221		288	Feb	205
River Avon: Fordingbridg/Ringwd	117	361	188	149	195	Feb	202
Blagdon Lake	400	75	146	160	(220)	Sep	200
Arun Valley	227	259	195	175	98	Dec	191
Walland Marsh	520	125	120	120	60	Dec	189
Rye Harbour and Pett Level	282	167	204	162	120	Jan	187
Walthamstow Reservoirs	179	135	212	265	142	Nov	187
Grafham Water	143	51	112	266	357	Mar	186 ▲
Medway Estuary	(280)	(20)	(26)	19 [10]	248	Jan	182 ▲
Hampton & Kempton Reservoirs	208	(88)	165	(134)	144	Feb	172
Tees Estuary	114	245	181	(145)	145	Sep	171
Middle Yare Marshes	(151)	(169)	(96)	(111)	(170)	Oct	(170)
Trinity Broads	(142)	55	137	304			165 ▲
Woolston Eyes	103	71	175	157	317	Sep	165 ▲
Malltraeth RSPB	145	186	124	173	147	Dec	155
Wraysbury Gravel Pits	260 [12]	221	97	172	21	Dec	154
Hickling Broad	153	307	81	108	100	Jan	150 ▲
Sites of all-Ireland importance in Northern Ireland							
Strangford Lough	182	199	201	119	147	Nov	170
Sites no longer meeting table qualifying levels in WeBS-Year 2005/2006							
Fen Drayton Gravel Pits	157	128	115	135	80	Mar	123
Other sites surpassing table qualifying levels in WeBS-Year 2005/2006 in Great Britain							
Belvide Reservoir	79		50	160	283	Sep	143
Swithland Reservoir	66 [12]	99	92	139	241	Sep	127
Tring Reservoirs	135	128	43	99	225	Sep	126
WWT Martin Mere	92	43	162	198	217	Oct	142
Llyn Traffwll	21	60	58	58	178	Apr	75
Holme Pierrepont Gravel Pits			14	69	176	Mar	86
Colne Valley Gravel Pits	120	38	96	154	173	Feb	116
Hollowell Reservoir	56	43	77	27	169	Oct	74
North West Solent	92	82	89	88	164	Feb	103
Cotswold Water Park (West)	88	(218)	91	126	163	Feb	137
Brent Reservoir	230	125	20	129	155	Oct	132

Ringed Teal
Callonetta leucophrys

Escape
Native Range: S America

GB max: 1   Jul
NI max: 0

Single Ringed Teal were present at Liden Lagoon in July and August, Vyne Floods in October and December and Coate Water in February and March.

Red-crested Pochard
Netta rufina

International threshold:	500
Great Britain threshold:	?[†]
All-Ireland threshold:	?[†]

GB max: 270   Jan
NI max: 1   Feb

The origin of Red-crested Pochards in the UK is mixed and, to a great extent, confused. Whilst wild birds presumably arrive from time to time from the continent, most birds seen here are resident and derive from escapes from captivity. Escaped birds in earlier years have formed the nucleus of a naturalised population, based largely in the upper Thames valley, whilst new birds still appear to be escaping from collections and adding to these numbers, with new nuclei appearing elsewhere around the country. Moreover, the picture is likely to be clouded further by the deliberate release of large numbers of birds in north Suffolk for shooting purposes from 2005 onwards (S. Piotrowski *pers. comm.*). The peak monthly British total increased dramatically in 2005/06,

by over 50% above the previous year. As usual, most of the birds (and indeed most of the increase) related to birds in the upper Thames Valley, with the combined total for both "halves" of the Cotswold Water Park reaching a record 188 in January 2006, and the Lower Windrush peak also reaching a new high. Away from there, a steady rise has been apparent at Ouse Fen & Pits and Bourton-on-the-Water Gravel Pits, but the count of 17 at Grimley New Workings in Worcestershire was a surprise. Overall, Red-crested Pochards were noted at 79 sites in Britain during 2005/06. There was just one record in Northern Ireland of a single bird at Upper Lough Erne in February.

	01/02	02/03	03/04	04/05	05/06	Mon	Mean
Sites with mean peak counts of 10 or more birds in Great Britain[†]							
Cotswold Water Park (West)	58	(74)	114	81	119	Oct	93
Cotswold Water Park (East)	72	40	33	48	70	Jan	53
Baston and Langtoft Gravel Pits	16	8	(23)				16
Lower Windrush Valley GPs	8	5	6	(19)	41	Dec	16
Hanningfield Reservoir	6	6	(43)	2	0		11
Other sites surpassing table qualifying levels in WeBS-Year 2005/2006 in Great Britain[†]							
Ouse Fen & Pits (Hanson/RSPB)		2	4	7	18	Jan	8
Grimley New Workings				0	(17)	Dec	9
Bourton-on-the-Water GPs	1	0	2	5	12	Mar	4
Sutton and Lound Gravel Pits	3	7	6	16	12	Apr	9

[†] *as no British or All-Ireland thresholds have been set a qualifying level of 10 has been chosen to select sites for presentation in this report*

Pochard
Aythya ferina

International threshold:	3,500
Great Britain threshold:	595
All-Ireland threshold:	400

GB max: 23,857   Dec
NI max: 8,664   Jan

Numbers in Northern Ireland showed tentative signs of revival following the steep declines recorded since 2000/01, although numbers were still below average through most of the year. The monthly maximum for the region was 20% higher than the previous year but similar to that of 2003/04. Figures shown in this report for Northern Ireland largely reflect those at the principal site for this species in the

UK, Loughs Neagh and Beg, which in 2005/06 held over 95% of all Pochard recorded in the province. Peak numbers at Loughs Neagh and Beg were higher than in the two preceding years but still remain just half those of 2001/02.

The British maximum fell by some 20% between 2004/05 and 2005/06. This was reflected in the annual index, which fell to its lowest ever level. The underlying trend is one of decline and a clear fall in numbers over the past ten years is illustrated. The monthly indices show how low numbers were recorded through the year and only in September, October and March were numbers within the range recorded during the previous five seasons. Numbers at the top two English sites each showed signs of decline and mean peak numbers at the Ouse Washes no longer surpass the international threshold for this species. Loch Leven remains the key site in Scotland, despite numbers falling to their lowest level

since 2000/01. Some sites, such as the Lower Derwent Valley and the Nene Washes, can clearly correlate numbers with the degree of winter flooding on the sites, and thus fluctuate greatly. The only sites that seem to be experiencing increases in recent years are Dungeness Gravel Pits, Chew Valley Lake and Wraysbury Gravel Pits. Apparent low numbers at the Middle Tame Valley Gravel Pits in recent years have resulted from a lack of counts from the favoured lakes within this complex.

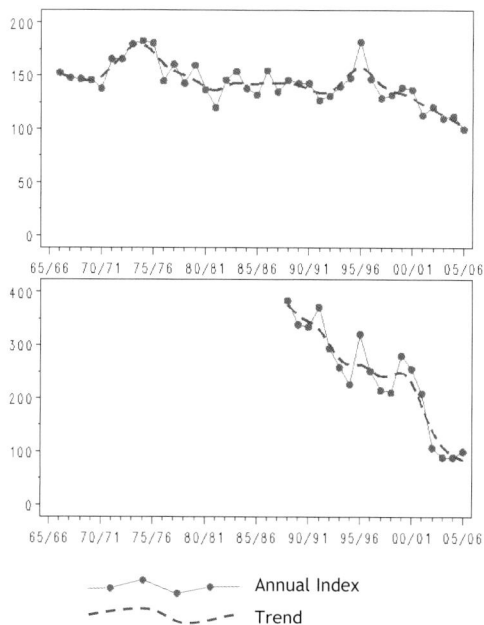

Figure 26.a, Annual indices & trend for Pochard for GB (above) & NI (below).

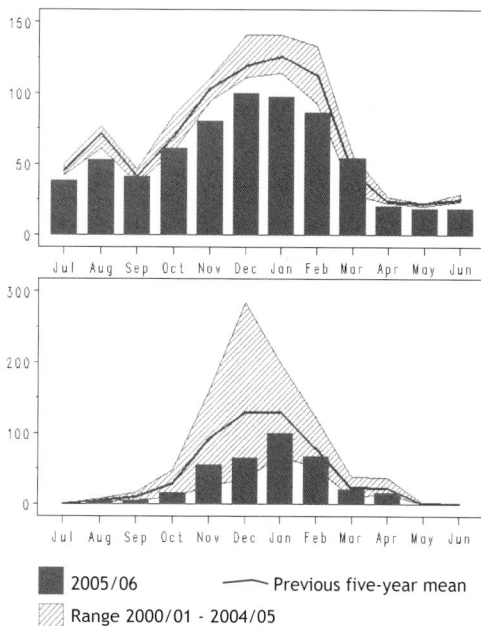

Figure 26.b, Monthly indices for Pochard for GB (above) & NI (below).

	01/02	02/03	03/04	04/05	05/06	Mon	Mean
Sites of international importance in the UK							
Loughs Neagh and Beg	16,168	9,082	7,835	6,764	8,256	Jan	9,621
Abberton Reservoir	3,125	4,325	5,290	3,188	2,852	Dec	3,756
Sites of national importance in Great Britain							
Ouse Washes	4,206	4,583	3,304 [12]	2,099	1,227	Feb	3,084 ▼
Loch Leven	4,074	2,934	2,548	2,193	1,715	Sep	2,693
Middle Tame Valley Gravel Pits	1,423	(442)	(203)	(56)	(12)	Oct	1,423
Hornsea Mere	1,115	1,415	1,325	1,150	1,150	Dec	1,231
Fleet and Wey	1,072	926	850	746	682	Dec	855
Severn Estuary	1,064	772	(905)	652	760	Feb	831
Dungeness Gravel Pits	595	765	855	788	1,053	Aug	811
Chew Valley Lake	735	475	480	635	1,580	Jan	781 ▲
Lower Derwent Ings	437	1,973	1,236	39	20	Jan	741
Loch of Boardhouse	822	605	705	770	709	Nov	722
Nene Washes	48	2,853	66	32	88	Mar	617
Sites of all-Ireland importance in Northern Ireland							
Upper Lough Erne	780	916	801	473	329	Feb	660
Sites no longer meeting table qualifying levels in WeBS-Year 2005/2006							
Cotswold Water Park (West)	512	(377)	499	(571)	573	Feb	539
Lower Windrush Valley GPs	600	(384)	505	(423)	(410)	Jan	553

	01/02	02/03	03/04	04/05	05/06	Mon	Mean
WWT Martin Mere	860	750	565	472	322	Jan	594
Woolston Eyes	570	637	663	620	321	Feb	562
Other sites surpassing table qualifying levels in WeBS-Year 2005/2006 in Great Britain							
Rutland Water	630	411	645	282	693	Sep	532
Wraysbury Gravel Pits	358	516	409	423	648	Dec	471

Ring-necked Duck
Aythya collaris

Vagrant
Native Range: N America

GB max: 7 Apr
NI max: 1 Dec

Ring-necked Ducks were noted at a total of 11 sites in England and one in Northern Ireland. Most records were of single birds although two were at Chew Valley Lake from December to April and the Tees Estuary from January to April. There were records in every month from August to April with birds present at a total of five sites in both March and April. The only record from Northern Ireland was of one at Upper Lough Erne in December.

Ferruginous Duck
Aythya nyroca

Vagrant and escape
Native Range: Europe, Africa & central Asia

GB max: 2 Feb
NI max: 0

Records of Ferruginous Duck were received from five sites and all but one were of single birds. Singles were at Fen Drayton Gravel Pits in July, Tittesworth Reservoir in December, Leybourne and New Hythe Gravel Pits in January and February and Grimley New Workings in February. The peak count was of two at Chew Valley Lake in June.

Tufted Duck
Aythya fuligula

International threshold: 12,000
Great Britain threshold: 901
All-Ireland threshold: 400

GB max: 55,148 Dec
NI max: 10,138 Jan

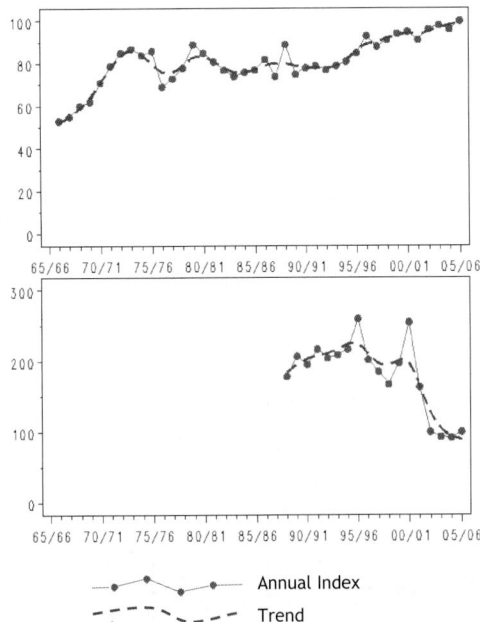

Figure 27.a, Annual indices & trend for Tufted Duck for GB (above) & NI (below).

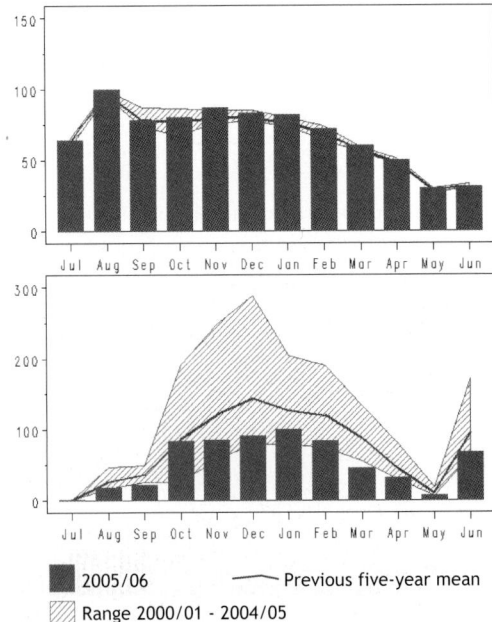

Figure 27.b, Monthly indices for Tufted Duck for GB (above) & NI (below).

The British long-term trend continues to rise, although the maximum count was slightly lower than for 2004/05 with the monthly index showing largely average counts throughout the year. Rutland Water became the first British site on record to hold in excess of 8000 Tufted Ducks (and for the first time, counts there exceeded those at Loughs Neagh and Beg), whilst Chew Valley Lake, Staines Reservoirs, Grafham Water and King George V Reservoir also supported notably high peak counts. Conversely, as with other species such as Gadwall, Shoveler and Smew, counts of Tufted Ducks from Wraysbury Gravel Pits were very much reduced.

In Northern Ireland, although the index increased slightly, there was little sign of recovery with the peak at the key site of Loughs Neagh and Beg, previously the only internationally important site for this species in the UK, now the lowest on record. Indeed this decline has been so marked that the site no longer qualifies for international importance; it was the only site to do so in recent years. Numbers on both Upper and Lower Lough Erne remained relatively stable however.

	01/02	02/03	03/04	04/05	05/06	Mon	Mean	
Sites of national importance in Great Britain								
Rutland Water	5,115	7,496	6,818	6,488	8,487	Oct	6,881	
Loch Leven	3,650	4,872	3,913	3,826	3,802	Sep	4,013	
Abberton Reservoir	1,418	2,487	2,067	5,112	(4,857)	Nov	3,188	
Middle Tame Valley Gravel Pits	2,164	(915)	(325)	(129)	(64)	Sep	2,164	
Pitsford Reservoir	1,263	2,441	2,226	2,506	2,066	Sep	2,100	
Walthamstow Reservoirs	1,838	1,867	1,772	1,771	1,828	Aug	1,815	
Hanningfield Reservoir	1,160	1,641	3,109	400	1,573	Aug	1,577	
Staines Reservoirs	1,026	1,971	1,133	792	2,844	Aug	1,553	
Chew Valley Lake	1,020	1,080	1,465	1,235	2,115	Oct	1,383	
Wraysbury Gravel Pits	2,091	2,422	846	1,015	465	Jan	1,368	
Lee Valley Gravel Pits	1,027	1,248	1,404	1,222	985	Dec	1,177	
Ouse Washes	1,395 [12]	1,192	973 [12]	1,330	966	Mar	1,171	
Thames Estuary	(537)	(436)	(461)	(584)	1,079	Dec	1,079	▲
Cotswold Water Park (West)	736	(792)	1,199	960	1,199	Feb	1,024	
Alton Water	961	815	1,440	644	1,063	Feb	985	
Hornsea Mere	866	1,225	1,050	900	840	Nov	976	▲
Sites of all-Ireland importance in Northern Ireland								
Loughs Neagh and Beg	13,318	9,771	8,999	9,277	7,871	Jan	9,847	▼
Upper Lough Erne	998	1,065	1,236	1,295	1,457	Feb	1,210	
Lower Lough Erne		635	580	674	575	Feb	616	
Other sites surpassing table qualifying levels in WeBS-Year 2005/2006 in Great Britain								
Grafham Water	(913)	473	815	824	1,337	Dec	872	
King George V Reservoirs	347	503	636	184	1,078	Nov	550	

Tufted Duck (John Harding)

Scaup
Aythya marila

International threshold: 3,100
Great Britain threshold: 76
All-Ireland threshold: 30*

GB max: 2,392 Feb
NI max: 6,495 Feb

*50 is normally used as a minimum threshold

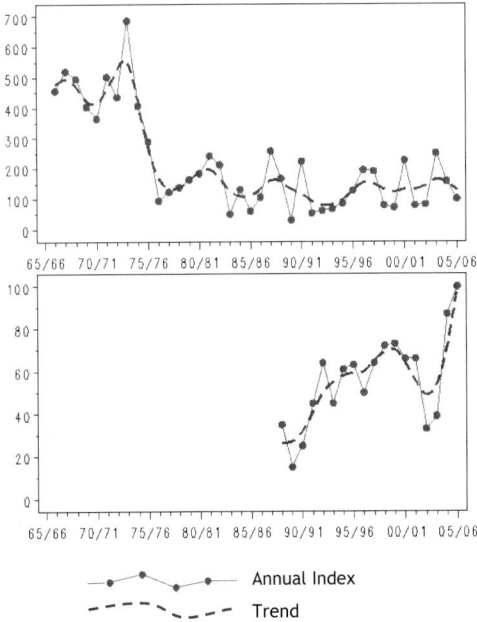

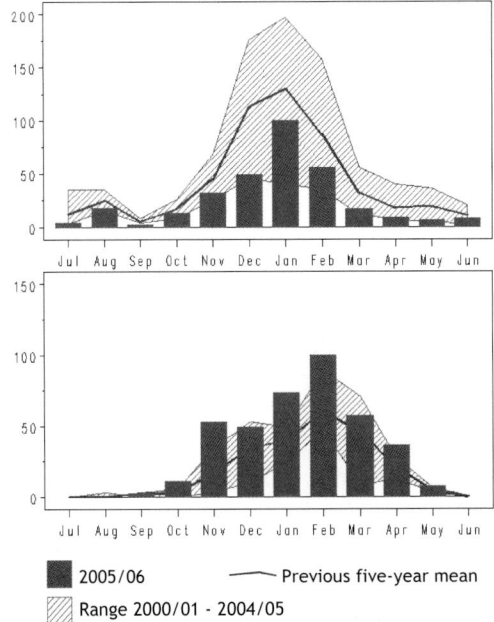

—●— Annual Index
- - - - - Trend

■ 2005/06   —— Previous five-year mean
▨ Range 2000/01 - 2004/05

Figure 28.a, Annual indices & trend for Scaup for GB (above) & NI (below).

Figure 28.b, Monthly indices for Scaup for GB (above) & NI (below).

Following a decline in the early 1970s, Scaup numbers in Great Britain have remained relatively stable ever since, albeit with substantial between-year fluctuations. These fluctuations are to be expected for a species, which can be difficult to count on its favoured estuarine and sea-loch habitats. The British index fell for the second consecutive year, whilst the counted maximum was at its lowest level since 2001/02. Conversely, the Northern Ireland trend continues to rise, and this is also reflected in the monthly index, which shows counts for most months to be much higher than the five year mean. This large rise in the past two years is largely due to an increase in peak numbers at the key site of Loughs Neagh and

Beg, which, in contrast to the low numbers of Pochard and Tufted Duck at the site in recent years, were again much higher than average. Numbers at Belfast Lough remained high, although were down slightly on the peak count of 2004/05. Despite a low count in 2005/06, the Solway Estuary becomes a site of international importance on the basis of its five-year mean. Numbers at several sites of national importance were also much lower than average; this was especially true on the Inner Moray and Inverness Firth although numbers there have fluctuated markedly in recent years; it may not be unconnected that counts on the nearby Cromarty Firth were at their highest for over five years.

	01/02	02/03	03/04	04/05	05/06	Mon	Mean
Sites of international importance in the UK							
Loughs Neagh and Beg	3,389	2,565	2,674	5,144	5,826	Feb	3,920
Solway Estuary	2,367 [10]	(1,077)	(1,782)	(4,610)	(185)	Dec	3,489 ▲
Sites of national importance in Great Britain							
Loch Ryan	766 [12]	907 [12]	986	1,577	1,020	Dec	1,051
Inner Moray and Inverness Firth	323	923	518	2,641 [1]	576	Feb	996
Inner Loch Indaal	241	755	1,003				666
Loch of Stenness	513	309	266	315	306	Nov	342
Loch of Harray	97	(185)	420	490	360	Feb	342
Cromarty Firth	353 [12]	160 [1]	13	47	400	Jan	195
Ayr to North Troon	200	120	(12)	(14)			160

	01/02	02/03	03/04	04/05	05/06	Mon	Mean
Auchenharvie Golf Course			145	107	97	Feb	116
Dornoch Firth	107	163	70	150 [12]	77	Feb	113
Rough Firth	88	0	107	204 [10]			100
Forth Estuary	189	130	14 [10]	22	(12)	Jan	89
Sites of all-Ireland importance in Northern Ireland							
Belfast Lough	270	642	669 [10]	1,224 [10]	833	Jan	728
Carlingford Lough	618	168	(158)	233	222	Jan	310
Other sites surpassing table qualifying levels in WeBS-Year 2005/2006 in Great Britain							
Alt Estuary	(65)	84	24	67	84	Feb	65

Lesser Scaup
Aythya affinis

Vagrant
Native Range: N America

GB max: 3 Feb
NI max: 0

Lesser Scaup were recorded from seven sites with all records being of single birds. Birds were at Drift Reservoir in November, December and February, Hornsea Mere in December, Reclamation Pond (Tees Estuary) in January, Grimley New Workings and Ouse Washes in February, College Reservoir in March and April and Rutland Water in April.

Eider
Somateria mollissima

International threshold: 12,850
Great Britain threshold: 730
All-Ireland threshold: 20*

GB max: 30,984 Feb
NI max: 2,103 Sep

*50 is normally used as a minimum threshold

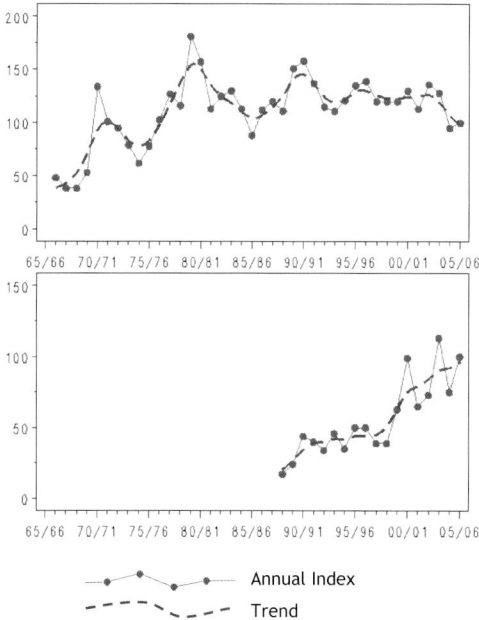

Annual Index
Trend

2005/06
Range 2000/01 - 2004/05
Previous five-year mean

Figure 29.a, Annual indices & trend for Eider for GB (above) & NI (below).

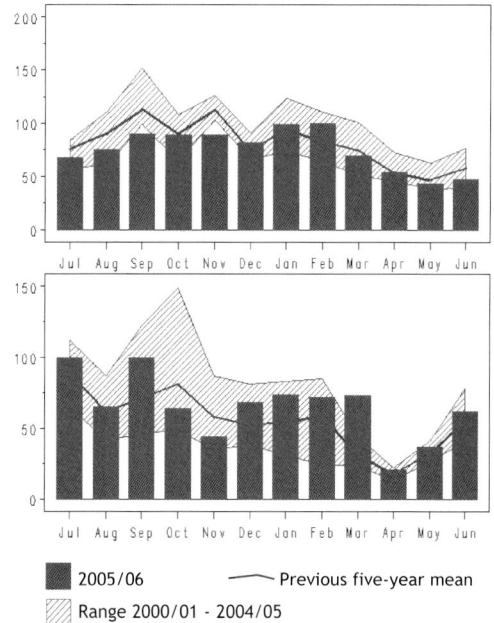

Figure 29.b, Monthly indices for Eider for GB (above) & NI (below).

The Great Britain index rose slightly, though the general trend since the early 1990s continues to be level or slightly declining. Conversely, the Northern Ireland trend continues to rise though the index has seen sharp fluctuations in recent years. Despite the index only rising slightly, the British maximum core count was at its highest level since 1991/92.

Dedicated counts from the greater Firth of Clyde area (again tabulated in whole and as its constituent parts) were incomplete along the Ayrshire coast but the long-term pattern of decline was still very evident, with all areas from Glasgow out to Arran and Kintyre seeing declines on the previous year. However, due to a re-evaluation of the international threshold level, the greater Clyde area now qualifies as a site of international importance.

Counts for the Tay Estuary were much higher than in recent years, although the main flock at this site can often be difficult to view satisfactorily. Counts on Montrose Basin and the Dee Estuary (Scotland) were also at their highest for many years and there was a high count from the dedicated boat-based survey of Hacosay/Bluemull/Colgrave Sounds in Shetland. The decline of Eiders along the Northumberland coast continues with the

Farne Islands no longer qualifying as a site of national importance, whilst the peak count from Lindisfarne, which came from low tide data, was also the lowest on record.

In Northern Ireland, many sites recorded higher numbers than in 2004/05, with the count from the key site at Belfast Lough, which was a low tide count, being one of the highest yet recorded at the site. Strangford Lough and Outer Ards Shoreline also supported notably high counts. Conversely, counts from Lough Foyle and Larne Lough were notably lower than their five year means.

Aerial surveys employing distance sampling

Area	Date	Counted	Estimate	Ref
Outer Hebrides	Jan	948	not available	Söhle *et al.* 2006
Scapa/Deer/Shapinsay	Feb	720	not available	Söhle *et al.* 2006
Sound of Gigha	Dec	623	not available	Söhle *et al.* 2006
Clyde/Loch Ryan	Mar	589	not available	Söhle *et al.* 2006
Moray Firth	Jan	393	not available	Söhle *et al.* 2006
Aberdeen Bay	May	283	not available	Söhle *et al.* 2006
Coll/Tiree/West Mull	Feb	267	not available	Söhle *et al.* 2006
Luce Bay	Feb	145	not available	Söhle *et al.* 2006

	01/02	02/03	03/04	04/05	05/06	Mon	Mean
Sites of international importance in the UK							
FIRTH OF CLYDE	15,692 [14]	14,297 [14]	15,276 [14]	13,042 [14]	(8,055) [14]	Sep	14,577
Sites of national importance in Great Britain							
Tay Estuary	7,500 [12]	6,000	4,700	5,636	11,500	Feb	7,067
Forth Estuary	5,684	7,616	7,014	4,750	5,047	Sep	6,022
Inner Firth of Clyde	3,901	4,730	6,194	4,152	3,837	Aug	4,563
Morecambe Bay	3,903	4,541	(3,940)	5,300 [14]	3,815	Apr	4,390
Aberdeen Bay offshore			1,756 [56]	6,003 [56]	5,302 [56]	Aug	4,354
Killantringan Bay				3,600 [14]			3,600
Ythan Estuary	3,531	2,082	3,417	(4,212)	2,239	Oct	3,096
Montrose Basin	3,013	3,051	2,075	1,754	4,322	Feb	2,843
Gare Loch	3,252 [14]	2,619 [14]	3,263 [14]	2,713 [14]	2,582 [14]	Sep	2,886
Irvine Bay				1,547 [14]			1,547
Lindisfarne	2,024	2,043	1,241	1,202 [10]	1,097 [10]	Nov	1,521
Loch Long and Loch Goil	1,299 [14]	1,459 [14]	1,390 [14]	1,614 [14]	1,458 [14]	Sep	1,444
Gourock to Largs	1,097 [14]	1,773 [14]	2,220 [14]	614 [14]	370 [14]	Sep	1,213
Girvan to Turnberry	(151)	1,198	(330)	1,500 [14]	(415)	Dec	1,349
Moray Firth	749	747	1,639	1,673	1,390	Jan	1,240
Holy Loch to Toward Point	615 [14]	1,146 [14]	1,114 [14]	2,225 [14]	766 [14]	Sep	1,173
Loch Ryan	(1,031)	1,188 [14]	1,803	1,150 [14]	539	Aug	1,170
Inner Loch Fyne	1,647 [14]	1,358 [14]	956 [14]	868 [14]	759 [14]	Sep	1,118
The Wash	1,344	2,546	703	91	557	Feb	1,048
Dee Estuary (Scotland)	408	874	852	865	1,673	Jul	934
Ayr to North Troon	1,203 [14]	458 [14]	1,064	225 [14]	(380)	Feb	738
Sites of all-Ireland importance in Northern Ireland							
Belfast Lough		906	1,016 [10]	1,813	1,490 [10]	1,839 [10]	Dec	1,413
Lough Foyle		344	551	645	431	164	Nov	427
Outer Ards Shoreline			428	256	271	335	Jan	323
Strangford Lough		283	165	259	282	480	Nov	294
Larne Lough		107	120	55	69	67	Sep	84
Port Stewart - Portrush					34 [14]			34
Ballycastle - Fair Head					26 [14]			26
Bann Estuary		32	21	10	26	11	Jul	20
Sites no longer meeting table qualifying levels in WeBS-Year 2005/2006							
Bute		1,143 [14]	944 [14]	457 [14]	451 [14]	490 [14]	Sep	697 ▼
Farne Islands		(671)	(293)	(183)	380	99	Jun	361
Other sites surpassing table qualifying levels in WeBS-Year 2005/2006 in Great Britain							
Hacosay, Bluemull & Colgrave Snds	183 [9]	631 [9]	790 [9]	855 [9]	992 [9]	Jan	690

King Eider

Somateria spectabilis

Vagrant
Native Range: Arctic

GB max: 1 Feb
NI max: 0

A single King Eider was present at Nairn Bar on the Inner Moray Firth in February. This is the first WeBS record from this area since one on the Moray Coast in October 1994.

Long-tailed Duck

Clangula hyemalis

International threshold: 20,000**
Great Britain threshold: 160[†]
All-Ireland threshold: +[†]

GB max: 12,094 Feb
NI max: 22 Mar

The counted British maximum far exceeded that of 2004/05 being the highest on record by some margin, although the total in Northern Ireland was typically very much smaller. The Moray Firth remains by far the most important site in the UK for Long-tailed Ducks with the huge count in January largely due to a flock of 9,000 birds off the Fort George-Delnies stretch, plus over 1,000 more off both Culbin Bar and Findhorn Bay. However, counts at most other sites were average or lower than average, with particularly low numbers at St Andrews Bay and Papa Westray. In Northern Ireland, Belfast Lough remains the key site though numbers here were again below the five-year average. It should be noted, however, that as with all seaducks, divers and grebes on inshore waters, counts are dependent on sea conditions on the day of the count, which can affect both visibility and location of birds.

Supplementary counts from Shetland on behalf of the Shetland Oil Terminal Environmental Advisory Group (SOTEAG) again provide a useful picture away from standard WeBS sections. In Shetland, South Yell Sound recorded the highest peak, though numbers in Hacosay, Bluemull and Colgrave Sounds were nearly half the total recorded in 2004/05.

Aerial surveys employing distance sampling

Area	Date	Counted	Estimate	Ref
Moray Firth	Jan	524	not available	Söhle *et al.* 2006
Scapa/Deer/Shapinsay	Jan	300	not available	Söhle *et al.* 2006
Outer Hebrides	Jan	75	not available	Söhle *et al.* 2006
Sound of Gigha	Dec	35	not available	Söhle *et al.* 2006

	01/02	02/03	03/04	04/05	05/06	Mon	Mean
Sites of national importance in Great Britain							
Moray Firth	1,501	3,585 [1]	5,446 [1]	6,402 [1]	11,565 [1]	Jan	5,700
South Uist West Coast		411 [49]	440 [49]	185 [49]			345
Forth Estuary	413	435	249	(240)	246	Feb	336
Sound of Harris	117 [49]		230 [49]	500 [49]			282
Branahuie Banks (Lewis)				196			196
Hacosay, Bluemull and Colgrave Sounds	201 [9]	59 [9]	249 [9]	303 [9]	160 [9]	Jan	194
Broad Bay (Lewis)		72 [49]	300 [49]				186
Sites with mean peak counts of 50 or more birds in Great Britain[†]							
West Whalsay and Sounds	152 [9]						152
South Yell Sound	136 [9]	108 [9]	201 [9]	91 [9]	169 [9]	Feb	141
Loch of Stenness	226	182	105 [12]	89	96	Mar	140
Loch Branahuie (Lewis)	4			272			138
Sound of Barra (Barra)	150 [49]		132 [49]	80 [49]			121
Burra and Trondra	109 [9]		97 [9]	117 [9]	99 [9]	Jan	106
Bressay Sound	130 [9]	176 [9]	66 [9]	90 [9]	44 [9]	Jan	101
Quendale to Virkie	117 [9]	122 [9]	103 [9]	100 [9]	57 [9]	Feb	100
Island of Papa Westray	4	182	184	102	10	Jan	96
St Andrews Bay	10	97	107	232	17	Nov	93
Traigh Luskentyre	126	50 [49]		100 [49]			92
Water Sound	68	155	80	60	37	Nov	80
West Coast (Benbecula)		63 [49]	92 [49]				78
Allasdale Bay to Borve (Barra)	30 [49]		112 [49]	68 [49]			70
Rova Head to Wadbister Ness	131 [9]	63 [9]	34 [9]	21 [9]	87 [9]	Dec	67
Kirkabister to Wadbister Ness	90 [9]	21 [9]	73 [9]	(4) [9]	78 [9]	Dec	66
Gulberwick Area					56 [9]	Jan	56
Other sites surpassing table qualifying levels in WeBS-Year 2005/06 in Great Britain[†]							
Deveron Estuary	0	0	0	4	50	Nov	11

[†] *as few sites exceed the British threshold and no All-Ireland threshold has been set, qualifying levels of 50 and 30 have been chosen to select sites, in Great Britain and Northern Ireland respectively.*

Common Scoter

Melanitta nigra

International threshold: 16,000
Great Britain threshold: 500
All-Ireland threshold: 40*

GB max: 13,055 Feb
NI max: 8 Oct

50 is normally used as a minimum threshold

The maximum counted in Britain during WeBS Core Counts was higher than in recent years, though as demonstrated by supplementary counts from offshore aerial surveys, this vastly underestimates the true total numbers present. Dedicated counts made in Carmarthen Bay found in excess of 20,000 birds for the fifth year running. On the North Norfolk Coast, the Core Count peak was the highest since 2001/02 whilst counts on the Moray Firth, Alt Estuary and Forth Estuary were also higher than in recent years. Counts in St Andrews Bay and Lavan Sands, however, were much lower than for 2004/05. As this species is present in substantial numbers offshore, annual fluctuations in land-based counts should be interpreted with caution, as they can be strongly influenced by the sea-conditions on the day of a count. The maximum for Northern Ireland was again very low with just a handful of birds recorded from the main coastal sites; no counts were received for the sea off Dundrum Bay which is traditionally the main site for this species in the province.

Aerial surveys employing distance sampling

Area	Date	Counted	Estimate (confidence intervals)	Ref
Liverpool Bay	Feb	23,439	60,215 (32,249-99,514)	Smith *et al.* 2007
Carmarthen Bay	Feb	14,652	28,456 (18,154-44,736)	Banks *et al.* 2007
Aberdeen Bay	May	455	not available	Söhle *et al.* 2006
Luce Bay	Feb	265	not available	Söhle *et al.* 2006
Moray Firth	Jan	205	not available	Söhle *et al.* 2006

	01/02	02/03	03/04	04/05	05/06	Mon	Mean
Sites of international importance in the UK							
Carmarthen Bay	20,078 [31]	23,288 [25]	20,271 [43]	24,460 [43]	20,287 [43]	Dec	21,677
Sites of national importance in Great Britain							
Colwyn Bay	(5,194) [31]	(7,436) [31]	(1,737)	(252)	(2,891)	Oct	(7,436)
Moray Firth	(3,072)	(8,351)	(7,987)	4,265	6,842	Jan	6,861
North Norfolk Coast	8,008	5,051	2,252	4,866	6,830	Jan	5,401
Cardigan Bay	4,045	(4,219)	(198)	(183)	(339)	Feb	4,132
Conwy Bay	(3,336) [31]	(1,424) [31]					(3,336)
Aberdeen Bay			2,992 [56]	3,475 [56]	3,514 [56]	Jan	3,327
Alt Estuary	1,900 [12]	1,818	2,169	3,000	4,300	Mar	2,637
Forth Estuary	2,557 [28]	3,255	1,349	(985)	1,495	Dec	2,164
St Andrews Bay	1,705	584	1,170	2,660	447	Sep	1,313
Dee Estuary (England and Wales)	4,000 [10]	5	26	17	40	Jan	818
Durham Coast	(30)	(151)	(0)	(40)	685	Nov	685 ▲
Sites of all-Ireland importance in Northern Ireland							
Dundrum Bay	828 [21]	(0)	(0)	(0)	(0)		828
Sites no longer meeting table qualifying levels in WeBS-Year 2005/2006							
The Wash	150	452	(15)	372	100	Dec	269

Black Scoter

Melanitta americana

Vagrant
Native Range: N America

GB max: 1 Jan
NI max: 0

The long-staying Black Scoter at Llanfairfechan Saltings, Lavan Sands was recorded during the January WeBS count. This individual has wintered in the area since 1999 but has only been recognised as a full species by WeBS since 2005 (Sangster *et al.* 2005).

Surf Scoter

Melanitta perspicillata

Vagrant
Native Range: N America

GB max: 3 Oct
NI max: 0

Apart from a single bird at Findhorn Bay in October all other records of Surf Scoter were from the Forth Estuary, two were recorded at Shell Bay to Ruddons Point in October, with singles at this site in November, December, February and March. Other singles were in Largo Bay in November and December.

Velvet Scoter
Melanitta fusca

International threshold: 10,000
Great Britain threshold: 30*
All-Ireland threshold: +

GB max: 1,869 Feb
NI max: 0

*50 is normally used as a minimum threshold

The British maximum was slightly lower than during the previous year and was recorded two months later. The majority of birds were noted in Scotland with a scattering throughout coastal sites in England and just two individuals recorded in Wales, at the Dysynni Estuary. The highest site total was typically at the Moray Firth, although the peak was lower than average; the majority of the birds there were found off Culbin Bar. The peak count on the Forth Estuary was the lowest since the 2000/01 winter, whilst counts at St Andrews Bay were also extremely low. At all sites, however, counts can be heavily influenced by visibility and sea-conditions on the dates of the counts. Away from Scotland, the North Norfolk Coast remains the only site with mean numbers in excess of 30 birds. Unusual inland records came from Chew Valley Lake with three there in November and one in December.

	01/02	02/03	03/04	04/05	05/06	Mon	Mean
Sites of national importance in Great Britain							
Moray Firth	610	4,398	2,103 [1]	1,169 [1]	1,261	Feb	1,908
Forth Estuary	1,923	1,487	1,008	1,007	775	Dec	1,240
St Andrews Bay	800	2	90	1,050	8	Sep	390
Lunan Bay	400	105	(300)	125	120	Sep	210
Aberdeen Bay offshore			17 [56]	50 [56]	89 [56]	Sep	52
North Norfolk Coast	41	55	14	45	25	Feb	36

Barrow's Goldeneye
Bucephala islandica

Vagrant and escape
Native Range: N Europe, N America

GB max: 0
NI max: 2 Feb

The first-ever records of Barrow's Goldeneye during WeBS were from Strangford Lough during 2005/06. This bird spent most of the winter at Quoile Pondage, although what was presumably the same bird was also reported from the north of the lough in February, leading to the apparent total of two birds for Northern Ireland.

Goldeneye
Bucephala clangula

International threshold: 11,500
Great Britain threshold: 249
All-Ireland threshold: 110

GB max: 12,763 Feb
NI max: 6,532 Feb

Since the peak in the late 1990s, the British indices have shown a pattern of sustained decline. In Northern Ireland, the decline has been steeper, with a slight drop in the index value following two years of relative stability. Due to a major increase in the international 1% threshold (from 4,000 to 11,500), the UK no longer has any sites that support internationally important numbers, though Loughs Neagh and Beg continue to hold far greater numbers than any other site. Numbers there were slightly down on the previous year, as were peaks at Lower Lough Erne and Belfast Lough, though increased peaks were noted from Strangford Lough, Larne Lough and the Outer Ards. No sites in Great Britain held peaks more than 1,000 birds recorded, with the peak for the Inner Moray and Inverness Firths peak being extremely low. At the Humber Estuary, the low total was largely due to no count being received from the key section of Goxhill to New Holland. However, the Firth of Clyde and Abberton Reservoir peaks were notably higher than in previous years.

Figure 30.a, Annual indices & trend for Goldeneye for GB (above) & NI (below).

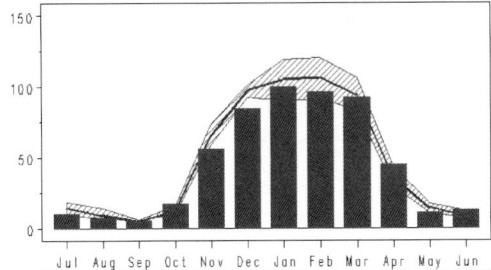

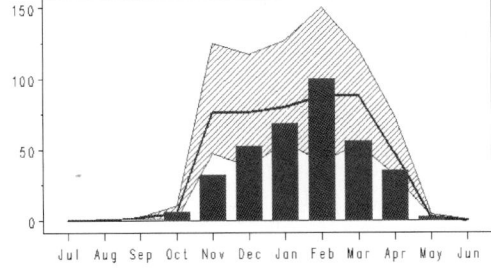

Figure 30.b, Monthly indices for Goldeneye for GB (above) & NI (below).

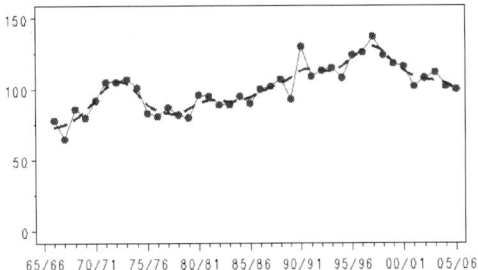

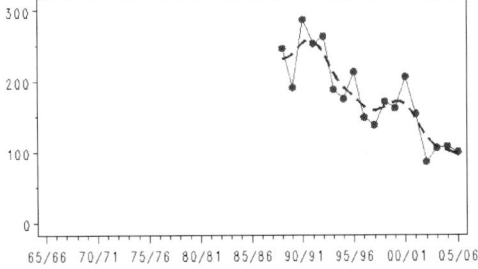

Legend:
- ●— Annual Index
- - - - Trend
- ■ 2005/06
- ⌇ Range 2000/01 - 2004/05
- ⌐ Previous five-year mean

	01/02	02/03	03/04	04/05	05/06	Mon	Mean
Sites of national importance in Great Britain							
Forth Estuary	1,113	1,241	(753)	879	(365)	Jan	1,078
Inner Moray and Inverness Firth	993	1,352 [10]	709 [1]	1,165 [1]	186	Feb	881
Abberton Reservoir	(619)	469	431	394	588	Mar	500
Rutland Water	450	428	511	420	521	Jan	466
Humber Estuary	208	618	296	595	(24)	Jan	429
Inner Firth of Clyde	321	264	(514)	159	636	Feb	379
Hornsea Mere	294	(480) [12]	235	325	280	Apr	323
Loch of Skene	270	(192)	298	207	334	Jan	277
Tweed Estuary	312	240	390	273	140	Mar	271
Solway Estuary	(122)	(254)	(188)	(126)	(76)	Mar	(254) ▲
Morecambe Bay	221	280	204	(297)	(247)	Jan	250
Sites of all-Ireland importance in Northern Ireland							
Loughs Neagh and Beg	6,454	3,661	4,497	5,787	5,688	Feb	5,217
Lower Lough Erne		218	(337)	319	254	Feb	282
Strangford Lough	256	295	253	161	187	Feb	230
Belfast Lough	140 [12]	249	242 [10]	164 [10]	103	Feb	180
Larne Lough	189	130	95	73	155	Feb	128
Other sites surpassing table qualifying levels in WeBS-Year 2005/2006 in Great Britain							
Loch Leven	249	153	86	385	289	Nov	232
Other sites surpassing table qualifying levels in WeBS-Year 2005/2006 in Northern Ireland							
Outer Ards Shoreline		63	65	2	181	Jan	78

Hooded Merganser

Lophodytes cucullatus

Escape

Native Range: N America

GB max: 1 Dec
NI max: 0

The only record was from Chilham & Chartham Gravel Pits in December. Hooded Mergansers, which generally are assumed to relate to escapes from captivity (although natural vagrancy cannot be ruled out), are only very rarely recorded during WeBS counts, this being the eighth location to host the species.

Smew
Mergellus albellus

International threshold: 400
Great Britain threshold: 4*
All-Ireland threshold: +

GB max: 214 Jan
NI max: 3 Feb

50 is normally used as a minimum threshold

Following the extremely low British total in 2004/05, the maximum rose somewhat this year, but was still low in comparison to other recent years. The peak count at the key site, Wraysbury Gravel Pits, was much lower than in recent years, as was the case with many other wildfowl species at this site, potentially due to increased disturbance there. Counts at several of the other regular sites for wintering Smew such as Lee Valley Gravel Pits, Rye Harbour and Pett Level, Abberton Reservoir and Fen Drayton Gravel Pits were also lower than in recent years. Conversely, Cotswold Water Park (West), Thorpe Water Park, Pitsford Reservoir and Rutland Water all supported higher than average counts. The maximum count of three in Northern Ireland was not unusually low.

	01/02	02/03	03/04	04/05	05/06	Mon	Mean
Sites of national importance in Great Britain							
Wraysbury Gravel Pits	66	63	55	68	38	Jan	58
Cotswold Water Park (West)	31	(32)	20	18	33	Feb	27
Dungeness Gravel Pits	32	18	33	14	17	Mar	23
Lee Valley Gravel Pits	23	29	23	8	9	Dec	18
Rye Harbour and Pett Level	20	28	19	10	8	Feb	17
Thorpe Water Park	6	11	18	10	20	Dec	13
Seaton Gravel Pits and River	11	7 [12]	14	8	11	Jan	10
Twyford Gravel Pits	7	12					10
Fen Drayton Gravel Pits	15	11	16	4	5	Dec	10
Sonning Eye & Henley Rd GPs	(3)	(9)	(0)	9			9
Rutland Water	12	8	8	4	14	Jan	9
Middle Tame Valley Gravel Pits	(8)	(5)	(1)	(0)			(8)
Abberton Reservoir	(9)	4	5	9	(1)	Nov	7
Thrapston Gravel Pits	(2)	2	11				7
Ouse Fen & Pits (Hanson/RSPB)		(0)	2	10	10	Jan	7
Little Paxton Gravel Pits	4	8	4	12	5	Jan	7
Marsh Lane GPs Hemingford Grey		6		9	5	Jan	7
Bedfont and Ashford Gravel Pits	6	(6)					6
Pitsford Reservoir	9	2	3	3	11	Jan	6
Belhus Woods Country Park	0	7	10	6	0		5
Grange Waters Complex	2	11	0	5	5	Dec	5
Colne Valley Gravel Pits	0	6	(8)	7	6	Jan	5
Cassington & Yarnton GPs	(0)	(0)	10	1	5	Feb	5
Tophill Low Reservoirs	3	5	6	7	5	Jan	5
Earls Barton Gravel Pits	0	7	7	0	6	Feb	4
Meadow Lane Gravel Pits St Ives	1	17	3	0	0		4
Eyebrook Reservoir	7	1	3	3	7	Mar	4
Fairburn Ings	6	6	4		1	Feb	4
Sites no longer meeting table qualifying levels in WeBS-Year 2005/2006							
Aston On Trent Gravel Pits			(2)	(0)	(2)	Dec	(2)
Chew Valley Lake	3	7	4	0	0		3
Leybourne and New Hythe GPs	7	3	0		0		3
Other sites surpassing table qualifying levels in WeBS-Year 2005/2006 in Great Britain							
Brandesburton Ponds West	1	1	0	0	6	Feb	2
Holme Pierrepont Gravel Pits			(1)	0	5	Jan	3
Minsmere	4	3	2	3	5	Feb	3
Linne Mhuirich		0	0	0	4	Nov	1
Ruislip Lido			(0)	2	4	Feb	3

Red-breasted Merganser
Mergus serrator

International threshold: 1,700
Great Britain threshold: 98
All-Ireland threshold: 20*

GB max: 3,277 Feb
NI max: 489 Oct

50 is normally used as a minimum threshold

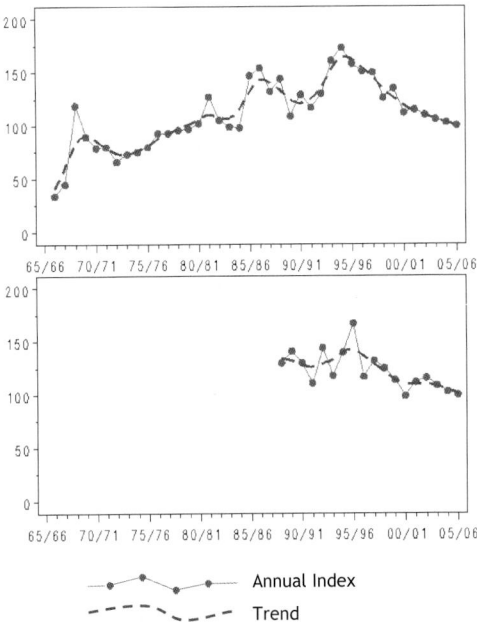

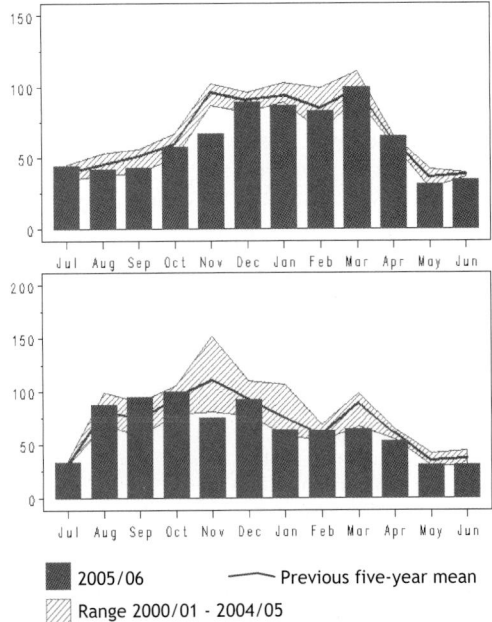

Annual Index — — — Trend

2005/06 — Previous five-year mean
Range 2000/01 - 2004/05

Figure 31.a, Annual indices & trend for Red-breasted Merganser for GB (above) & NI (below).

Figure 31.b, Monthly indices for Red-breasted Merganser for GB (above) & NI (below).

The decline in the British index for Red-breasted Merganser since the mid 1990s continues, though numbers still remain higher than during the 1970s. The Northern Ireland index also continued to fall. The monthly indices suggest that the peak count for Northern Ireland was much earlier than for Britain, coming in October as opposed to February.

In Britain, notably high counts were recorded from Fleet and Wey, Morecambe Bay, Chichester Harbour, Inner Firth of Clyde,

Langstone Harbour, Loch Lomond, Tay Estuary, Whiteness and Skelda Ness and Montrose Basin. However, the peak count from the Forth Estuary was the lowest since 2000/01, whilst the Moray Firth and the Exe Estuary also supported lower numbers than usual. In Northern Ireland, counts at Strangford Lough and Lough Foyle were higher than in recent years. The high count from Carlingford Lough in 2004/05 was not sustained, though the count was still higher than other recent years.

	01/02	02/03	03/04	04/05	05/06	Mon	Mean
Sites of national importance in Great Britain							
Forth Estuary	599	769	791	544	489	Oct	638
Fleet and Wey	366	358	425	413	438	Mar	400
Poole Harbour	(417)	469	(392)	315	(250)	Dec	398
Moray Firth	234	355	338	300	254	Oct	296
Morecambe Bay	229	(265)	(170)	167	263	Dec	231
Lavan Sands	164	170	264	(211)	196	Aug	201
Chichester Harbour	(159)	(184)	191	194	(212)	Mar	199
Inner Firth of Clyde	(196)	141	(164)	107	252	Jul	174
Duddon Estuary	136	220	167	152	(121)	Feb	169
Langstone Harbour	192	158	127	128	187	Dec	158
Loch Ryan	(113)	133 [12]	74	179	180	Oct	142
Loch Lomond	(14)	(4)	(14)	(4)	(129)	Jul	(129) ▲
Inner Loch Indaal	40	172	138				117

	01/02	02/03	03/04	04/05	05/06	Mon	Mean
North Norfolk Coast	102	109	105	126	132	Dec	115
Loch of Tankerness				222	1	Sep	112
Goring	186	35	(35)	(102)			111 ▲
Arran	(94)	126	103	90	113	Nov	108
Exe Estuary	94	112	(132)	82	78	Jan	100
Sites of all-Ireland importance in Northern Ireland							
Strangford Lough	342	187	188	189	263	Sep	234
Larne Lough	176	123	135	211	151	Oct	159
Belfast Lough	162	228	216	75	104	Dec	157
Lough Foyle	73	37	122	52	169	Dec	91
Carlingford Lough	24	106	40	154	118	Aug	88
Outer Ards Shoreline		62	48	54	31	Jan	49
Sites no longer meeting table qualifying levels in WeBS-Year 2005/2006							
Solway Firth	(58)	(55)	92	(84)	(53)	Oct	92
Other sites surpassing table qualifying levels in WeBS-Year 2005/2006 in Great Britain							
Tay Estuary	30	39	98	60	172	Aug	80
Montrose Basin	16	33	139	(39)	163	Jul	88
Whiteness to Skelda Ness	47 [9]	82 [9]	91 [9]	68 [9]	145 [9]	Feb	87
Kirkabister to Wadbister Ness	79 [9]	47 [9]	100 [9]	(52) [9]	110 [9]	Dec	84

Goosander

Mergus merganser

International threshold: 2,700
Great Britain threshold: 161[†]
All-Ireland threshold: +

GB max: 3,087 Dec
NI max: 1 Nov

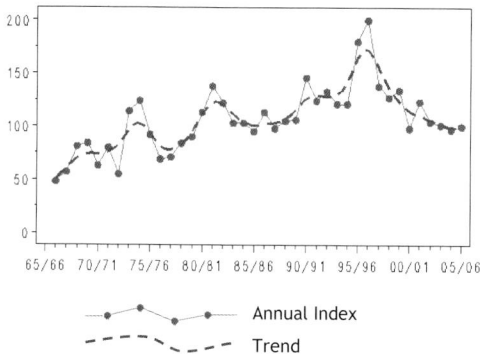

Annual Index
Trend

Figure 32.a, Annual indices & trend for Goosander for GB.

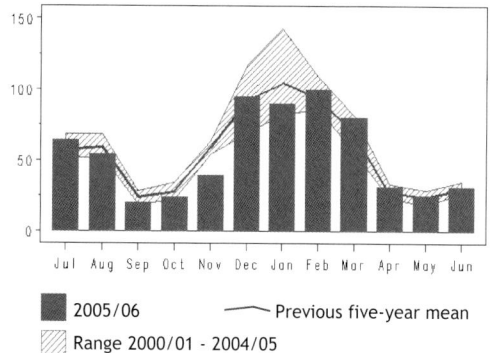

2005/06
Range 2000/01 - 2004/05
Previous five-year mean

Figure 32.b, Monthly indices for Goosander for GB.

The British annual index showed a very slight increase in 2005/06, although the overall trend continues to be one of decline. As in previous years, the monthly indices show a distinct dip between September and November, as many birds vacate the country at this time to moult in Norway. Typically the species was again very scarce in Northern Ireland with single birds recorded at Belfast, Larne and Strangford Loughs, all in different months. Only two sites now support nationally important numbers with mean peak numbers at Tyninghame Estuary dropping below the qualifying level following a particularly low count in 2005/06. Following the high count on the Tay Estuary in 2004/05, the subsequent peak for 2005/06 was the lowest for over five years. The zero count at Castle Loch (Lochmaben) was due to an absence of winter counts at the site. Yetholm Loch, Hirsel Lake, and Windermere all supported higher counts than normal, and the peak at the Severn/Vyrnwy Confluence was exceptionally high for a site which generally holds only single figures. As so few sites now support nationally important numbers an arbitrary level of 70 birds has been set for the purposes of the table in this report.

Goosander present a problem for WeBS counts in that in many areas they disperse from large water bodies during the day, only returning at dusk to roost, spending time on rivers which may not be covered by WeBS. Increased submissions of roost counts and/or counts of riverine stretches would give a more complete picture of numbers present during the winter months.

	01/02	02/03	03/04	04/05	05/06	Mon	Mean
Sites of national importance in Great Britain							
Tay Estuary	245	248	192 [12]	263	153	Aug	220
River Tweed - Kelso/Coldstream	371	179	61	112	113	Oct	167
Sites with mean peak counts of 70 or more birds in Great Britain[†]							
Tyninghame Estuary	161	97	177	189	69	Jun	139
Eccup Reservoir	103	95	137	94	115	Dec	109
Yetholm Loch	16	(13)	(16)	(54)	167	Feb	92
Hirsel Lake			6	15	240	Dec	87
Solway Estuary	(27)	72	(105)	84	(47)	Aug	87
Castle Loch (Lochmaben)	113	82	137	88	0		84
Loch Lomond	(37)	(84)	(23)	(15)	(19)	Sep	(84)
Inner Moray and Inverness Firth	191 [12]	85 [12]	137 [12]	2	1	Oct	83
Forth Estuary	54	(89)	53	81	119	Jul	79
Other sites surpassing table qualifying levels in WeBS-Year 2005/2006 in Great Britain[†]							
River Severn/Vyrnwy Confluence	4	7	2	8	175	Dec	39
Windermere	19	24	(70)	48	127	Aug	58
Eden Estuary	38	58	56	60	(75)	Aug	57
Spittal to Cocklawburn	(0)	7	92	86	72	Aug	64

[†] *as few sites exceed the British threshold a qualifying level of 70 has been chosen to select sites for presentation in this report*

Ruddy Duck
Oxyura jamaicensis

Naturalised introduction[†]
Native Range: N & S America

GB max: 2,991 Dec
NI max: 36 Nov

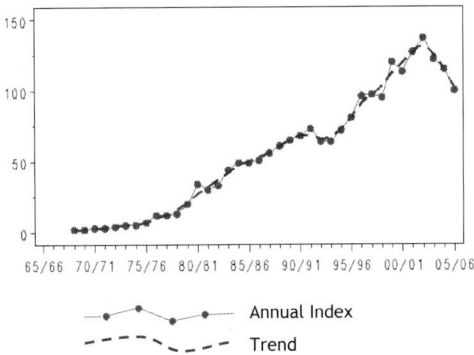

Figure 33.a, Annual indices & trend for Ruddy Duck for GB.

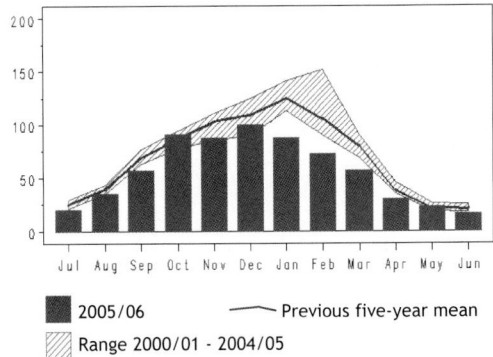

Figure 33.b, Monthly indices for Ruddy Duck for GB.

The annual British index continues to show a steep decline since the peak in 2002/03, with many monthly indices being well below their five-year average. This pattern can clearly be attributed to the control programme intended to reduce numbers of this naturalised introduction with the aim of aiding the conservation of the White-headed Duck in southern Europe.

Peak counts at many of the top sites for Ruddy Duck were below the five-year mean, with notably low counts coming from Rutland Water, Blithfield Reservoir, King George V Reservoirs, Stanford Reservoir, Cotswold Water Park (West), Sutton and Lound Gravel Pits, Eyebrook Reservoir and Hollowell Reservoir. Conversely, some sites showed slightly increased numbers from 2004/05, such as Abberton Reservoir, Chew Valley Lake, Blagdon Lake, Pitsford Reservoir and King George VI Reservoir, whilst counts from Colne Valley Gravel Pits, Carsington Water, Hurworth Burn Reservoir and Holme Pierrepont Gravel Pits were at their highest for over five years. As usual, the peak count for Northern Ireland was comprised entirely of birds at Loughs Neagh and Beg.

	01/02	02/03	03/04	04/05	05/06	Mon	Mean
Sites with mean peak counts of 30 or more birds in Great Britain[†]							
Staines Reservoirs	444	(696)	694	695 [41]	521 [41]	Jan	610
Abberton Reservoir	456	493	678	403	455	Mar	497
Hanningfield Reservoir	553	(664)	285 [41]	412 [41]	246 [41]	Jan	432
Rutland Water	911	482	200	251	57	Mar	380
Chew Valley Lake	491 [12]	427 [12]	488 [41]	220 [41]	257 [41]	Jan	377
Blagdon Lake	463	394	249	151 [41]	172	Feb	286
Dungeness Gravel Pits	224	264	222	287	250	Dec	249
Pitsford Reservoir	293	358	103	178	311 [41]	Jan	249
Blithfield Reservoir	265	187	180 [41]	401	59 [41]	Jan	218
Hilfield Park Reservoir	159	125	187	241 [41]	176	Dec	178
Holme Pierrepont Gravel Pits			115	189	202	Feb	169
King George V Reservoirs	156	135	268	(23)	83	Dec	161
Stanford Reservoir	274	97	277 [41]	76	29	Oct	151
Tophill Low Reservoirs	173	89	110 [41]	124	131	Nov	125
Anglers Country Park Lake	58	76	78	180 [41]	185	Nov	115
Middle Tame Valley Gravel Pits	146 [41]	(120)	(96)	58 [41]	126 [41]	Jan	113
Cotswold Water Park (West)	128	(60)	127	125	59	Dec	110
Carsington Water	141 [41]	132	0	82 [41]	182	Jan	107
King George VI Reservoir	283 [12]	0	2	0	145	Oct	86
Walthamstow Reservoirs	40 [41]	(67)	118	90	86 [41]	Jan	84
Bolton-on-Swale Gravel Pits	97	108	118 [41]	55 [41]	37	Sep	83
Brent Reservoir	73	104	25	133	77	Sep	82
Sutton and Lound Gravel Pits	132	26	46	175	13	Apr	78
Thames Estuary	34	106	(82)	85	(85)	Jan	78
Blackwater Estuary	152	53	69	71	39	Nov	77
Colwick Country Park	69 [41]		88 [41]	100 [41]	51	Feb	77
Eyebrook Reservoir	236	56	84	5	2	Nov	77
Hollowell Reservoir	76	39	191	53	19	Dec	76
Humber Estuary	45	55	116	84	(15)	Nov	75
Wigan Flashes	78	49	60	86			68
Llyn Traffwll	36 [41]	80	83	78	52	Oct	66
Clumber Park Lake	122	72	76	16 [41]	41	Mar	65
Fairburn Ings	69	94	115	5 [41]	40	May	65
Tees Estuary	(40)	77	70	37	63	Apr	62
Colne Valley Gravel Pits	9	12	16	33	215 [41]	Jan	57
Llynnau Y Fali	32	86	57	70	39	Apr	57
Knight & Bessborough Res	122	29	23	46	45	Mar	53
Hurworth Burn Reservoir	23	34		9	130	Sep	49
Llyn Alaw	54 [41]	44	2	45	95	Feb	48
Barn Elms Reservoirs	39	36	43	59			45
London Wetland Centre					49 [41]	Jan	45
Hampton & Kempton Reservoirs	47	(30)	39	14	76	Jan	44
Newsham Park			42 [41]				42
Old Moor	40	47	28	24	71	May	42
Pugneys Country Park Lakes	57	7	63	27	50 [41]	Jan	41
Thoresby Lake	30 [41]	3 [41]	69 [41]	46 [41]	42 [41]	Jan	38
Great Pool Westwood Park	59	57	22	35	14	Nov	37
Swithland Reservoir	16	10	61	62	38	Dec	37
Grafham Water	121	10	10	28	12	Dec	36
Houghton Green Pool	36	36	40	40	29	Dec	36
Barrow Gurney Reservoir	59	38	49	19	3	Nov	34
Church Wilne Reservoir	68 [41]				0 [41]		34
Loch Gelly	58	24 [12]	66	6	8	Nov	32
Sites with mean peak counts of 30 or more birds in Northern Ireland[†]							
Loughs Neagh and Beg	59	67	56	33	36	Nov	50
Other sites surpassing table qualifying levels in WeBS-Year 2005/2006 in Great Britain[†]							
Skelton Lake	9	16	32	14	54	Sep	25
Cropston Reservoir		0	4		51 [41]	Jan	18
Gailey Pools	5 [41]		15	18	49	Aug	22
Rostherne Mere	7 [41]	26	5	28	48	Jan	23
Crookfoot Reservoir	3	2	2	14	47	Dec	14
Attenborough Gravel Pits	39		4	22	43	Feb	27
Nosterfield Gravel Pits	10	(22)	(24)	14	43	Sep	23

	01/02	02/03	03/04	04/05	05/06	Mon	Mean
Bardney Pits	0	2			37	Sep	13
Farmwood Pool	21 [41]	8	4	1	34	Jan	14
Durkar Sand Quarry	0	0	0	9	31	Feb	8
Stoke Newington Reservoirs	10	7	(26)	(25)	31	Jan	20
Kilconquhar Loch	30	42	12	5	30	Sep	24

[†] as no British or All-Ireland thresholds have been set a qualifying level of 30 has been chosen to select sites for presentation in this report

Argentine Bluebill
Oxyura vittata

Escape
Native Range: S America

GB max: 1 Jul
NI max: 0

A single Argentine Bluebill was reported from Netherfield Gravel Pits, Nottinghamshire, in July and October. This bird has been recorded regularly at this site since October 2001. The only other records of this species have been of singles at Colwick Country Park, Nottinghamshire, in March 2000 and Melton Country Park, Leicestershire, in September 2000. It is conceivable that all these records refer to the same individual.

White-headed Duck
Oxyura leucocephala

Escape and potential vagrant
Native Range: S Europe, W Asia

GB max: 1 Aug
NI max: 0

The only report of a White-headed Duck was of one from August through to November at Hilfield Park Reservoir. This species was previously recorded at this site in the 2004/05 winter.

Red-throated Diver
Gavia stellata

International threshold: 3,000
Great Britain threshold: 49*
All-Ireland threshold: 10*

GB max: 472 Feb
NI max: 115 Dec

50 is normally used as a minimum threshold

Following a record low count in 2004/05, the maximum British total rose to its highest level since 2002/03, whilst the Northern Ireland maximum count was the second highest ever recorded by WeBS. Three-figure counts were recorded at two sites: the Inner Firth of Clyde, with its highest count to date following a particularly low count in 2004/05; and Glyne Gap in East Sussex with a record high count for the site. Of potential concern was the absence of this species on core counts from Lade Sands for the second consecutive year. Five years ago this site supported a peak of 800 birds but this number has dramatically decreased since then. These reduced counts may be due to sea conditions on the days on which counts were made, or to birds being too far offshore to be seen, although Great Crested Grebes are still being recorded in high numbers which may indeed indicate a lack of birds. Numbers were again low in Cardigan Bay, although birds often feed well offshore here, whilst the Moray Firth and Loch Ryan also yielded lower than average counts. The highest count for Great Britain again came from a supplementary count for Aberdeen Bay where the count was lower than for 2004/05 but average for the three years for which counts have been made there. The notably high maximum count in Northern Ireland was due largely to Lough Foyle where counters recorded their second highest count to date.

Aerial surveys employing distance sampling

Area	Date	Counted	Estimate (confidence intervals)	Ref
Greater Thames	Feb	1,360	7,998 (6,609-9,661)	WWT data
Liverpool Bay	Dec	187	1,518 (981-2,163)	WWT data
Moray Firth	Jan	81	not available	Söhle *et al.* 2006
Aberdeen Bay	May	39	not available	Söhle *et al.* 2006

	01/02	02/03	03/04	04/05	05/06	Mon	Mean
Sites of national importance in Great Britain							
Aberdeen Bay offshore			225 [28]	423 [56]	352 [56]	Apr	333
Cardigan Bay	732	32	22	30	(67)	Dec	204
Lade Sands	800	100	10	0	0		182
Inner Firth of Clyde	112	151	126	34	202	Apr	125
Moray Firth	74	126	166	117	81	Jan	113
Thames Estuary	40	(344) [29]	23	32	66	Feb	101
Forth Estuary	93	106	61	132	87	Nov	96
Loch Ryan	47	111 [12]	89	81	49	Feb	75
Pegwell Bay	54	215	0	10	5	Dec	57
Lavan Sands	13	202 [12]	30	22	8	Dec	55
Sites of all-Ireland importance in Northern Ireland							
Lough Foyle	7	29	147	21	98	Dec	60
Belfast Lough	41	31	13	14	30	Mar	26
Strangford Lough	57 [10]	2	0	2	1	Feb	12
Carlingford Lough	5	19	(4)	4	15	Jan	11 ▲
Sites no longer meeting table qualifying levels in WeBS-Year 2005/2006							
Outer Ards Shoreline		1	6	14	8	Mar	7
Dengie Flats	51	114	50	15	2	Feb	46
Minsmere	4	3	57	3			17
Other sites surpassing table qualifying levels in WeBS-Year 2005/2006 in Great Britain							
Glyne Gap	3	(0)	35	6	103	Feb	37

Black-throated Diver
Gavia arctica

International threshold: 3,750
Great Britain threshold: 7*
All-Ireland threshold: 1*

GB max: 164 Feb
NI max: 2 Mar

*50 is normally used as a minimum threshold

The maximum British count of 164 in February was the highest ever recorded by WeBS counters, being more than double the 2004/05 peak, thanks in part to the efforts of the RAF Ornithological Society expedition to northwest Scotland. Moreover, the peak count of 70 birds at the key British site of Gerrans Bay in Cornwall was the highest site total recorded under WeBS. The counts from Loch Gairloch were very low, although the count at nearby Gruinard Bay was the highest recorded there since 2000/01. Supplementary counts from the Sound of Barra in the Western Isles again illustrate the importance of this area as a wintering ground for this species. Black-throated Divers remain scarce in Northern Ireland however, with a peak of two on the Outer Ards Shoreline and singles at Belfast Lough and Loughs Neagh & Beg.

	01/02	02/03	03/04	04/05	05/06	Mon	Mean
Sites of national importance in Great Britain							
Gerrans Bay	33	53	37	47	70	Mar	48
Sound of Barra (Barra)	40 [49]	37 [49]	31 [49]		35 [49]	Oct	36
Loch Slapin			21 [45]	26 [45]			24
Moray Firth	9	18	48	6	19	Feb	20
Loch Gairloch				28	6	Feb	17
Loch Roag	18 [49]	13 [49]					16
Broad Bay (Lewis)	7 [49]		21 [49]				14
Girvan to Turnberry	8	20	19	(9)	5	Mar	13
Forth Estuary	24	9	5	3	10	May	10
Applecross Bay				5	14	Feb	10 ▲
Little Loch Broom				3	(10)	Feb	7
Sites of all-Ireland importance in Northern Ireland							
Outer Ards Shoreline		(0)	(0)	1	2	Mar	2
Strangford Lough	0	8	0	4	0		2
Belfast Lough	2	1	4 [10]	(0)	1	Oct	2
Sites no longer meeting table qualifying levels in WeBS-Year 2005/2006							
Loch Ewe			0	3		Feb	2
Other sites surpassing table qualifying levels in WeBS-Year 2005/2006 in Great Britain							
Gruinard Bay			0	9		Feb	5
Glyne Gap	0	(0)	0	0	8	Feb	2
Other sites surpassing table qualifying levels in WeBS-Year 2005/2006 in Northern Ireland							
Loughs Neagh and Beg		0	0	0	1	Sep	0

Great Northern Diver
Gavia immer

International threshold: **50**
Great Britain threshold: **30*†**
All-Ireland threshold: **?†**

GB max: 334 Feb
NI max: 63 Dec

*50 is normally used as a minimum threshold

The British maximum total was the highest recorded to date whilst the Northern Ireland total was the highest recorded for WeBS for the province. The highest Core Count in Britain came from Outer Loch Indaal, whilst in Northern Ireland, the peak count at Lough Foyle was the highest WeBS count there on record. The peak count at Moray Firth however was the lowest there since 1999/00. Again, the large amount of supplementary data received has helped to build a truer picture of the numbers and distribution of Great Northern Divers around our coastlines in areas otherwise not covered by WeBS. The count of 203 in the Sound of Gigha off the coast of Kintyre was the second highest count anywhere in the last five years, whilst in the Western Isles, 188 were recorded from aerial surveys offshore between Scarp and Vatersay; moreover, both of these counts were made during distance sampling aerial surveys and the true numbers present will have been much higher than these figures.

Aerial surveys employing distance sampling

Area	Date	Counted	Estimate (confidence intervals)	Ref
Sound of Gigha	Dec	203	not available	Söhle *et al.* 2006
Outer Hebrides	Jan	188	not available	Söhle *et al.* 2006
Scapa/Shapinsay/Deer	Jan	85	not available	Söhle *et al.* 2006
Coll/Tiree/West Mull	Feb	51	not available	Söhle *et al.* 2006
Luce Bay	Feb	29	not available	Söhle *et al.* 2006
Moray Firth	Jan	14	not available	Söhle *et al.* 2006
Firth of Clyde/Loch Ryan	Mar	11	not available	Söhle *et al.* 2006

	01/02	02/03	03/04	04/05	05/06	Mon	Mean	
Sites of international importance in the UK								
South Uist West Coast	(148)[49]	(57)[49]	(48)[49]	(63)[49]			(148)	
Sound of Barra (Barra)	97[49]	142[49]	96[49]	94[49]			107	
Outer Loch Indaal				20	108	Feb	64	▲
Moray Firth	54	60	(109)	37	14	Dec	55	
Inner Loch Indaal	74	68	18				53	▲
Loch Slapin			44[45]	59[45]			52	
Sites with mean peak counts of 10 or more birds in Great Britain†								
Kirkabister to Wadbister Ness	49[9]	22[9]	50[9]	(2)[9]	37[9]	Dec	40	
Traigh Luskentyre	8	60[49]	70[49]	22			40	
East Unst	37[9]						37	
Broadford Bay	(25)	35	(43)	24	(20)	Nov	34	
Gruinard Bay				26	40	Feb	33	
Scousburgh to Maywick					32[9]	Dec	32	
Sound of Harris	32[49]	35[49]	20[49]	42[49]			32	
Pontllyfni to Aberdesach			28[12]				28	
Scalloway Islands	58[9]		19[9]	13[9]	19[9]	Jan	27	
Whiteness to Skelda Ness	34[9]	34[9]	27[9]	30[9]	11[9]	Feb	27	
Loch Ewe				19	33	Feb	26	
Quendale to Virkie	30[9]	22[9]	24[9]	22[9]	27[9]	Dec	25	
Gualan and Balgarva			23[49]				23	
Rova Head to Wadbister Ness	38[9]	19[9]	30[9]	4[9]	17[9]	Dec	22	
Red Point to Port Henderson				17	12	Feb	15	
Gerrans Bay	6	17	15	14	16	Feb	14	
Island of Papa Westray	6	20	22	(1)	5	Feb	13	
Callakille				1	19	Feb	10	
Island of Egilsay	4	3	(0)	10	21	Feb	10	
Sites with mean peak counts of 10 or more birds in Northern Ireland†								
Lough Foyle	4	26	24	5	60	Dec	24	
Carlingford Lough	3	(15)	25	2	17	Nov	12	

† *as few sites exceed the British threshold and no All-Ireland threshold has been set, qualifying levels of 10 have been chosen to select sites for presentation in this report*

Little Grebe
Tachybaptus ruficollis

International threshold: 4,000
Great Britain threshold: 78
All-Ireland threshold: ?[†]

GB max: 5,237 Oct
NI max: 535 Nov

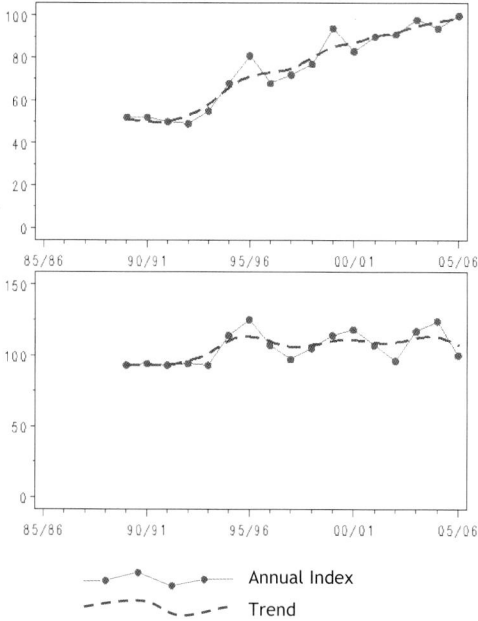

Figure 34.a, Annual indices & trend for Little
Grebe for GB (above) & NI (below).

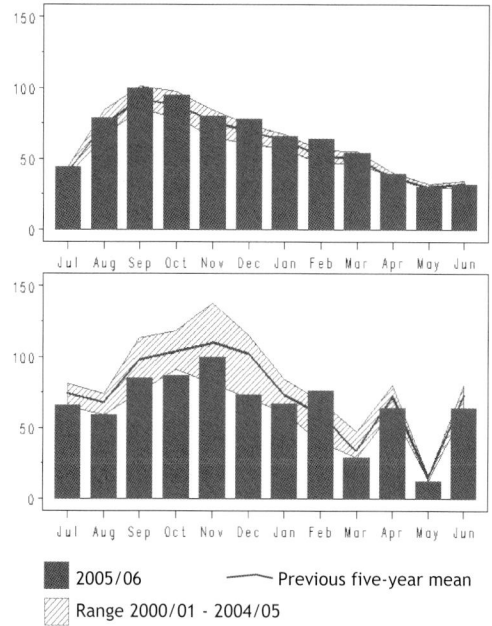

Figure 34.b, Monthly indices for Little Grebe for
GB (above) & NI (below).

Following a small fall in the British annual index in 2004/05, in 2005/06 there was once again a rise in line with the trend seen over the past ten years, with the Great Britain maximum being the highest recorded by WeBS to date. The British monthly index values averaged higher than the five-year mean and mirrored the pattern seen in previous years with a peak in September. It is interesting to note that despite the record British maximum being recorded, peak counts at the top five sites were actually lower than the 2004/05 peaks. Counters at Rutland Water recorded

their highest peak for six years and along with Lee Valley Gravel Pits these sites now qualify as nationally important. The Swale Estuary, however, no longer qualifies despite two consecutively above average years. After two successive years when the index rose, the Northern Ireland index in 2005/06 declined although the overall trend for the province is one of stability. This was reflected in the peak counts from the sites, with Loughs Neagh and Beg supporting its lowest peak for 15 years, with a decrease of 30% on the 2004/05 peak.

	01/02	02/03	03/04	04/05	05/06	Mon	Mean	
Sites of national importance in Great Britain								
Thames Estuary	351	378	(198)	444	377	Dec	388	
Chichester Harbour	150 [10]	111	125	135	95 [10]	Dec	123	
Chew Valley Lake	100	70	110	110	95	Sep	97	
Bewl Water	132	136	94	80	43	Sep	97	
Holme Pierrepont Gravel Pits			55	120	114	Sep	96	
Hamford Water	105	68	(92)	89	114	Feb	94	
Blagdon Lake	127	18	127	98	(69)	Sep	93	
Alde Complex	(76)	54	(47)	109	112	Dec	92	
Tees Estuary	(70)	104	70	54	88	Sep	79	
Lee Valley Gravel Pits	59	71	83	102	77	Oct	78	▲
Rutland Water	77	58	87	70	96	Sep	78	▲
Sites no longer meeting table qualifying levels in WeBS-Year 2005/2006[†]								
Swale Estuary	64 [10]	43	49	89	86	Dec	66	

	01/02	02/03	03/04	04/05	05/06	Mon	Mean
Sites with mean peak counts of 30 or more birds in Northern Ireland[†]							
Loughs Neagh and Beg	418	438	433	466	330	Nov	417
Upper Lough Erne	122	75	131	104	78	Feb	102
Strangford Lough	103	113	83	76	75	Feb	90
Lower Lough Erne		(39)	(57)	(53)	(54)	Dec	(57)
Larne Lough	27	32	65	77	52	Oct	51
Lough Money	53	41	39	51	48	Oct	46
Hillsborough Main Lake	45	37	27	28	21	Sep	32
Other sites surpassing table qualifying levels in WeBS-Year 2005/2006 in Great Britain[†]							
Dungeness Gravel Pits	(75)	48	65	56	99	Sep	69
Deben Estuary	70	73	76	74	90	Mar	77
Carsington Water	51	51	43	45	89	Jan	56
Severn Estuary	41	68	54	52	87	Oct	60
Pitsford Reservoir	74	37	57	50	86	Oct	61
Swale Estuary	64 [10]	43	49	89	86	Dec	66
Other sites surpassing table qualifying levels in WeBS-Year 2005/2006 in Northern Ireland[†]							
Upper Quoile River	19	13	28	35	33	Sep	26
Lough Foyle	12	20	31	31	32	Sep	25

[†] as no All-Ireland threshold has been set a qualifying level of 30 has been chosen to select sites for presentation in this report

Great Crested Grebe
Podiceps cristatus

International threshold: 3,600
Great Britain threshold: 159
All-Ireland threshold: 30*

GB max: 8,459 Sep
NI max: 2,503 Feb

50 is normally used as a minimum threshold

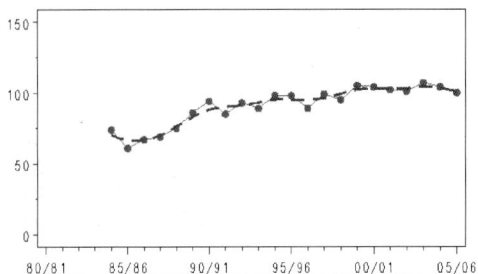

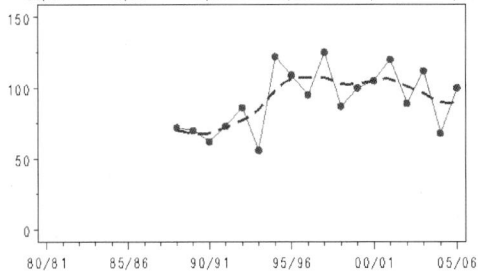

Figure 35.a, Annual indices & trend for Great Crested Grebe for GB (above) & NI (below).

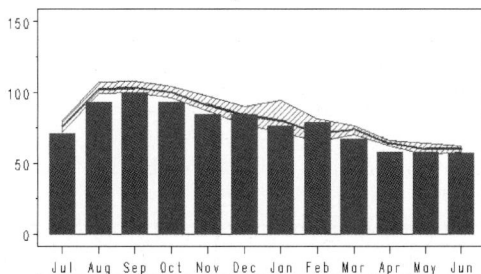

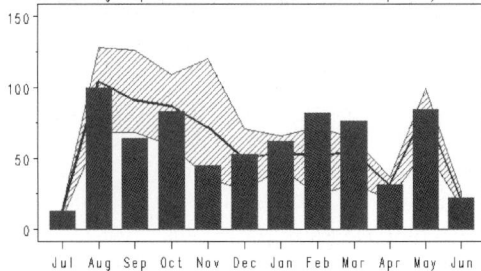

Figure 35.b, Monthly indices for Great Crested Grebe for GB (above) & NI (below).

The annual index for Great Crested Grebe in Great Britain shows a slight decline in 2005/06, with the underlying trend now perhaps levelling off after a period of growth. This slight decline is reflected in the peak counts at the top five sites, which were below the 2004/05 peaks. Notably low counts were recorded at Queen Mary Reservoir, Forth Estuary, Lavan Sands and Pegwell Bay, yet a record high count for the site was noted at Rye Harbour & Pett Level and also at King George V Reservoir and Swansea Bay. For the second consecutive year, there were no four figure counts recorded in Great Britain, the peak count of 771 coming from Rutland Water in October.

The index for Northern Ireland continues to fluctuate with an increase on 2004/05 when numbers were at their lowest for a decade. Belfast Lough numbers exceeded 2,000 for the first time since 1997 and four other sites in Northern Ireland supported three figure counts. At Bough Beech Reservoir, Poole Harbour, Outer Ards Shoreline and Minsmere, peak counts decreased to such an extent that the sites no longer support nationally important numbers. However, it is always important to remember that offshore counts of this species at open coast sites can be strongly affected by sea conditions on the day of the count.

	01/02	02/03	03/04	04/05	05/06	Mon	Mean
Sites of national importance in Great Britain							
Lade Sands	1,315 [44]	1,600	1,080 [12]	860	700	Jan	1,111
Rutland Water	600	607	619	815	771	Oct	682
Grafham Water	619	311	463	526	463	Jan	476
Queen Mary Reservoir	671	267	495	262	126	Nov	364
Chew Valley Lake	480	320	330	330	275	Jul	347
Cotswold Water Park (West)	258	(188)	(245)	283	354	Sep	298
Pitsford Reservoir	268	203	341	309	308	Oct	286
Rye Harbour and Pett Level	160	48	365	186	621	Feb	276
Swale Estuary	446 [44]	(42)	316	63	(52)	Oct	275
Bewl Water	292	356	190	330	204	Aug	274
Forth Estuary	224	(389)	295	(313)	123	Sep	269
Loch Ryan	(121)	(300)	210	299	193	Aug	251
Pegwell Bay	354 [44]	604	20	233	38	Dec	250
Morecambe Bay	222	187	218	(91)	(138)	Dec	209
Lavan Sands	113	308	114	(446)	57	Dec	208
Solway Firth	164	119	88	333	230	Dec	187
Lee Valley Gravel Pits	181	169	204	(147)	175	Dec	182
Draycote Water	221	255	151	98			181
Loch Leven	222	127	204	127	150	Nov	166
Sites of all-Ireland importance in Northern Ireland							
Belfast Lough	1,995	1,214	1,832	1,577	2,095	Feb	1,743
Loughs Neagh and Beg	336	930	1,695	518	449	Sep	786
Lough Foyle	278	782	1,030	50	169	Dec	462
Carlingford Lough	284	174	184	232	246	Dec	224
Upper Lough Erne	190	110	112	191	147	Feb	150
Lower Lough Erne		71	(66)	(117)	(48)	Feb	94
Larne Lough	80	105	115	50	56	Sep	81
Strangford Lough	50	(36)	(43)	(64)	(82)	Dec	65
Sites no longer meeting table qualifying levels in WeBS-Year 2005/2006							
Bough Beech Reservoir	146 [44]	149	196	142	121	Oct	151
Poole Harbour	171	127	202	(87)	74	Dec	144
Outer Ards Shoreline		9	(7)	2	11	Mar	7
Minsmere	5	19	463	30	2	Jul	104
Other sites surpassing table qualifying levels in WeBS-Year 2005/2006 in Great Britain							
Swansea Bay	71	67	57	26	205	Feb	85
King George V Reservoirs	67	26	44	52	193	Nov	76
Blagdon Lake	110	113	161	176	(176)	Sep	147

Red-necked Grebe
Podiceps grisegena

International threshold:	510
Great Britain threshold:	2*
All-Ireland threshold:	?

GB max: 34 Aug
NI max: 1 Mar

50 is normally used as a minimum threshold

The maximum count for Red-necked Grebes in August was relatively low. Typically, the Forth Estuary remains the key site with the highest monthly peak there since 2002/03. Counts from other sites of national importance were generally average with North Norfolk Coast and Lindisfarne the only other sites holding mean peaks of more than two birds. Gerrans Bay is well known as a site supporting high numbers of divers, but this was the highest count of Red-necked Grebes at this site to date for WeBS. As well as the nine sites listed here where two birds were noted, single Red-necked Grebes were also recorded at a further 44 sites. The low peak of one bird in Northern Ireland at Belfast Lough was typical for the province.

	01/02	02/03	03/04	04/05	05/06	Mon	Mean
Sites of national importance in Great Britain							
Forth Estuary	39	44	16	24	32	Aug	31
North Norfolk Coast	9	2	2	1	6	Feb	4
Lindisfarne	5	4 [10]	0	2	3 [10]	Nov	3
Poole Harbour	(0)	(2)	(0)	(0)	(0)		(2)
Loch Ryan	(1)	(0)	3	0	2	Oct	2 ▲
Moray Firth	2	1	1	2	2	Jan	2 ▲
Other sites surpassing table qualifying levels in WeBS-Year 2005/2006 in Great Britain							
Gerrans Bay	1	0	1	1	4	Mar	1
Medway Estuary	(0)	(0)	0	0	3	Oct	1
Belvide Reservoir	0		0	0	2	Feb	1
Cambois to Newbiggin	0	0	0	0	(2)	Oct	0
Fleet and Wey	1	1	0	0	2	Dec	1
Hamford Water	1 [10]	0	0	2	2	Dec	1
Kilconquhar Loch	0	1	0	0	2	Aug	1
Par Sands Pools & St Andrews Rd	0	0	0	0	2	Dec	0
Tring Reservoirs	0	0	0	0	2	Oct	0

Slavonian Grebe

Podiceps auritus

International threshold:	55
Great Britain threshold:	7*
All-Ireland threshold:	?[†]

GB max: 224 Feb
NI max: 42 Dec

50 is normally used as a minimum threshold

The maximum count in Great Britain was about average for recent years, being slightly up on 2004/05, as was the Northern Ireland maximum. The Forth Estuary remains the key site though counts there were much lower than in recent years, as were counts at the only other internationally important site, the Moray Firth. Counts at Loch Ryan and the Blackwater Estuary were also lower than in previous years, whilst counts at the majority of other sites were about average. The count from Gerrans Bay in Cornwall was well above average, and a number of good counts were also recorded by the RAF Ornithological Society expedition to northwest Scotland in February. Supplementary counts provided by the Shetland Oil Terminal Environmental Advisory Group showed the highest recent counts for Kirkabister to Wadbister Ness and Rova Head to Wadbister Ness, whilst counts from Whiteness to Skelda Ness were slightly lower than 2004/05 but still about average. In Northern Ireland, Lough Foyle was typically the key site, with the peak of 42 birds representing the second highest count there in the last decade.

	01/02	02/03	03/04	04/05	05/06	Mon	Mean
Sites of international importance in the UK							
Forth Estuary	61	80	110	73	55	Nov	76
Moray Firth	75	69	62	55	42	Oct	61
Sites of national importance in Great Britain							
Whiteness to Skelda Ness	29 [9]	55 [9]	55 [9]	59 [9]	52 [9]	Feb	50
Traigh Luskentyre	19	50 [49]	44 [49]	31			36
Sound of Gigha		51 [11]		20 [11]	30 [11]	Nov	34
Loch Ryan	31	31 [12]	32	42	23	Feb	32
Jersey Shore	(31)						(31)
Loch of Harray	25	25	23	49	24	Oct	29
Inner Firth of Clyde	10	45	(20)	16	35	Feb	27
Inner Loch Indaal	11	31	30				24
Lindisfarne	14	23 [10]	(2)	30 [10]	22 [10]	Nov	22
Blackwater Estuary	22	9 [10]	41	11	2 [10]	Nov	17
Loch Na Keal		12 [11]	18 [11]				15
Kirkabister to Wadbister Ness	5 [9]	13 [9]	17 [9]	(16) [9]	20 [9]	Dec	14
Rova Head to Wadbister Ness	15 [9]	11 [9]	6 [9]	18 [9]	22 [9]	Dec	14
Gualan and Balgarva		12 [49]		11 [49]			12
Pagham Harbour	6	6	28	8	8	Dec	11
Loch of Swannay	4	10	11	19	10	Jan	11
Upper Loch Torridon				0	17	Feb	9 ▲
Broadford Bay	8	10	6	10	(7)	Nov	9

	01/02	02/03	03/04	04/05	05/06	Mon	Mean
Vaila Sound and Gruting Voe	9 [9]						9
Gerrans Bay	0	3	4	5	26	Mar	8 ▲
Sound of Harris		8 [49]	5 [49]	10 [49]			8
Sullom Voe	8 [9]	6 [9]	6 [9]	13 [9]	7 [9]	Feb	8
Lavan Sands	(4)	15 [12]	6	5	2	Jan	7
Loch Ewe				0	13	Feb	7 ▲
South Yell Sound	5 [9]		9 [9]				7 ▲
Sites with mean peak counts of 4 or more birds in Northern Ireland[†]							
Lough Foyle	6	13	61	10	42	Dec	26
Other sites surpassing table qualifying levels in WeBS-Year 2005/2006 in Great Britain							
Exe Estuary	(3)	4	(9)	1	8	Feb	6
North Norfolk Coast	6	4	(4)	4	8	Jan	6
Kinghorn Loch	(0)	0	0	0	7	Jan	2
Other sites surpassing table qualifying levels in WeBS-Year 2005/2006 in Northern Ireland[†]							
Strangford Lough	0	5	0	4	5	Sep	3

[†] as no All-Ireland threshold has been set a qualifying level of 4 has been chosen to select sites for presentation in this report

Black-necked Grebe
Podiceps nigricollis

International threshold: 2,200
Great Britain threshold: 1*[†]
All-Ireland threshold: ?[†]

GB max: 98 Feb
NI max: 0

*50 is normally used as a minimum threshold

During February Black-necked Grebes were noted at 19 sites, the total from which was the second-highest ever record by WeBS counters. The highest count during this month was of 45 on the Fal Complex; however, this rose to 56 in March and this flock, all at Carrick Roads, was the highest single-site total ever recorded during a WeBS count. Throughout the year birds were present at 67 sites, compared to 64 the year before. Most records were from England although birds were noted at Inner Moray & Inverness Firth and Gruinard Bay in Scotland and St Davids Pools, Llys-y-fran Reservoir and Llyn Alaw in Wales.

The winter peak at Langstone Harbour was the highest for ten years and new all-time site peaks were recorded at Thames Estuary, Barrow Gurney Reservoir and Pitsford Reservoir. Several areas featured in the key sites table refer to breeding colonies so some of these have been kept confidential following the advice of the Rare Breeding Birds Panel and/or local counters.

	01/02	02/03	03/04	04/05	05/06	Mon	Mean
Sites with mean peak counts of 5 or more birds in Great Britain[†]							
Woolston Eyes	41	6	23	35	13	Apr	24
Fal Complex	16	15	7	19	56	Mar	23
William Girling Reservoir	16	16	21	27	21	Sep	20
Langstone Harbour	15 [10]	15	11	16 [10]	20	Feb	15
Confidential Hertfordshire Site	8	7	17	12	10	Apr	11
Teignmouth to Berry Head			4	18			11
Confidential Northumberland Site	2	11	10	11	16	May	10
Lower Derwent Ings	2	0	1	47	0		10
Other sites surpassing table qualifying levels in WeBS-Year 2005/2006 in Great Britain[†]							
Thames Estuary	1	(3)	2	4	9	Apr	4
Barrow Gurney Reservoir	0	0	0	0	5	Mar	1
Pitsford Reservoir	1	0	1	1	5	Oct	2

[†] as the British threshold is so low, and as no All-Ireland threshold has been set a qualifying level of 5 has been chosen to select sites for presentation in this report

Cormorant
Phalacrocorax carbo

International threshold: 1,200
Great Britain threshold: 230
All-Ireland threshold: ?[†]

GB max: 16,124 Oct
NI max: 2,684 Sep

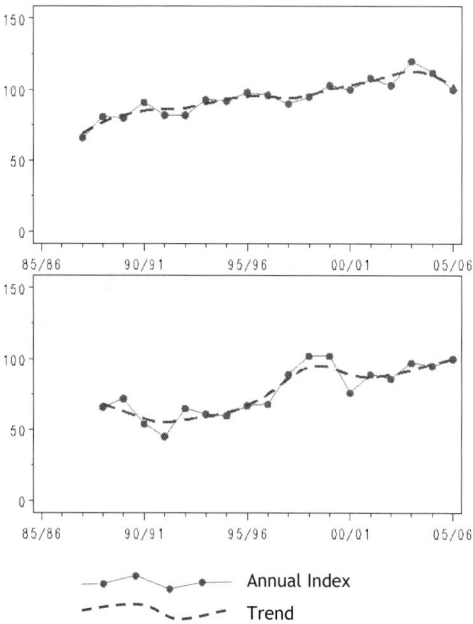

Figure 36.a, Annual indices & trend for Cormorant for GB (above) & NI (below).

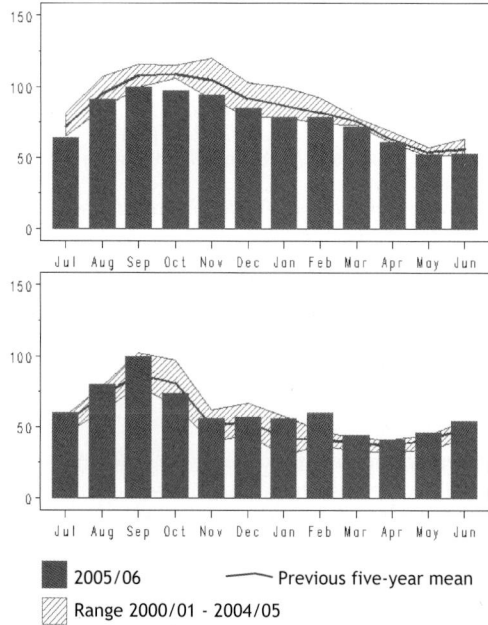

Figure 36.b, Monthly indices for Cormorant for GB (above) & NI (below).

National maxima for both Britain and Northern Ireland were only slightly lower than during 2004/05. The British index for Cormorant fell for the second year running and if this decline continues in future seasons, may lead to the reversal of the trend of steady increase seen since the late 1980s. The monthly indices suggest that this reduction was apparent throughout most of the year. Although perhaps too early to be sure, the increase in the number of licences for culling Cormorants introduced in the 2004/05 winter is one potential factor which could explain the decline in the British index. Conversely, the Northern Irish index has shown that numbers of Cormorants have increased slightly over the past few years and in 2005/06 continued to do so. Numbers in the region were above average for much of the year.

The principal site in the UK, and the only site holding internationally important numbers in the five years up to 2005/06, was Loughs Neagh and Beg. Peak numbers during 2005/06 were the fourth highest recorded at the site. A total of 40 sites held nationally important numbers of Cormorant in the five years to 2005/06, one more than previously. The November count at the Alt Estuary was the highest ever at the site. A new site record was also recorded at Rutland Water. Low peak counts were noted at Abberton Reservoir following a continuing a run of declines at the site, Walthamstow Reservoirs, Ribble Estuary, which was perhaps due in part to a local redistribution to the Alt Estuary, and, for the second year running, at Queen Mary Reservoir, although a number of the other west London sites saw increases following the low counts that were noted in 2004/05.

	01/02	02/03	03/04	04/05	05/06	Mon	Mean
Sites of international importance in the UK							
Loughs Neagh and Beg	723	1,383	1,468	1,591	1,490	Sep	1,331
Sites of national importance in Great Britain							
Alt Estuary	960	569	739	984	1,079	Nov	866
Forth Estuary	(761)	(982)	631	669	713	Sep	751
Dee Estuary (England & Wales)	(692)	668	718	780	623	Oct	697
Rutland Water	520	529	788	697	825	Sep	672
Thames Estuary	578	736	596	654	526	Dec	618
Morecambe Bay	398	(657)	(539)	(681)	655	Aug	586
Tees Estuary	432	438	773	471	511	Aug	525
Abberton Reservoir	780	600	480	450	306	Mar	523
Dungeness Gravel Pits	625	235 [33]	251	870	622	Sep	521
Poole Harbour	585	558	(412)	431	408	Oct	496
Inner Firth of Clyde	528	553	425	452	(500)	Sep	492
Solway Estuary	378	500	(594)	(454)	(270)	Nov	482
Walthamstow Reservoirs	(531)	570	505	453	306	Sep	473
The Wash	233	502	449	538	371	Sep	419
Ribble Estuary	358	398	(456)	543	293	Nov	410
Besthorpe & Girton GPs & Fleet	386	415	372	336	363	May	374
Rye Harbour and Pett Level	218	340	382	446	466	Aug	370
Queen Mary Reservoir	580	342	768	44	85	Aug	364
Little Paxton Gravel Pits		362 [33]					362
North Norfolk Coast	(268)	581	276	242	272	Aug	343
Grafham Water	204	349	193	344	531	Feb	324
Hanningfield Reservoir	585	189 [33]	411	109	318	Oct	322
Wraysbury Gravel Pits	306	181	607	119	306	Dec	304
Wraysbury Reservoir	93	132	899	83			302
Blackwater Estuary	450	104	473	191	224	Dec	288
Queen Elizabeth II Reservoir	115	308 [33]	340	295	360	Sep	284
Staines Reservoirs	77	41	773	21	436	Aug	270
Colne Estuary	(151)	(29)	423	297	81	Oct	267
Pagham Harbour	247	240	303	225	308	Jan	265
Alde Complex	318 [10]	84	(106)	549	99	Oct	263
Ouse Washes	213 [11]	347	252 [11]	294	182 [11]	Jan	258
Queen Mother Reservoir	50	91	850	25	252	Sep	254
Middle Tame Valley Gravel Pits	213	293	(168)	(256)	(93)	Oct	254 ▲
Ranworth and Cockshoot Broads	398 [11]	270 [11]	324 [11]	115	108	Feb	243
Loch Leven	421	68	310	222	180	Oct	240
Medway Estuary	167	(136)	305	(68)	(93)	Dec	236
Durham Coast	(8)	(13)	(2)	(52)	236	May	236 ▲
Rostherne Mere	31	293	306	256	273	Jan	232
South Stoke	266 [11]	248 [33]	180 [11]				231 ▲
Ayr to North Troon	(190)	169	(110)	292	(97)	Dec	231
Sites no longer meeting table qualifying levels in WeBS-Year 2005/2006							
Lee Valley Gravel Pits	271	231	286	153	169	Dec	222
Moray Firth		261 [1]	246 [1]	151 [1]	Nov	219
Sites with mean peak counts of 130 or more birds in Northern Ireland[†]							
Outer Ards Shoreline		652	563	350	455	Oct	505
Belfast Lough	528	388	348	234	378	Oct	375
Strangford Lough	245	358	400	405	455	Sep	373
Carlingford Lough	208	206	154	221	238	Sep	205
Upper Lough Erne	199	124	225	125	107	Dec	156
Other sites surpassing table qualifying levels in WeBS-Year 2005/2006 in Great Britain							
Clwyd Estuary	122	135	99	(210)	339	Nov	181
Fairburn Ings	140 [11]	219	187		265	May	203
Other sites surpassing table qualifying levels in WeBS-Year 2005/2006 in Northern Ireland[†]							
Larne Lough	84	137	140	88	135	Sep	117

[†] *as no All-Ireland threshold has been set a qualifying level of 130 has been chosen to select sites for presentation in this report*

Shag
Phalacrocorax aristotelis

International threshold: 2,000
Great Britain threshold: ?[†]
All-Ireland threshold: ?[†]

GB max: 1,755 Feb
NI max: 526 Jan

The British maximum count of Shag fell by over 20% from that of 2004/05. However, given that the number counted for WeBS represents a small proportion of those wintering around the UK it would be unwise, under current understanding, to assume that this decline is representative of the overall wintering population.

During 2005/06 Shags were noted at 179 sites, 11 more than in the previous year. The Forth Estuary remains the key site in terms of mean numbers, although the peak count was extremely low, being less than a third of the five-year mean. Other low counts came from the Inner Moray Firth, Moray Coast, Lindisfarne, Thurso Bay and Island of Egilsay. Notably high core counts were recorded at

Arran, Red Point to Port Henderson and Gerrans Bay.

The Northern Ireland maximum was at it highest since WeBS has recorded the species with record counts from Strangford Lough, Dundrum Bay and Larne Lough, although counts at Outer Ards Shoreline, Belfast Lough and Carlingford Lough were lower than for 2004/05.

Supplementary counts provided by the Shetland Oil Terminal Environmental Advisory Group (SOTEAG) found numbers to be high in South Yell Sound, Quendale to Virkie, Rova Head to Wadbister Ness, Hacosay, Bluemull and Colgrave Sounds but lower than average at North Bressay, Burra, Trondra and Sullom Voe.

	01/02	02/03	03/04	04/05	05/06	Mon	Mean
Sites with mean peak counts of 100 or more birds in Great Britain[†]							
Forth Estuary		2,315	(1,664)	(760)	420	Nov	1,466
South Yell Sound	1,690 [9]	710 [9]	893 [9]	558 [9]	790 [9]	Nov	928
Hacosay, Bluemull & Colgrave Snds	1,132 [9]	423 [9]	709 [9]	232 [9]	625 [9]	Jan	624
Moray Firth			413 [1]	995 [1]	308 [1]	Jan	572
North-west Yell Sound	495 [9]						495
Scalloway Islands	760 [9]		424 [9]	255 [9]	448 [9]	Jan	472
Burra and Trondra	478 [9]		476 [9]	441 [9]	287 [9]	Jan	421
West Whalsay and Sounds	383 [9]						383
Inner Moray and Inverness Firth		636 [1]	108 [1]	663 [1]	31	Feb	360
West Yell	349 [9]						349
Kirkabister to Wadbister Ness	73 [9]	172 [9]	778 [9]	(97) [9]	198 [9]	Dec	305
North Bressay		53 [9]		728 [9]	128 [9]	Jan	303
Quendale to Virkie	605 [9]	123 [9]	176 [9]	97 [9]	503 [9]	Dec	301
Bressay Sound	657 [9]	114 [9]	100 [9]	272 [9]	97 [9]	Jan	248
East Unst	246 [9]						246
Scousburgh to Maywick					245 [9]	Dec	245
Widewall Bay		68	580	140	150	Jan	235
South Unst	339 [9]		206 [9]	63 [9]	246 [9]	Jan	214
South Havra	428 [9]			59 [9]	125 [9]	Jan	204
Gulberwick Area					189 [9]	Jan	189
Red Point to Port Henderson				92	246	Feb	169
Easter Ross Coast			214 [1]	122 [1]			168
Inner Firth of Clyde	139	(213)	(159)	190	115	Dec	163
Rova Head to Wadbister Ness	83 [9]	166 [9]	132 [9]	126 [9]	299 [9]	Dec	161
Anstruther Harbour	639	64	8	45	27	Feb	157
South Yell	157 [9]						157
Loch Ewe				197	115	Feb	156
Arran	86	100	(151)	131	304	Nov	155
Broadford Bay		150	(100)	150	152	Jan	151
Moray Coast (Consolidated)		121	180	251	33	Jan	146
Whiteness to Skelda Ness	142 [9]	149 [9]	169 [9]	138 [9]	115 [9]	Feb	143
Linga Beach	133 [12]						133
Ayr to North Troon	63	184	(26)	(30)	(6)	Jan	124
South-east Yell	120 [9]						120

106

	01/02	02/03	03/04	04/05	05/06	Mon	Mean
Loch Ryan	(90)	(110)	79	144	127	Oct	117
Island of Papa Westray	47	107	210	50	150	Dec	113
Girvan to Turnberry	111	80	117	115	128	Dec	110
Helmsdale to Lothbeg				103 [1]			103
Sites with mean peak counts of 100 or more birds in Northern Ireland[†]							
Strangford Lough	(166)	189	226	(218)	295	Nov	237
Outer Ards Shoreline		227	187	280	236	Jan	233
Belfast Lough	30 [10]	215	194 [10]	90	49	Nov	116
Carlingford Lough	294	48	(37)	60	55	Jan	114
Other sites surpassing table qualifying levels in WeBS-Year 2005/2006 in Great Britain							
Gerrans Bay	0	0	18	25	101	May	29
Other sites surpassing table qualifying levels in WeBS-Year 2005/2006 in Northern Ireland[†]							
Dundrum Bay	20	0	0	0	153	Jan	35

[†] as no British or All-Ireland thresholds have been set a qualifying level of 100 has been chosen to select sites for presentation in this report

Bittern

Botaurus stellaris

International threshold:	65
Great Britain threshold:	?
All-Ireland threshold:	?

GB max: 25   Dec
NI max: 0

Bitterns were recorded at 50 sites across the country, 44 in England, five in Wales and one in Scotland. The British maximum reached double figures throughout November to March and peaked at 25 in December; this was very similar to the previous year. The highest single-site count was of four at Minsmere during August and three were recorded at Potterick Carr in both December and January.

Night Heron

Nycticorax nycticorax

Vagrant and escape
Native Range: Europe, America, Asia

GB max: 1   Mar
NI max: 0

A single Night Heron was at Esso Pools, Dyfed in March. This is only the second time this species has been recorded during a WeBS count; the first was at Bainton Pits, Cambridgeshire in September 2002.

Cattle Egret

Bubulcus ibis

Vagrant and escape
Native Range: Worldwide

GB max: 6   Mar
NI max: 0

The total of six Cattle Egrets in March was made up of two groups of three, at Avon Valley (Salisbury to Fordingbridge) and Pagham Harbour. The three birds on the Avon Valley remained from January to April, after which time only a single bird remained into May. Other records included two on the North Norfolk Coast in September, this fell to one in October, and one at Chichester Harbour in January. Prior to 2005/06, Cattle Egrets had been recorded during WeBS counts on 14 occasions, all involving single birds. As a species which has spread northwards rapidly in recent years, and which is now breeding on the other side of the English Channel, it seems very likely that this potential colonist will continue to increase in numbers in the UK.

Little Egret
Egretta garzetta

International threshold: 1,300
Great Britain threshold: ?[†]
All-Ireland threshold: ?[†]

GB max: 3,105 Sep
NI max: 10 Feb

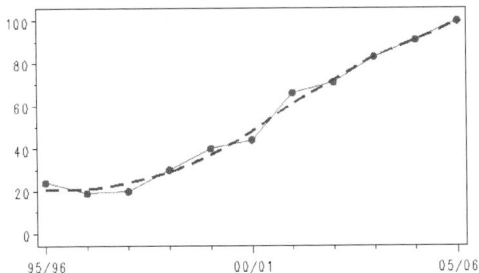

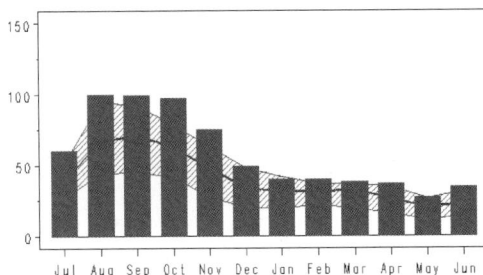

— Annual Index
- - - Trend

■ 2005/06 — Previous five-year mean
▨ Range 2000/01 - 2004/05

Figure 37.a, Annual indices & trend for Little Egret for GB.

Figure 37.b, Monthly indices for Little Egret for GB.

The increase in Little Egret shows no sign of abating with the annual index again rising sharply in 2005/06 and an increase also seen in the British counted maximum. The monthly indices again show peaks above the five-year mean for all months, with the standard late summer/early autumn peak still very apparent and a low point during the breeding season. Whilst there is still no official 1% threshold value for this species, for the purposes of this report the table qualifying level remains at a mean of 30 birds for clarity; 70 sites supported means in excess of ten birds in 2005/06.

Although peak numbers remain in the south and east, from the counts on individual sites, it

is noticeable that there has been an increase in numbers in the west. This is especially true in Wales, with many Estuaries such as the Cleddau Estuary, Burry Inlet, Lavan Sands, Severn Estuary and Dee Estuary showing notable increases. Many estuaries in the southwest of England have shown an increase too with the Taw-Torridge Estuary and the Camel Estuary showing record peak counts. There also appears to be an increase on the northern limit of the range with increased peaks on both the Humber Estuary and in Morecambe Bay. The increase is also apparent in Northern Ireland where the maximum peak count of ten was also higher than for 2004/05.

	01/02	02/03	03/04	04/05	05/06	Mon	Mean
Sites with mean peak counts of 30 or more birds in Great Britain[†]							
Medway Estuary	106	(125)	413 [11]	(76)	(62)	Oct	260
Thames Estuary	132	201	(262)	295	260	Aug	230
Chichester Harbour	255 [11]	218	228	129	(206)	Oct	208
Poole Harbour	197 [11]	(140)	(179)	(116)	(112)	Oct	197
Blackwater Estuary	(35)	(51)	(66)	(159)	133	Oct	146
North Norfolk Coast	50 [10]	81	149 [11]	228 [11]	164	Aug	134
Tamar Complex	141	129	143	120	(129)	Oct	133
Jersey Shore	126						126
Exe Estuary	149	67	131	93	107	Nov	109
Longueville Marsh	132	145	105	102	60	Nov	109
Burry Inlet	99	87	(141)	103	108	Nov	108
Kingsbridge Estuary	100	105	(99)	86	85	Sep	95
Swale Estuary	44	(59)	131	95	(100)	Oct	93
Langstone Harbour	99	88	90	87	91	Oct	91
The Wash	(6)	29	72	92	139	Aug	83
Portsmouth Harbour	123 [11]	11 [110]	(34)	51	45	Nov	82
Pagham Harbour	81 [11]	76	63	(60)	94	Sep	79
Taw-Torridge Estuary	64	(60)	(74)	(56)	93	Aug	79
Colne Estuary	118 [11]	(2)	(35)	46	47	Oct	70
Camel Estuary	48	64	65	71	96	Sep	69
Fal Complex	(30)	55	(52)	89	60	Jul	68
Severn Estuary	59	41	47	66	104	Sep	63
Cleddau Estuary	66	48	36	71	83	Sep	61
Stour Estuary	29	32	57	87	102	Aug	61
Lavan Sands	6	15	67	71	107	Aug	53

108

	01/02	02/03	03/04	04/05	05/06	Mon	Mean
Hamford Water	31	20	53	81	72	Sep	51
Avon Valley: Salisbury/Fordingbr	49	(79)	19	57	48	Nov	50
North West Solent	(44)	(25)	42	(51)	(56)	Oct	48
Crouch-Roach Estuary	24	42	43	73 [10]	(35)	Oct	46
Dee Estuary (England & Wales)	18 [10]	20	32 [11]	50 [11]	109 [11]	Sep	46
Fowey Estuary	79	48	35	33	37	Feb	46
Guernsey Shore	50	(48)	(51)	34	39	Dec	44
Fleet and Wey	(37)	38	25	46	56	Apr	41
Helford Estuary	33	47	35	46	41	Jul	40
Southampton Water	45 [11]	19	(51)	(39)	(44)	Oct	40
Carmarthen Bay	(13)	(9)	(23)	35	41	Sep	38
Newtown Estuary	44 [11]	22	41	(21)	(30)	Oct	36
Alde Complex	15	(20)	(23)	45	44	Nov	35
Pegwell Bay	20	23	26	(48)	56	Sep	35
Orwell Estuary	43 [11]	11 [37]	56 [11]	9 [10]	27	Sep	34
Beaulieu Estuary	6	42	22	30	49	Aug	30
Breydon Wtr & Berney Marshes	7	19	22	42	61	Jul	30

[†] as no British or All-Ireland thresholds have been set a qualifying level of 30 has been chosen to select sites for presentation in this report

Great White Egret
Ardea alba

Vagrant
Native Range: Europe, Africa, Asia & America

GB max: 4   Oct
NI max: 0

A total of three Great White Egrets was recorded at Elmley Marshes on the Swale Estuary during October, the first time that more than one has been recorded on a WeBS count, and perhaps included the single bird recorded during July at South Allhallows Marsh, Thames Estuary. Additionally, the long-staying individual around Mockbeggar Lake by the River Avon (Fordingbridge to Ringwood) was noted on WeBS counts between September and November.

Grey Heron
Ardea cinerea

International threshold: 2,700
Great Britain threshold: ?[†]
All-Ireland threshold: ?[†]

GB max: 4,002   Sep
NI max: 422   Sep

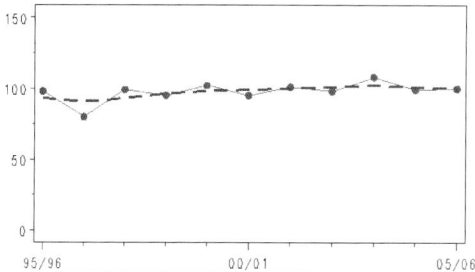

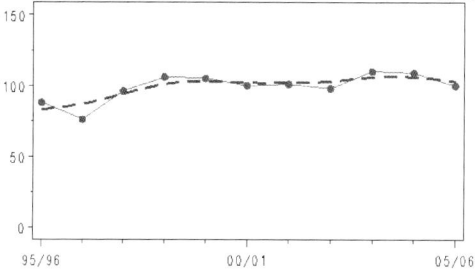

Figure 38.a, Annual indices & trend for Grey Heron for GB (above) & NI (below).

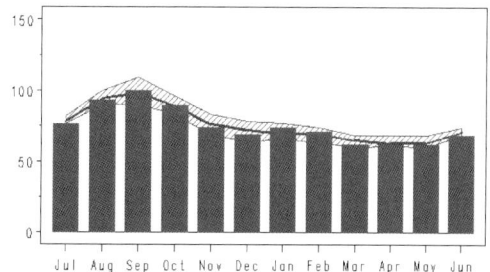

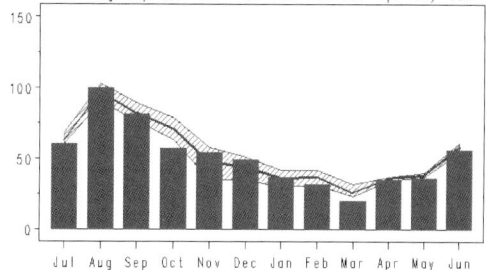

Figure 38.b, Monthly indices for Grey Heron for GB (above) & NI (below).

The numbers of Grey Herons recorded for WeBS rarely vary significantly from year to year. The national maxima recorded during 2005/06 were similar to those of the past ten years in both Great Britain and Northern Ireland. Grey Herons are widely distributed, in both range and habitat, and were recorded on over three quarters of all WeBS sites counted during 2005/06. Counts in excess of 100 birds were noted from seven sites, the highest single-site count being 202 at Loughs Neagh and Beg. The highest British count was at Coombe Country Park in Warwickshire, although peak counts at this site have declined somewhat. The peak at Walthamstow Reservoirs was very low in 2005/06, mirroring the low counts of Cormorants at the same site.

	01/02	02/03	03/04	04/05	05/06	Mon	Mean
Sites with mean peak counts of 50 or more birds in Great Britain[†]							
Avon Valley: Salisbury-Fordingbr	(100)	(83)	150	(80)	(106)	Mar	150
Coombe Country Park		159	159	105	120	Apr	136
Somerset Levels	121	136	130	151	119	Feb	131
Thames Estuary	(129)	(124)	(94)	100	117	Sep	118
Inner Firth of Clyde	90	87	81	90	93	Aug	88
Dee Estuary (England & Wales)	63 [10]	(111)	87	67	(48)	Sep	82
Morecambe Bay	51	101	91	68	88	Sep	80
Forth Estuary	47	62	10 [78]	104	106	Jan	79
Ouse Washes	100	104 [12]	78	66	42	Mar	78
Severn Estuary	69	104 [10]	73	69	55	Sep	74
Walthamstow Reservoirs	91	133	64	60	16	Nov	73
Inner Moray and Inverness Firth	56	91	67	55	68	Jan	67
Colne Valley Gravel Pits	(36)	(33)	68	(47)	62	Apr	65
Tees Estuary	58	66	64	56	62	Sep	61
Solway Estuary	28 [10]	(69)	(70)	72	(28)	May	60
The Wash	49	54	76	50	52	Sep	56
Cromarty Firth	48	44	73	47	58	Oct	54
Tamar Complex	44	53 [10]	52	49	67	Aug	53
Sites with mean peak counts of 50 or more birds in Northern Ireland[†]							
Loughs Neagh and Beg	87	226	208	172	202	Sep	179
Strangford Lough	113	102	102	90	121 [10]	Nov	106
Other sites surpassing table qualifying levels in WeBS-Year 2005/2006 in Great Britain[†]							
Wraysbury Gravel Pits	32	34	22	58	96	Mar	48
River Avon: Fordingbr-Ringwood	19	46	28	56	73	Nov	44
Lower Derwent Ings	14	21	20	19	55	Mar	26

[†] *as no British or All-Ireland thresholds have been set a qualifying level of 50 has been chosen to select sites for presentation in this report*

Purple Heron
Ardea purpurea

Vagrant
Native Range: Europe, Africa, Asia

GB max:	2	May
NI max:	0	

Purple Herons were recorded at three sites during 2005/06. One was at Doxey Marshes SSSI in August and singles were recorded at both New Grounds Slimbridge and College Reservoir in May. Prior to 2005/06 there had only been seven individuals recorded by WeBS counters since the first at Slapton Ley in July 1996.

White Stork
Ciconia ciconia

Vagrant and escape
Native Range: Europe, Africa, Asia

GB max:	3	Sep
NI max:	0	

Up to two free-flying White Storks were present at Harewood Lake, where they have been recorded in almost every month since April 1997. The only other report of was of a single bird at Poole Harbour in September. Reports of a colour-ringed bird in the area around this time suggest that this is unlikely to be of wild origin.

Sacred Ibis
Threskiornis aethiopicus

Escape
Native Range: Africa & Middle East

GB max:	1	Aug
NI max:	0	

The regular Sacred Ibis was recorded at Outwood Swan Sanctuary from August to November. A later record, presumably of the same individual, was received from Gatton Park in February. This bird has been seen in this area regularly since the end of 1998.

Spoonbill
Platalea leucorodia

International threshold:	110
Great Britain threshold:	?
All-Ireland threshold:	?

GB max:	18	Oct
NI max:	0	

Spoonbills were recorded in every month during 2005/06 and, with the exception of three at Goldcliff Saline Lagoons in August and one at Llyn Hafodol in April, all were recorded at sites within England. Birds were noted at 18 sites, one more than in the previous year. The summed site maximum of 40 was also slightly higher than the 2004/05 total of 31. Typical winter records from the southwest included up to eight birds on the Tamar Complex between October and January, whilst the Taw-Torridge Estuary held up to five between October and February.

Water Rail
Rallus aquaticus

International threshold:	10,000
Great Britain threshold:	?[†]
All-Ireland threshold:	?[†]

GB max:	540	Dec
NI max:	4	Oct

Water Rails, being difficult to detect, are under-recorded by WeBS, with the total number recorded representing only a fraction of the birds present in the country. However, the numbers recorded for WeBS at the majority of key sites remains fairly consistent. The British maximum was only slightly higher than in the previous year. In total, 19 sites held average counts of ten or more birds, four less than during 2004/05. At individual sites, the peak count from Stodmarsh NNR & Collards Lagoon was lower than usual, but counts at Swanpool and the Firth of Clyde were notably high. The Northern Ireland maximum was just four birds, of which three were at Upper Lough Erne.

	01/02	02/03	03/04	04/05	05/06	Mon	Mean
Sites with mean peak counts of 10 or more birds in Great Britain[†]							
Somerset Levels	45	(45)	45	63	50	Jan	51
Grouville Marsh	30	25	20	20	30	Nov	25
Kenfig Pool	30	39	27	17	12	Mar	25
Poole Harbour	15	24	(10)	(12)	(6)	Oct	20
Stodmarsh NNR & Collards Lagn	27	28	20	15	9	Oct	20
Middle Yare Marshes	23	17	18	(4)	(5)	Sep	19
Longueville Marsh	15	15	15	20	20	Dec	17
Southampton Water	20	18	(7)	11	20	Nov	17
Burry Inlet	16	10	18	16	(0)		15
Lee Valley Gravel Pits	18	18	12	(7)	11	Dec	15
North Norfolk Coast	22	(10)	10	7	15	Dec	14
Chichester Harbour	14	16	6	13	14	Dec	13
River Cam - Kingfishers Bridge	6	22 [12]	7	8	22	Nov	13
Severn Estuary	(5)	15	5	6	(25)	Jan	13
Thames Estuary	12	21	(8)	11	9	Oct	13
Loe Pool	1	19	(16)	10	12	Dec	12
Ingrebourne Valley	10	14	12	7	10	Aug	11
Nth Warren & Thorpeness Mere	4	2 [12]	44	1	3	Dec	11
Other sites surpassing table qualifying levels in WeBS-Year 2005/2006 in Great Britain[†]							
Brading Harbour	6	8	4	3	13	Dec	7
Swanpool (Falmouth)	1	0	1	2	11	Feb	3
Inner Firth of Clyde	4	2	3	3	10	Feb	4

[†] *as no British or All-Ireland thresholds have been set a qualifying level of 10 has been chosen to select sites for presentation in this report*

Spotted Crake
Porzana porzana

Scarce

GB max: 3 Sep
NI max: 0

All seven records of Spotted Crake were from July to October and all were of single birds. The earliest record was at Inner Marsh Farm, Dee Estuary (England & Wales) in July and the latest was at Old Moor in October. The bird at Inner Marsh Farm remained into August with another recorded at South Huish Marsh in the same month. All remaining records were in September and were at Baron's Haugh, Blacktoft Sands and at Salthouse on the North Norfolk Coast.

Sora
Porzana carolina

Vagrant
Native Range: N America

GB max: 1 Oct
NI max: 0

The presence of a Sora on Lower Moors Pool on the Isles of Scilly in October inspired one recorder to carry out a WeBS count of the pool at the time. This is the first record of the species within the WeBS database.

Little Crake
Porzana parva

Vagrant
Native Range: Europe, Africa, Asia

GB max: 1 Sep
NI max: 0

The first ever Little Crake recorded during a WeBS count was a single at New Grounds Slimbridge in September.

Corncrake
Crex crex

Scarce

GB max: 3 Jun
NI max: 0

Although not exactly a waterbird, single Corncrakes were recorded at Loch na Meallaird in April and May. Three were recorded at Loch a` Chinn Uacraich, also known as Coot Loch, on Benbecula in June. Only four Corncrakes had been recorded on WeBS counts prior to 2005/06, the first of which was in August 2003.

Moorhen
Gallinula chloropus

International threshold: 20,000**
Great Britain threshold: 7,500[†]
All-Ireland threshold: ?[†]

GB max: 14,530 Dec
NI max: 260 Jan

The peak total for this under-recorded species was higher than for 2004/05 though these numbers are assumed to be a small proportion of the true total. The peak for Northern Ireland was also higher than the previous winter, which was reflected in the counts from the three main sites, which were higher than in 2004/05. The Severn Estuary remained the top site based on the five-year mean although the peak count for the winter was recorded from WWT Martin Mere with 490 in October. High counts were recorded from the North Norfolk Coast, Chichester Gravel Pits and Pitsford Reservoir and particularly low counts were recorded at Bewl Water and Thanet Coast.

	01/02	02/03	03/04	04/05	05/06	Mon	Mean
Sites with mean peak counts of 100 or more birds in Great Britain[†]							
Severn Estuary	557	476	443	409	465	Nov	470
WWT Martin Mere	485	490	440	420	490	Oct	465
Lower Derwent Ings	412	463	444	321	366	Nov	401
Thames Estuary	345	472	324	371	314	Dec	365
Somerset Levels	308	325	276	327	410	Oct	329

	01/02	02/03	03/04	04/05	05/06	Mon	Mean
Lee Valley Gravel Pits	357	312	340	301	292	Dec	320
North Norfolk Coast	334	243	280	192	281	Feb	266
Pitsford Reservoir	267	209	326	133	266	Oct	240
Dee Estuary (England & Wales)	(199)	(116)	(121)	(86)	(97)	Feb	(199)
Rutland Water	211	189	191	192	188	Sep	194
Burry Inlet	209	175	169	202	(104)	Aug	189
Humber Estuary	101	215	224	(170)	(96)	Dec	180
Bewl Water	200	254	215	165	61	Aug	179
Durham Coast	225	160	(0)	(158)	133	Dec	173
Chichester Gravel Pits	157	149	161	167	228	Feb	172
Arun Valley	148	172	163	128	162	Dec	155
Avon Valley: Salisbury/Fordingbri	(79)	(56)	143	112	178	Sep	144
Chew Valley Lake	165	105	245	125	80	Oct	144
Barn Elms Reservoirs	170	131	137	135			143
Hamford Water	(72)	134	(156)	90	(91)	Jan	127
Sutton and Lound Gravel Pits	(160)	118	112	94	108	Apr	118
Thanet Coast	95	123	169	133	65	Jan	117
Colne Valley Gravel Pits	(149)	(110)	(58)	(120)	86	Mar	116
Orwell Estuary	117 [10]	100 [10]	164 [10]	109 [10]	90 [10]	Dec	116
Tring Reservoirs	110	106	115	135	110	Mar	115
River Wye - Bakewell to Haddon	89	131	126	109	104	Feb	112
Rye Harbour and Pett Level	107	71	116	117	151	Dec	112
Southampton Water	(81)	(81)	125	83	114	Feb	107
Grnd W Cnl: G'wy Br/N Dvn Rd	73	80	103	132	137	Jan	105
Old Moor	122	(131)	116	45	(80)	Dec	104
R. Cam: Owlstone Rd/Baits B Lk	99	93	76	117	126	Feb	102
Ingrebourne Valley	122	116	121	77	71	Dec	101
Tees Estuary	73	115	110	103	102	Sep	101
Sites with mean peak counts of 30 or more birds in Northern Ireland[†]							
Loughs Neagh and Beg	380	211	177	124	143	Sep	207
Upper Lough Erne	60	46	32	46	60	Dec	49
Belfast Lough	47	62	27	51	54	Jan	48
Other sites surpassing table qualifying levels in WeBS-Year 2005/2006 in Great Britain[†]							
Cotswold Water Park (West)	61	(49)	(89)	73	132	Sep	89
Chichester Harbour	74	78	85	98	127	Mar	92
Clumber Park Lake	76	104	60		122	Feb	91
Ouse Washes	61	70	95	92	111	Feb	86
Blagdon Lake	129	63	132	46	(105)	Sep	95
Langstone Harbour	10	6	73	45	104	Mar	48
Other sites surpassing table qualifying levels in WeBS-Year 2005/2006 in Northern Ireland[†]							
Lough McNean Lower				1	37	Jan	19

[†] as few sites exceed the British threshold and no All-Ireland threshold has been set, qualifying levels of 100 and 30 have been chosen to select sites, in Great Britain and Northern Ireland respectively, for presentation in this report

---

Coot
Fulica atra

International threshold: 17,500
Great Britain threshold: 1,730
All-Ireland threshold: 250

GB max: 106,790 Dec
NI max: 4,139 Dec

The Great Britain index for Coot continues to show a slight increase, with monthly peaks being in line with the five-year means. Northern Ireland did show a slight rise following the record low count in 2004/05, largely due to increased peaks at Loughs Neagh & Beg and Upper Lough Erne. Abberton Reservoir remained by far the most important site in the UK and the peak count was a significant increase on 2004/05. Other proportionately high peaks were noted at Cotswold Water Park (West) and Blagdon Lake yet significantly lower than expected peaks were noted at several sites, in particular Rutland Water, Chew Valley Lake and Fleet & Wey. Following an exceptionally low peak in 2004/05, the peak for Hanningfield Reservoir was back up to around the mean five year total. Both Blagdon Lake and Chichester Harbour now hold nationally important numbers of Coot but Little Paxton Gravel Pits no longer meets the qualifying level.

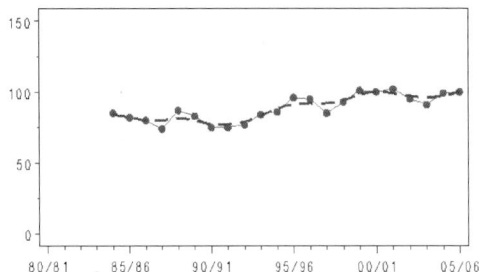

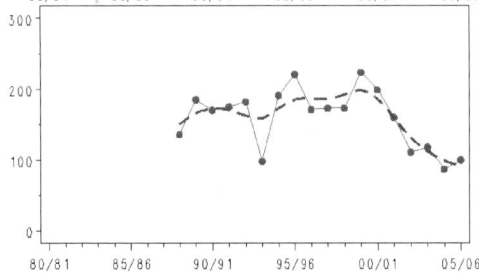

Annual Index
Trend

2005/06
Previous five-year mean

Range 2000/01 - 2004/05

Figure 39.a, Annual indices & trend for Coot for GB (above) & NI (below).

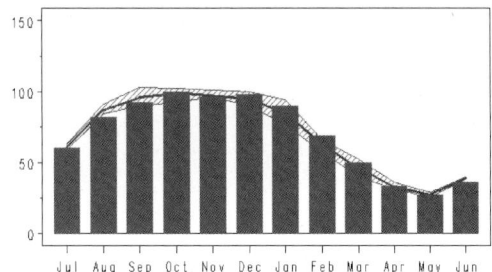

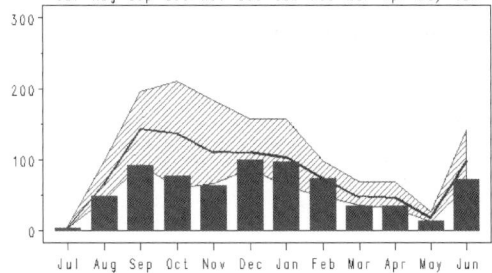

Figure 39.b, Monthly indices for Coot for GB (above) & NI (below).

	01/02	02/03	03/04	04/05	05/06	Mon	Mean
Sites of national importance in Great Britain							
Abberton Reservoir	7,610	6,885	6,166	9,697	10,965	Oct	8,265
Cotswold Water Park (West)	4,161	(2,528)	4,042	4,077	4,548	Dec	4,207
Rutland Water	3,283	3,969	4,021	4,733	3,490	Jan	3,899
Lee Valley Gravel Pits	3,245	3,250	3,213	3,435	3,459	Dec	3,320
Cheddar Reservoir	2,950	2,975	3,100	3,873	3,140	Dec	3,208
Chew Valley Lake	2,360	3,715	3,285	3,335	2,205	Oct	2,980
Fleet and Wey	3,418	2,353	(2,923)	3,275	2,699	Dec	2,936
Ouse Washes	2,488 [12]	1,349	2,039	4,229	4,194	Mar	2,860
Loch Leven	1,818	3,205	2,650 [12]	2,375	1,610	Sep	2,332
Cotswold Water Park (East)	2,634	2,365	2,296	1,850	2,045	Jan	2,238
Pitsford Reservoir	2,746	1,949	1,823	2,354	2,212	Oct	2,217
Hanningfield Reservoir	1,369	3,426	3,791	463	2,000	Dec	2,210
Blagdon Lake	2,846	628	1,993	2,080	3,151	Aug	2,140 ▲
Middle Tame Valley Gravel Pits	(2,106)	(1,284)	(559)	(393)	(15)	Oct	(2,106)
Lower Windrush Valley GPs	1,720	2,016	2,341	2,075	(1,338)	Jan	2,038
Chichester Gravel Pits	2,545	2,213	1,250	1,393	1,266	Dec	1,733 ▲
Sites of all-Ireland importance in Northern Ireland							
Loughs Neagh and Beg	2,555	4,344	4,124	1,890	2,506	Sep	3,084
Upper Lough Erne	1,660	1,447	2,062	1,462	2,023	Jan	1,731
Strangford Lough	581	420	230	223	378	Nov	366
Lower Lough Erne		272	197	308	411	Feb	297
Sites no longer meeting table qualifying levels in WeBS-Year 2005/2006							
Little Paxton Gravel Pits	1,679	1,831	1,334	1,422	1,289	Jan	1,511
Other sites surpassing table qualifying levels in WeBS-Year 2005/2006 in Great Britain							
River Avon: Fordingbridg/Ringwd	1,628	1,069	1,494	1,765	1,800	Oct	1,551

American Coot

Fulica americana

Vagrant

Native Range: N America

GB max: 1 Sep
NI max: 0

The single American Coot recorded at Benston Loch in September is almost certainly the same bird that was recorded in the area during the past two winters. This bird was first recorded at Loch of Clickimin in December 2003, at which time it was the first record of this species during WeBS.

Crane

Grus grus

GB max: 2 Sep
NI max: 0

The only record of Crane was of two at Loch Scarmclate in September. This is the first time that this species has been recorded at this site during WeBS counts.

Oystercatcher

Haematopus ostralegus

International threshold: 10,200
Great Britain threshold: 3,200
All-Ireland threshold: 500

GB max: 206,275 Nov
NI max: 15,138 Nov

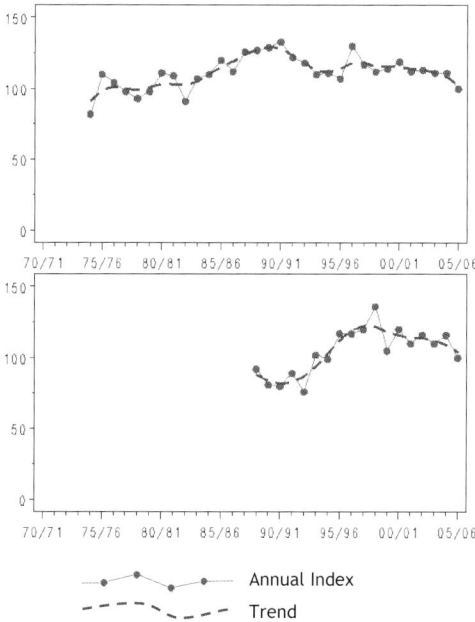

Figure 40.a, Annual indices & trend for Oystercatcher for GB (above) & NI (below).

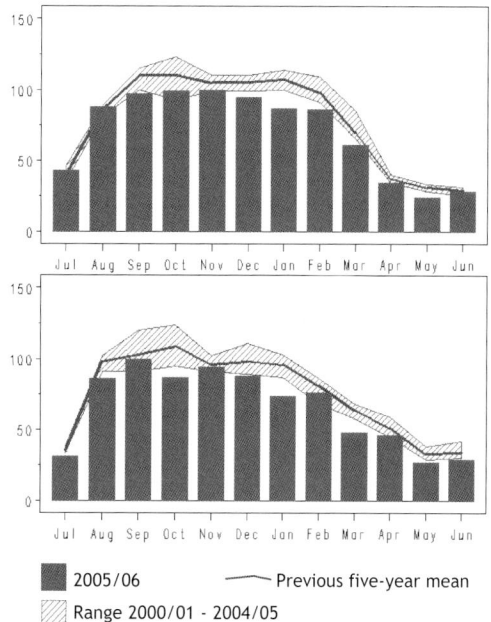

Annual Index
Trend

2005/06
Previous five-year mean
Range 2000/01 - 2004/05

Figure 40.b, Monthly indices for Oystercatcher for GB (above) & NI (below).

The Oystercatcher is one of the UK's most numerous and widespread waders. Over the past 30 years the number of Oystercatchers wintering in Britain has remained fairly stable. However, the index fell quite sharply during 2005/06 and reached its lowest level for over 20 years; it will be interesting to see whether this decline continues to be seen in 2006/07. The British counted maximum also fell slightly compared to 2004/05. Monthly indices indicate that numbers were below average from September to March and were particularly low from December onwards. The Northern Ireland index also showed a decline, continuing a trend seen over the past seven years. There were no changes in the sites holding internationally or nationally important numbers during 2005/06. Morecambe Bay remains the key site for this species both during winter and passage periods. The number of wintering birds on the Burry Inlet was the lowest for eight years and relatively low peak counts were also recorded from the Forth, Duddon and Swale Estuaries. In general, numbers at key sites across Northern Ireland were similar to the previous year, the exception being Lough Foyle where numbers were the lowest for over ten years.

	01/02	02/03	03/04	04/05	05/06	Mon	Mean
Sites of international importance in the UK							
Morecambe Bay	47,286	48,752	(48,600)	55,072	(43,923)	Nov	50,370
Solway Estuary	(35,035)	(47,415)	(34,099)	(30,397)	(9,647)	Jan	(47,415)
Dee Estuary (England & Wales)	31,851 [10]	20,373	23,906	25,956	22,847	Nov	24,987
Thames Estuary	18,814	26,803	23,858	(14,907)	22,368	Nov	22,961
Ribble Estuary	23,072	(12,395)	19,915	12,953	(6,378)	Dec	18,647
The Wash	13,371	16,760	14,684	16,395	14,705	Dec	15,183

	01/02	02/03	03/04	04/05	05/06	Mon	Mean
Burry Inlet	15,253	14,570	13,831 [10]	16,219	11,728	Nov	14,320
Sites of national importance in Great Britain							
Forth Estuary	(6,631)	9,279	7,834	6,569	4,364	Feb	7,012
Lavan Sands	7,831	7,612	6,796	5,718	5,926	Nov	6,777
Duddon Estuary	6,907	(6,476)	8,683	5,272	5,577	Nov	6,610
Carmarthen Bay	(5,575)	(4,530)	4,597	6,736	7,754	Nov	6,362
Inner Moray and Inverness Firth	5,153	6,087	7,624	4,681	4,930	Jan	5,695
Swale Estuary	6,270	5,058	5,858	5,225	4,646	Nov	5,411
Inner Firth of Clyde	5,488	5,386	4,627	4,737	5,756	Nov	5,199
North Norfolk Coast	3,990	3,011	3,858	3,778	3,707	Dec	3,669
Humber Estuary	(3,318)	(2,963)	3,305 [10]	(4,582)	(3,468)	Dec	3,668
Swansea Bay	3,563	3,797	2,857 [10]	4,605	3,511	Feb	3,667
Dengie Flats	7,061 [10]	3,034	(1,450)	1,865	1,595	Nov	3,389
Sites of all-Ireland importance in Northern Ireland							
Strangford Lough	8,296	8,557	7,412	6,454	6,861	Nov	7,516
Belfast Lough	4,276 [10]	5,542 [10]	4,248 [10]	3,909 [10]	3,889	Nov	4,373
Lough Foyle	2,294	2,326	2,231	3,095	1,805	Feb	2,350
Outer Ards Shoreline		1,968	1,812	1,740	1,565	Jan	1,771
Dundrum Bay	(1,428)	(1,250)	(1,425)	(1,252)	(1,389)	Dec	(1,428)
Carlingford Lough	986	1,289	(1,414)	1,410	1,442	Jan	1,308

Sites surpassing international passage threshold in the UK in 2005/2006

Morecambe Bay	46,760	Aug	The Wash	18,677	Aug
Thames Estuary	22,956	Sep	Solway Estuary	15,802	Oct
Dee Estuary (England & Wales)	19,164	Oct			

Sites surpassing national passage threshold in Great Britain in 2005/2006

Burry Inlet	8,850	Oct	Lavan Sands	5,353	Oct
Carmarthen Bay	7,295	Oct	Swale Estuary	5,011	Oct
Forth Estuary	6,598	Sep	Inner Moray and Inverness Firth	4,325	Oct
Ribble Estuary	6,071	Aug	Duddon Estuary	4,019	Oct
Inner Firth of Clyde	5,880	Sep	Humber Estuary	3,266	Aug

Sites surpassing national passage threshold in Northern Ireland in 2005/2006

Strangford Lough	6,232	Sep	Lough Foyle	1,707	Sep
Belfast Lough	4,756	Sep	Carlingford Lough	1,342	Sep
Outer Ards Shoreline	1,747	Oct	Dundrum Bay	1,078	Aug

Black-winged Stilt

Himantopus himantopus

Vagrant
Native Range: Worldwide

GB max: 3   May
NI max: 0

Three Black-winged Stilts were recorded at WWT Martin Mere in May and two remained into June. This was the first time this species had been recorded at the site during WeBS. The only other record was at Thurlestone Marsh, where one was present in April.

Avocet

Recurvirostra avosetta

International threshold: 730
Great Britain threshold: 35*

GB max: 5,575   Dec
NI max: 0

*50 is normally used as a minimum threshold

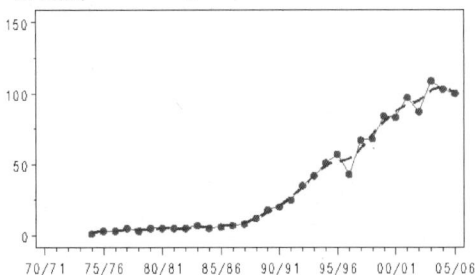

Figure 41.a, Annual indices & trend for Avocet for GB.

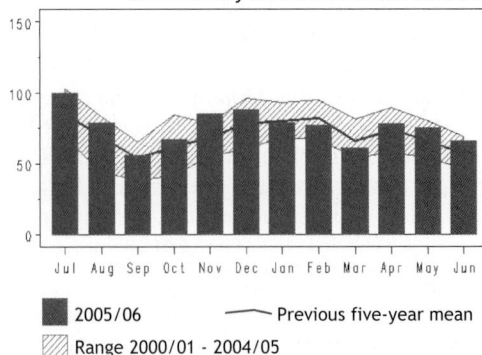

Figure 41.b, Monthly indices for Avocet for GB.

The British index for Avocet has been one of continuous increase, with the rate of increase accelerating rapidly since the late 1980s. However, during the past two years this trend has been reversed, and whilst index values do fluctuate year-to-year this is the first time that the index has fallen in two consecutive years. This said, the British maximum was the fourth highest on record and was only ten percent lower than the all time high in February 2002. Furthermore, there was little change in the key sites from recent years. As usual there were no Avocets reported from Northern Ireland, the only previous record there being of three in July 1993. Monthly indices reveal relatively stable numbers throughout the year, although in reality, there is a large amount of redistribution throughout the year, both within and outwith Britain.

Three sites held mean numbers of international importance in the five years up to 2005/06. As well as being the highest count during 2005/06, the November count for the Alde Complex was also the highest ever recorded at this site. Conversely, the Thames held its lowest numbers for seven years. A further 17 sites are nationally important for Avocet, two more than in the previous year. The inclusion of the Orwell Estuary was helped by an unusually high count that matched the site record of July 2002. During passage, four sites held internationally important numbers and a further 12 nationally important numbers. The highest count was at the Thames Estuary, which was three times the peak of the previous year. Numbers at Breydon Water & Berney Marshes were similar to recent years.

	01/02	02/03	03/04	04/05	05/06	Mon	Mean
Sites of international importance in the UK							
Poole Harbour	1,893	1,007	(1,493)	1,480 [10]	1,387	Feb	1,452
Alde Complex	1,174	1,089	1,073	1,058	1,392	Nov	1,157
Thames Estuary	1,447	839	658	1,153	542	Nov	928
Sites of national importance in Great Britain							
Medway Estuary	860	(650)	(615)	309 [10]	(557)	Nov	609
Swale Estuary	532	318	451	1,290	320	Mar	582
Hamford Water	485	(406)	433	520	488	Nov	482
Blyth Estuary	463						463
Exe Estuary	528	436	353	297	(500)	Dec	423
Tamar Complex	277	317 [10]	394	438	494	Feb	384
Colne Estuary	465	(383)	205 [12]	(90)	(285)	Dec	351
North Norfolk Coast	228	334	508	283	337	Mar	338
Blackwater Estuary	125	151 [10]	295	428	445	Nov	289
Deben Estuary	193	170	353	323	236	Dec	255
Breydon Wtr & Berney Marshes	172	224	268	232	228	Mar	225
Humber Estuary	121	281	(271)	215	194	Nov	216
Crouch-Roach Estuary	(43)	(9)	(17)	288 [10]	26	Mar	157
The Wash	347	130	180	37	82	Mar	155
Orwell Estuary	3 [10]	(36)	(63)	3	162	Mar	58 ▲
Minsmere	10	1	107	86	(1)	Jan	51
Stour Estuary	1	(0)	(0)	25	(89)	Nov	38 ▲
Sites surpassing international passage threshold in the UK in 2005/2006							
Thames Estuary	1,663	Oct	Alde Complex	983	Oct		
Breydon Wtr & Berney Marshes	1,044	Aug	The Wash	760	Jul		
Sites surpassing national passage threshold in Great Britain in 2005/2006							
Blackwater Estuary	622	Oct	Swale Estuary	273	Oct		
North Norfolk Coast	617	Apr	Deben Estuary	161	Oct		
Poole Harbour	562	Sep	Minsmere	65	Jul		
Medway Estuary	481	Oct	Rye Harbour and Pett Level	45	Jun		
Hamford Water	394	Oct	Ribble Estuary	38	Jun		
Humber Estuary	374	Aug	Stour Estuary	35	Oct		

Stone-curlew
Burhinus oedicnemus

Scarce

GB max:	4	Jun
NI max:	0	

Stone-curlews were recorded at just one site, which is close to a known breeding location in eastern England. Birds were present from April to June, the peak being four birds in June.

Little Ringed Plover
Charadrius dubius

International threshold: 2,500
Great Britain threshold: ?[†]
All-Ireland threshold: ?[†]

GB max: 189 Jul
NI max: 0

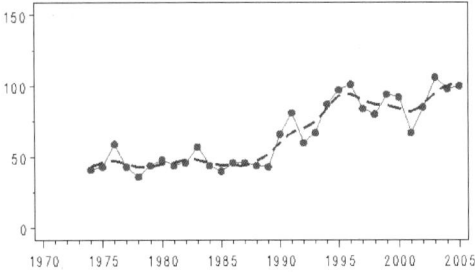

— Annual Index
---- Trend

Figure 42.a, Annual indices & trend for Little Ringed Plover for GB.

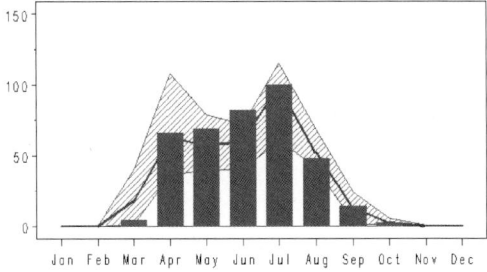

■ 2005/06   ⌒ Previous five-year mean
▨ Range 2000/01 - 2004/05

Figure 42.b, Monthly indices for Little Ringed Plover for GB.

The British peak of 189 was somewhat lower than the previous year and indeed was the lowest total since 2001. As Little Ringed Plover are a summering species the data presented here are for the calendar year 2005 rather than the usual WeBS year. The species was noted at 123 sites across Britain with, as usual, none recorded in Northern Ireland. The first birds of the year were recorded in March; these were one at Hamer Warren Lake and five at River Avon - Fordingbridge to Ringwood. Birds were then noted in Britain in every month through the summer until the last records in September. Most records were single-figure counts, with double figures only being recorded at four sites (compared to ten sites in 2004).

Sites with 10 or more birds during 2005[†]

Old Moor	14	Apr	Nosterfield Gravel Pits	11	May
Rutland Water	14	Jul	Belvide Reservoir	11	Jul

[†] *as no British or All-Ireland thresholds have been set a qualifying level of 10 has been chosen to select sites for presentation in this report*

Ringed Plover
Charadrius hiaticula

International threshold: 730
Great Britain threshold: 330
Great Britain passage threshold: 300
All-Ireland threshold: 125

GB max: 15,485 Aug
NI max: 717 Nov

Although the British index for Ringed Plover went up slightly for the first time in over five years, the underlying trend continues to be one of a decline, and monthly indices for most months were lower than average. The Northern Ireland index appears to be showing something of a recovery since the low point in the late 1990s, although the counted maximum was lower than for 2004/05. A notable difference between the British and Northern Ireland monthly indices is that the spring passage seems to peak earlier in Northern Ireland, coming in April as opposed to May in Britain, whilst the pulse of passage in August apparent in Britain is far less obvious in Northern Ireland.

The Thames Estuary remains the key site in Britain for wintering birds, with Hamford Water the only other site supporting internationally important numbers in the UK during this season. Record peak counts at both Tiree and on the Duddon Estuary, the latter being a low tide count, have seen these sites qualify as sites of national importance, whilst Morecambe Bay and Swansea Bay both supported their highest peak numbers for over five years.

At six sites, peak passage numbers surpassed the international passage threshold, with counts at all these sites being higher than wintering counts. A further fifteen sites held nationally important passage numbers in Britain with a further three in Northern Ireland.

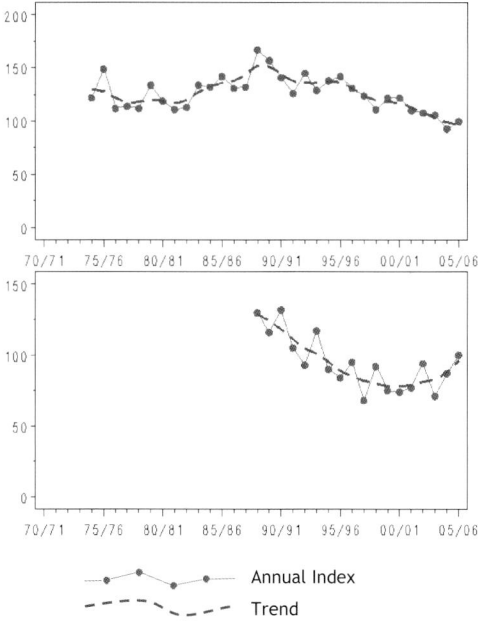

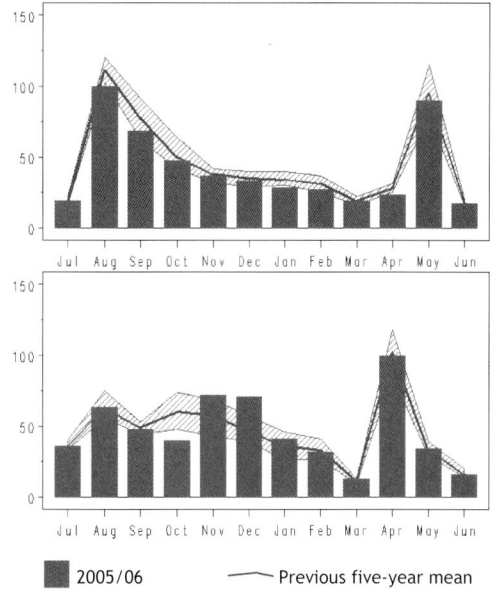

| | | Annual Index | | 2005/06 | | Previous five-year mean |
| | | Trend | | Range 2000/01 - 2004/05 | | |

Figure 43.a, Annual indices & trend for Ringed Plover for GB (above) & NI (below).

Figure 43.b, Monthly indices for Ringed Plover for GB (above) & NI (below).

	01/02	02/03	03/04	04/05	05/06	Mon	Mean	
Sites of international importance in the UK								
Thames Estuary	765	794	(654)	872	846	Dec	819	
Hamford Water	1,302 [10]	201	(576)	(333)	(361)	Nov	752	
Sites of national importance in Great Britain								
Tiree					648 [48]	Feb	648	▲
Solway Estuary	(289)	(599)	(286)	(305)	(157)	Mar	(599)	
Duddon Estuary	232	(227)	222	350 [10]	757 [10]	Nov	390	▲
Swansea Bay	436	269	330	431	453	Feb	384	
North Norfolk Coast	(471)	262	464	411	231	Mar	368	
Morecambe Bay	298	246	303	357	587	Dec	358	
South Ford		373	250	400			341	
Medway Estuary	(89)	(249)	(136)	332 [10]	(94)	Dec	332	
Humber Estuary	350	225	418 [10]	(194)	(223)	Jan	331	
Sites of all-Ireland importance in Northern Ireland								
Strangford Lough	618 [10]	236 [10]	277 [10]	342	449	Dec	384	
Outer Ards Shoreline		315	(198)	142	308	Jan	255	
Carlingford Lough	(203)	(240)	161	223	247	Nov	218	
Belfast Lough	188	189	234 [10]	109 [10]	168 [10]	Nov	178	
Sites no longer meeting table qualifying levels in Winter 2005/2006								
Thanet Coast	407	412	389	123	245	Nov	315	
Other sites surpassing table qualifying levels in Winter 2005/2006 in Great Britain								
Orwell Estuary	181 [10]	234	291 [10]	160 [10]	330 [10]	Feb	239	

Sites surpassing international passage threshold in the UK in 2005/2006					
North Norfolk Coast	2,310	Aug	The Wash	1,416	May
Ribble Estuary	1,950	May	Thames Estuary	1,262	Oct
Humber Estuary	1,447	May	Morecambe Bay	1,000	May
Sites surpassing national passage threshold in Great Britain in 2005/2006					
Solway Estuary	665	Aug	Dee Estuary (England & Wales)	392	Aug
Severn Estuary	662	Aug	Swale Estuary	392	Aug
Stour Estuary	610	Aug	Blackwater Estuary	367	Aug
Lindisfarne	415	May	Forth Estuary	348	Sep
Alt Estuary	404	Aug	Dengie Flats	331	Aug
Chichester Harbour	400	Aug	Colne Estuary	322	Sep
Mersey Estuary	400	Aug	Pool of Virkie	302	Aug
Taw-Torridge Estuary	395	Aug			
Sites surpassing national passage threshold in Northern Ireland in 2005/2006					
Outer Ards Shoreline	204	Oct	Belfast Lough	164	Sep
Carlingford Lough	180	Oct			

Kentish Plover

Charadrius alexandrinus

Scarce

GB max:	1	May
NI max:	0	

A single Kentish Plover was at Pegwell Bay in May. This was the 16th recorded for WeBS at this site, the last being in May 2002.

Golden Plover

Pluvialis apricaria

International threshold:	9,300
Great Britain threshold:	2,500
All-Ireland threshold:	2,000

GB max:	227,182	Jan
NI max:	24,388	Jan

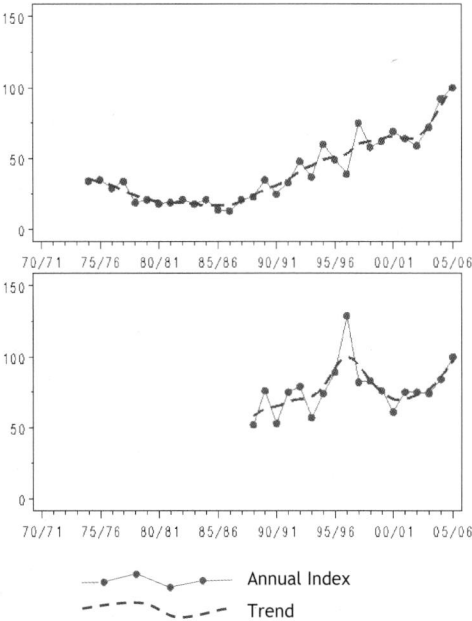

Figure 44.a, Annual indices & trend for Golden Plover for GB (above) & NI (below).

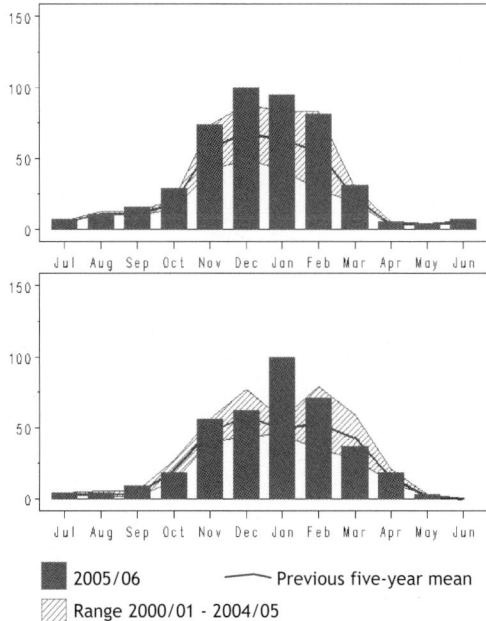

Figure 44.b, Monthly indices for Golden Plover for GB (above) & NI (below).

The recent sharp increase in the numbers of Golden Plover on WeBS sites continues to be seen with the British maximum again rising by 8% from the previous record level. Monthly indices from November to February were again much higher than the previous five year mean, peaking in December. Many of the largest counts came from east coast sites with the highest number again at the Humber Estuary again. However, the peak recorded on the Wash was well down on the previous year. Five sites in Great Britain maintained their internationally important status though Carmarthen Bay now only qualifies as nationally important. The peak count at Dengie Flats was the highest ever recorded at the site whilst counters at the Thames Estuary, Ouse Washes, Hamford Water, Lindisfarne, Lower Derwent Ings, Camel Estuary, Colne Estuary and Rutland Water also recorded notably high numbers.

Similarly, numbers in Northern Ireland continue to increase with the index at its highest peak since the one-off peak in the 1996/97 winter. Strangford Lough remains the only site supporting internationally important numbers of Golden Plovers in the province although the peak here, though high, was below the five-year mean.

Again only the Humber Estuary and the Wash surpassed the international threshold during the passage months, when the peak for the Wash was higher than for the winter months.

	01/02	02/03	03/04	04/05	05/06	Mon	Mean
Sites of international importance in the UK							
Humber Estuary	29,607	40,585	(50,662)	43,217	(41,608)	Jan	41,136
The Wash	14,109	19,089	25,817	34,900	17,418	Nov	22,267
Breydon Wtr & Berney Marshes	10,200	8,900	10,464	30,940	28,220	Jan	17,745
Blackwater Estuary	(8,082)	12,455 [10]	(6,986)	12,747	11,949	Dec	12,384
Swale Estuary	13,898	3,282	10,935	(6,560)	12,014	Jan	10,032
Strangford Lough	11,726 [10]	8,766	15,988 [10]	4,578	7,970	Jan	9,806
Sites of national importance in Great Britain							
Carmarthen Bay	(800)	(500)	(9,832)	(7,661)	4,047	Jan	7,180 ▼
Pegwell Bay	7,000	7,229 [10]	8,000	5,330	7,000	Dec	6,912
Old Moor	5,500	(7,700)	7,000		(1,600)	Nov	6,733
Morecambe Bay	(5,649)	(3,349)	(7,304)	(4,431)	5,088	Nov	6,014
Thames Estuary	3,538	(3,268)	(1,823)	6,440	7,401	Jan	5,793
Lynemouth Ash Lagoons			5,700				5,700
Somerset Levels	5,169	1,260	8,609	8,136	5,018	Dec	5,638
Nene Washes	4,440	4,320	650	13,000	4,500	Feb	5,382
Solway Estuary	(3,333)	(3,708)	4,459	6,145 [10]	(3,762)	Mar	5,302
Dengie Flats	1,900	3,288	2,275	3,660	12,678	Dec	4,760 ▲
North Norfolk Coast	(4,917)	1,919	5,039	5,975	5,315	Jan	4,633
Ouse Washes	4,035	2,828 [12]	2,844	3,456	9,990	Jan	4,631
Hamford Water	2,464	2,384	3,204	5,606	8,859	Jan	4,503
Lindisfarne	2,881	(3,383)	3,822 [10]	3,920	(7,081)	Dec	4,426
Taw-Torridge Estuary	(4,500)	(2,612)	3,300	(6,000)	2,550	Dec	4,088
Cleddau Estuary	2,240	1,060	(2,664)	(4,273)	8,630	Dec	4,051 ▲
Durham Coast				(2,000)	(3,704)	Jan	(3,704) ▲
Forth Estuary	2,419	(4,632)	6,940 [10]	2,658	1,588	Nov	3,647
Blyth Estuary	3,510						3,510
Lower Derwent Ings	3,471	890	2,005	4,130	6,776	Jan	3,454
Camel Estuary	800	727	515	4,750 [10]	9,000	Jan	3,158 ▲
St Mary`s Island	(2,000)	3,000	3,200	3,000			3,067
Ribble Estuary	3,075	(2,671)	(3,300)	1,705	3,829	Nov	2,977
Wigtown Bay	2,000	(602)	(3,604)	(2,500)	3,175	Dec	2,926 ▲
Chichester Harbour	(2,436)	2,237	(2,822)	3,048	(3,586)	Dec	2,923
Crouch-Roach Estuary	2,602	2,165	1,354	4,771 [10]	3,718	Jan	2,922
Walland Marsh	2,600	500	600	6,500	3,200	Dec	2,680 ▲
Clifford Hill Gravel Pits	3,560	2,500	2,740	1,000	3,000	Jan	2,560
Sites of all-Ireland importance in Northern Ireland							
Lough Foyle	4,100	3,320	5,719	7,372	7,640	Jan	5,630
Loughs Neagh and Beg	(2,817)	4,631	7,091	3,447	6,537	Feb	5,427
Bann Estuary	1,660 [12]	1,400	2,265	2,100	2,610	Feb	2,007 ▲
Sites no longer meeting table qualifying levels in Winter 2005/2006							
Colne Estuary	1,820	(82)	(1,480)	(1,450)	2,840	Dec	2,330
Stanwick Gravel Pits	4,504		880		2,000	Nov	2,461
Other sites surpassing table qualifying levels in Winter 2005/2006 in Great Britain							
Severn Estuary	806 [12]	1,215 [10]	2,060	3,100	4,370	Feb	2,310
The Ouse and Lairo Water	90	320	140	150	4,000	Nov	940
Ouse Fen & Pits (Hanson/RSPB)		(2,118)	150	20	3,000	Jan	1,322
Colne Estuary	1,820	(82)	(1,480)	(1,450)	2,840	Dec	2,330
Alde Complex	793	1,444	(696)	3,346	2,765	Jan	2,087
Rutland Water	428	2,588	2,442	654	2,700	Jan	1,762
Middle Yare Marshes	1,945	85	(96)	4,400	2,597	Mar	2,257
Priory Water	168	2,750	1,500	450	2,536	Jan	1,481
Other sites surpassing table qualifying levels in Winter 2005/2006 in Northern Ireland							
Killough Harbour	560 [10]	1,101	0	1,850	2,100	Mar	1,122

Sites surpassing international passage threshold in the UK in 2005/2006

The Wash	26,996	Oct	Humber Estuary			16,045	Oct

Sites surpassing national passage threshold in Great Britain in 2005/2006

Breydon Wtr & Berney Marshes	8,930	Oct	Ribble Estuary	3,628	Oct
Old Moor	6,200	Oct	North Norfolk Coast	3,444	Oct
Morecambe Bay	5,768	Oct	Forth Estuary	3,326	Oct
Blackwater Estuary	5,755	Oct	Lower Derwent Ings	3,250	Oct
Dengie Flats	4,278	Oct	Carmarthen Bay	2,923	Oct
Lindisfarne	3,852	Oct			

Sites surpassing national passage threshold in Northern Ireland in 2005/2006

Lough Foyle	4,612	Oct
Loughs Neagh and Beg	2,441	Oct

Grey Plover
Pluvialis squatarola

International threshold: 2,500
Great Britain threshold: 530
All-Ireland threshold: 40*

GB max: 40,306 Nov
NI max: 203 Feb

50 is normally used as a minimum threshold

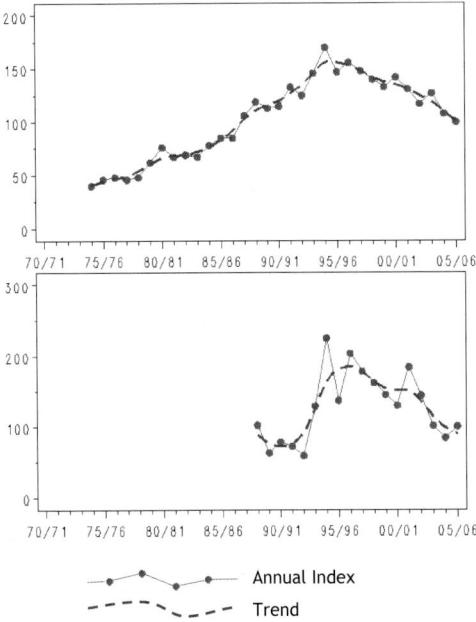

Annual Index
Trend

Figure 45.a, Annual indices & trend for Grey Plover for GB (above) & NI (below).

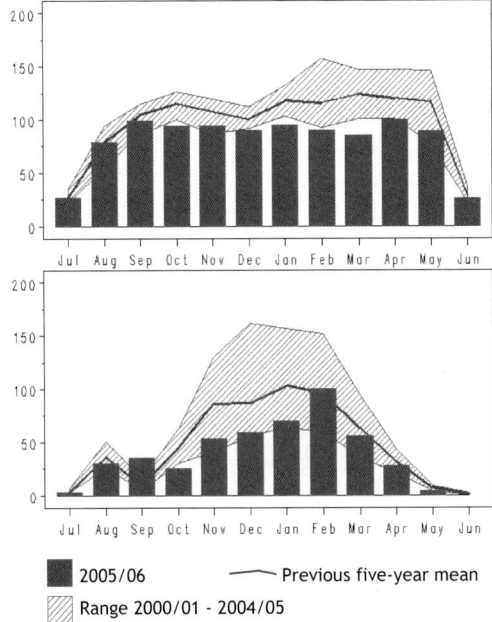

2005/06
Range 2000/01 - 2004/05
Previous five-year mean

Figure 45.b, Monthly indices for Grey Plover for GB (above) & NI (below).

The British counted maximum was only slightly lower than during the previous year and was similar to the average of the previous five-years. However, the British index has shown that numbers of wintering Grey Plovers have declined steadily after the zenith of the mid-1990s and it numbers are now similar to those during the late 1980s. The monthly indices show that although numbers were about normal during July-September, they were then low for the remainder of the winter. Grey Plovers are much scarcer in Northern Ireland. The counted maximum was the second lowest since 1992/93 and was less than a quarter of that in recorded 2004/05. The index for the region shows a similar pattern to that of Britain, being one of sustained decline since the mid-1990s.

Numbers on key sites were variable. The November total for the Thames Estuary was the highest ever recorded at the site, and for the first time in recent years this site becomes the top British site in terms of mean peak numbers. This was aided by numbers at the Wash being the lowest there for over twenty years. The high count at Dengie Flats and the low one at the adjacent Blackwater Estuary in November may well refer to local redistribution. Numbers on the Swale were at their lowest for 30 years. In Northern Ireland, however, the peak count at Carlingford Lough was the highest for 20 years. In line with wintering numbers, passage counts at the Wash were also lower than usual.

	01/02	02/03	03/04	04/05	05/06	Mon	Mean
Sites of international importance in the UK							
Thames Estuary	(5,160)	3,195	(3,812)	(2,681)	13,028	Nov	8,112
The Wash	8,395	7,778	10,447	6,605	3,642	Mar	7,373
Ribble Estuary	6,285	1,658	5,568	3,529	(1,524)	Mar	4,260
Dengie Flats	3,640	3,768	2,943	2,466	4,909	Nov	3,545
Hamford Water	3,267	2,984	(1,746)	(2,915)	(2,198)	Jan	3,126
Blackwater Estuary	2,228	3,230	2,011	4,043	1,697	Nov	2,642

	01/02	02/03	03/04	04/05	05/06	Mon	Mean
Sites of national importance in Great Britain							
Stour Estuary	1,612	(2,717)	1,553	2,128	3,263	Dec	2,255
Chichester Harbour	(3,180)	1,700	1,515	1,420	2,017	Dec	1,966
Humber Estuary	1,567	(1,300)	2,285 [10]	(964)	(1,639)	Jan	1,926
Alt Estuary	2,500	1,099	3,098	1,501	1,174	Nov	1,874
Swale Estuary	1,745	2,181	1,892	1,389	840	Dec	1,609
North Norfolk Coast	1,720	1,374	1,316	1,386	1,093	Feb	1,378
Dee Estuary (England & Wales)	2,201 [10]	966	(1,851)	758	1,091	Feb	1,373
Colne Estuary	1,357	(141)	(705)	(623)	(288)	Dec	1,357
Medway Estuary	1,616	938	1,544	733	989	Dec	1,164
Lindisfarne	1,016	(635)	(656)	775	1,361	Feb	1,051
Pagham Harbour	713	704	1,348	873	1,067	Mar	941
Morecambe Bay	1,043	657	778	1,001	1,074	Jan	911
Langstone Harbour	504	982	1,119	701	879	Mar	837
Deben Estuary	340	344	656	1,037	(719)	Jan	619
Eden Estuary	812	690	371	450	(356)	Jan	581
Solway Estuary	(482)	(466)	509	602 [10]	(474)	Mar	556
Sites of all-Ireland importance in Northern Ireland							
Strangford Lough	273	398 [10]	137	114	249 [10]	Dec	234
Dundrum Bay	(19)	(72)	(27)	(32)	(10)	Nov	(72)
Carlingford Lough	45	52	(57)	33	100	Feb	58
Other sites surpassing table qualifying levels in Winter 2005/2006 in Great Britain							
Mersey Estuary	260	201	205	202	597 [10]	Dec	293
Crouch-Roach Estuary	200	282	212	518	595	Feb	361
Orwell Estuary	323 [10]	(413)	710	350 [10]	(568)	Dec	488
Severn Estuary	359	555 [10]	(275)	222	561	Feb	424
Cromarty Firth	0	14	21	5	554	Dec	119

Sites surpassing international passage threshold in the UK in 2005/2006					
The Wash	8,604	Aug	Humber Estuary	2,789	May
Dengie Flats	4,200	Oct	Thames Estuary	2,724	Aug
Ribble Estuary	3,813	Sep	Blackwater Estuary	2,650	Oct
Alt Estuary	2,837	Apr			
Sites surpassing national passage threshold in Great Britain in 2005/2006					
Stour Estuary	1,628	Sep	Morecambe Bay	904	Oct
North Norfolk Coast	1,483	Aug	Colne Estuary	800	Sep
Swale Estuary	1,244	Sep	Pagham Harbour	755	Oct
Hamford Water	1,057	Oct	Langstone Harbour	584	Oct
Chichester Harbour	1,048	Oct	Medway Estuary	578	Oct
Lindisfarne	955	Apr			
Sites surpassing national passage threshold in Northern Ireland in 2005/2006					
Strangford Lough	79	Sep			

Lapwing
Vanellus vanellus

International threshold: 20,000**
Great Britain threshold: 20,000**[†]
All-Ireland threshold: 2,500

GB max: 475,236 Jan
NI max: 21,989 Jan

Both the British index and the counted maximum rose to their highest levels since 1999/00. The monthly indices suggest this increase was confined to January and February. However it is important to remember that this represents numbers of birds on wetland sites and does not take into account birds using agricultural land. Looking to the future, the picture may be further enhanced by the 2006/07 Winter Plover Survey, which covered a range of habitats including those not covered by WeBS. Following a steady decline since the mid 1990s, the Northern Ireland index has shown some stability in the last few years; the monthly indices here do suggest particularly low numbers present in November.

The Wash regained its status as the key site for the species, whilst numbers at Breydon Water & Berney Marshes were the second highest for over five years, leading to this site surpassing the international importance qualifying level. Counts at most sites in the UK were roughly in line with the five-year average, although Ouse Washes, North Norfolk Coast and Severn Estuary did support notably high peak numbers. Peak counts at the top three sites, the Wash, Somerset Levels and Ribble Estuary, however, were slightly lower than for 2004/05.

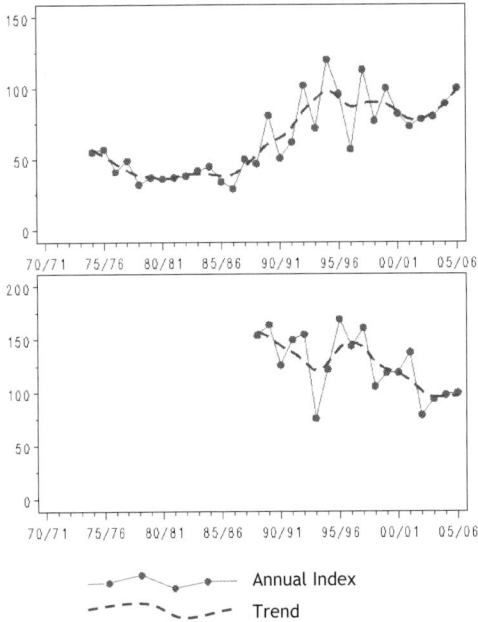

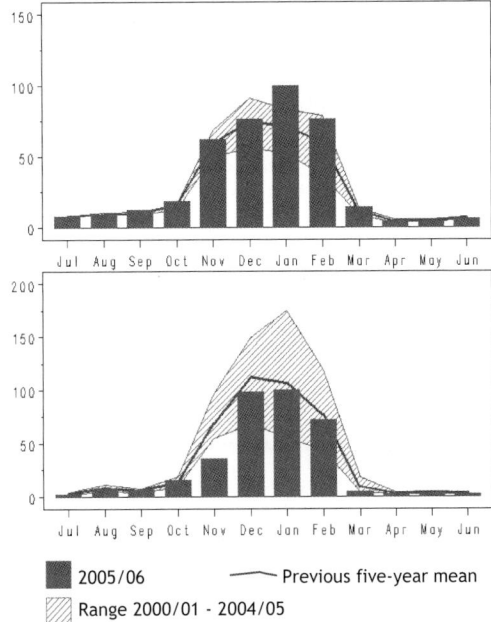

| | Annual Index |
| | Trend |

| ■ | 2005/06 | — Previous five-year mean |
| ▨ | Range 2000/01 - 2004/05 |

Figure 46.a, Annual indices & trend for Lapwing for GB (above) & NI (below).

Figure 46.b, Monthly indices for Lapwing for GB (above) & NI (below).

	01/02	02/03	03/04	04/05	05/06	Mon	Mean
Sites of international importance in the UK							
The Wash	43,558	43,672	29,350	43,822	36,327	Jan	39,346
Somerset Levels	41,675	16,053	23,641	60,834	48,116	Jan	38,064
Ribble Estuary	(9,579)	(14,500)	(15,374)	25,991	24,265	Jan	25,128
Humber Estuary	10,719	(36,309)	(39,865)	(16,856)	(20,559)	Jan	24,862
Breydon Wtr & Berney Marshes	19,380	15,230	15,890	29,136	25,140	Feb	20,955 ▲
Sites of all-Ireland importance in Northern Ireland							
Strangford Lough	10,527	6,977	8,884 [10]	5,792	6,635	Dec	7,763
Loughs Neagh and Beg	(4,264)	3,090	6,282	7,584	6,684	Jan	5,910
Lough Foyle	3,320	2,629	4,240	3,606	4,745	Dec	3,708
Sites with mean peak counts of 5,000 or more birds in Great Britain[†]							
Morecambe Bay	13,504	(13,714)	(20,750)	16,701	19,192	Jan	17,537
Ouse Washes	19,219 [12]	8,125 [12]	13,577	12,240	25,835	Feb	15,799
Swale Estuary	14,804	14,974	16,523	(13,270)	14,913	Jan	15,304
Severn Estuary	(7,439)	12,129 [10]	(6,889)	11,312	19,434	Feb	14,292
Thames Estuary	(10,282)	16,036	10,229	14,657	(12,390)	Jan	13,641
Mersey Estuary	(5,284)	(5,675)	(12,150)	(9,370)	10,098 [10]	Feb	11,124
Nene Washes	4,230	21,016	3,870	7,050	6,070	Feb	8,447
North Norfolk Coast	7,830	5,124	7,358	7,833	13,305	Jan	8,290
Solway Estuary	(5,211)	(7,340)	8,218	(5,989)	(8,036)	Dec	8,218
Blackwater Estuary	(9,005)	11,053 [10]	7,472	6,785	6,766	Dec	8,216
Dee Estuary (England & Wales)	9,206	6,470	7,853	7,512	8,800	Feb	7,968
Pegwell Bay	6,000	10,282 [10]	10,000	5,420	(8,100)	Jan	7,960
Crouch-Roach Estuary	3,697	4,939	5,386	11,288 [10]	8,464	Jan	6,755
Walland Marsh	11,000	1,800	1,700	10,000	5,000	Jan	5,900
Lower Derwent Ings	4,119	3,986	5,119	7,920	7,520	Jan	5,733
Tees Estuary	(3,196)	6,017	6,623	4,571	5,334	Dec	5,636
Other sites surpassing table qualifying levels in Winter 2005/2006 in Great Britain[†]							
Alde Complex	2,241 [10]	4,358	(3,841)	5,472	7,843	Jan	4,979
Taw-Torridge Estuary	(3,237)	(2,339)	1,270	(2,507)	(7,013)	Feb	3,273
Middle Yare Marshes	1,649	1,947	1,622	2,560	6,520	Jan	2,860
Malltraeth Cob and Pools			950	30	6,000	Feb	2,327
Lindisfarne	1,742 [10]	(1,483)	1,836	1,795 [10]	5,371	Nov	2,686
Other sites surpassing table qualifying levels in Winter 2005/2006 in Northern Ireland							
Upper Quoile River	300	160	730	322	2,502	Feb	803

[†] as few sites exceed the British threshold a qualifying levels of 5,000 has been chosen to select sites for presentation in this report

Knot
Calidris canutus

International threshold: 4,500
Great Britain threshold: 2,800
All-Ireland threshold: 375

GB max: 269,154 Nov
NI max: 6,653 Dec

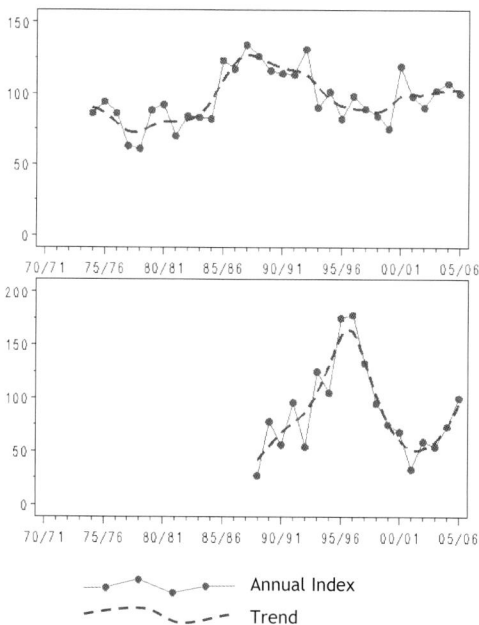

Annual Index

Trend

Figure 47.a, Annual indices & trend for Knot for GB (above) & NI (below).

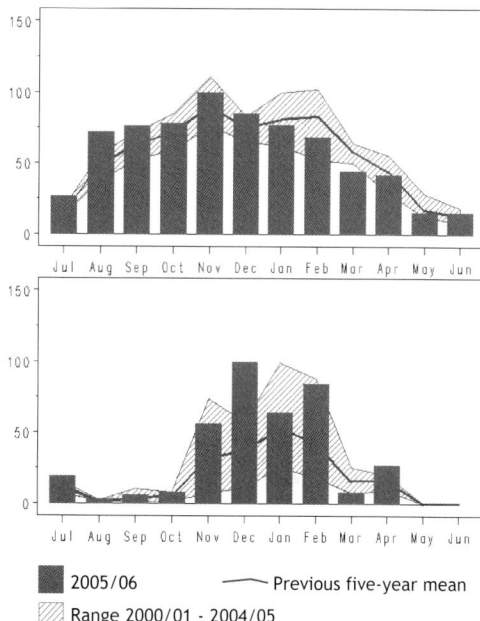

2005/06 — Previous five-year mean

Range 2000/01 - 2004/05

Figure 47.b, Monthly indices for Knot for GB (above) & NI (below).

The British annual index remained similar to the past two years with the underlying trend showing signs of recovery following a steady decline during the 1990s. Numbers were above average through the early part of 2005/06 with above average numbers from August to December, after which time they fell to below the average of the past five years. As in Britain, numbers in Northern Ireland also showed signs of recovery, although the increase there was somewhat clearer. The Northern Ireland maximum was similar to the average for the past five years although numbers in the province are largely dependent on those at the principal site, Strangford

Lough. Peak Core Count numbers there were the highest for over five years, however peak totals from Low Tide Counts have surpassed these figures in several years.

Numbers on the Wash peaked at their highest level since 1992/93 with a count that was almost three times that of two years ago. Totals from many of the other key sites were below average, however, with those on the North Norfolk Coast and the Forth Estuary being the lowest of the past five years. Passage numbers on the Wash were the highest for 15 years whereas those at most other key sites were lower than during the previous year.

	01/02	02/03	03/04	04/05	05/06	Mon	Mean
Sites of international importance in the UK							
The Wash	80,452	51,642	48,372	105,912	139,270	Nov	85,130
Morecambe Bay	66,031	(61,968)	67,959	(24,749)	31,207	Jan	56,791
Ribble Estuary	36,202	(23,691)	44,947	(21,540)	(13,970)	Mar	40,575
Humber Estuary	49,991	18,936	50,557 [10]	(37,015)	(24,035)	Dec	39,828
Thames Estuary	27,425	30,060	43,873	33,024	24,233	Feb	31,723
Dee Estuary (England & Wales)	52,792	26,769	38,070	10,243	24,505	Nov	30,476
Alt Estuary	44,012	25,045	30,000	19,006	12,454	Feb	26,103
Dengie Flats	13,600	10,550	8,000	22,700	15,650	Nov	14,100
Solway Estuary	(3,784)	(9,620)	8,725	13,142	(7,596)	Mar	10,934
North Norfolk Coast	16,214	9,224	7,523	6,735	4,462	Jan	8,832

	01/02	02/03	03/04	04/05	05/06	Mon	Mean
Forth Estuary	7,232	8,936	6,907 [10]	5,077	4,685	Feb	6,567
Strangford Lough	4,000 [10]	10,340 [10]	4,058	5,730	8,014 [10]	Jan	6,428
Blackwater Estuary	(2,495)	1,700 [10]	(5,982)	6,273	(5,326)	Dec	4,820
Sites of national importance in Great Britain							
Lindisfarne	2,858	(4,512)	(6,751)	4,197	(4,172)	Jan	4,498
Burry Inlet	2,000	3,800	3,500	8,259	4,301	Dec	4,372
Montrose Basin	5,000	5,800	(2,562)	1,990	3,360	Feb	4,038
Cromarty Firth	2,621	3,132	4,932	5,000	3,132	Dec	3,763
Inner Moray and Inverness Firth	1,980	1,873	3,663	3,446	5,146	Jan	3,222
Hamford Water	1,957	2,935	4,160	(2,481)	3,185	Dec	3,059
Swale Estuary	2,900	1,500	4,050	2,538	4,060	Dec	3,010
Orwell Estuary	1,601 [10]	3,172 [10]	4,021	2,115 [10]	3,569 [10]	Feb	2,896 ▲
Dornoch Firth	3,113 [10]	2,960	1,500	2,680	4,215	Jan	2,894 ▲
Medway Estuary	1,950	4,085	1,817	3,024 [10]	3,574	Dec	2,890 ▲
Sites of all-Ireland importance in Northern Ireland							
Lough Foyle	20	345	942	470	470	Jan	449
Dundrum Bay	(555)	(603)	320	(475)	270	Nov	445
Other sites surpassing table qualifying levels in Winter 2005/2006 in Great Britain							
Stour Estuary	160	41	84	1,470	6,701	Jan	1,691

Sites surpassing international passage threshold in the UK in 2005/2006

The Wash	112,438	Sep	Alt Estuary	8,651	Aug
Humber Estuary	35,004	Aug	Dengie Flats	8,500	Oct
Ribble Estuary	26,106	Apr	Thames Estuary	8,017	Oct
North Norfolk Coast	25,551	Sep	Dee Estuary (England & Wales)	7,121	Oct
Morecambe Bay	9,916	Oct	Solway Estuary	6,412	Oct

Sites surpassing national passage threshold in Northern Ireland in 2005/2006

Lough Foyle	404	Oct

Sanderling
Calidris alba

GB max: 10,626 Aug
NI max: 268 Apr

International threshold: 1,200
Great Britain threshold: 210
Great Britain passage threshold: 300
All-Ireland threshold: 35*

*50 is normally used as a minimum threshold

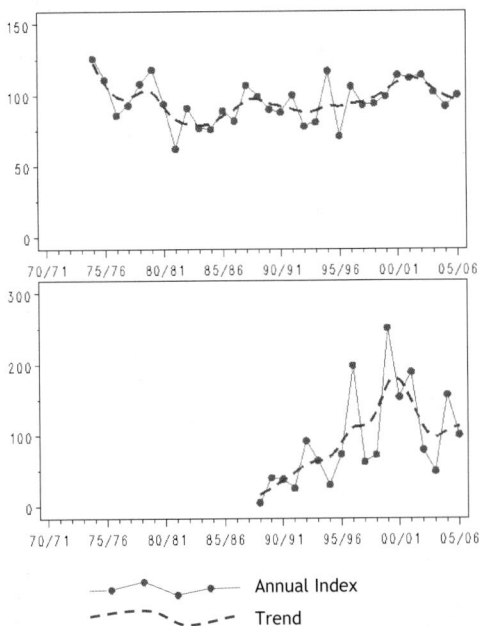

Figure 48.a, Annual indices & trend for Sanderling for GB (above) & NI (below).

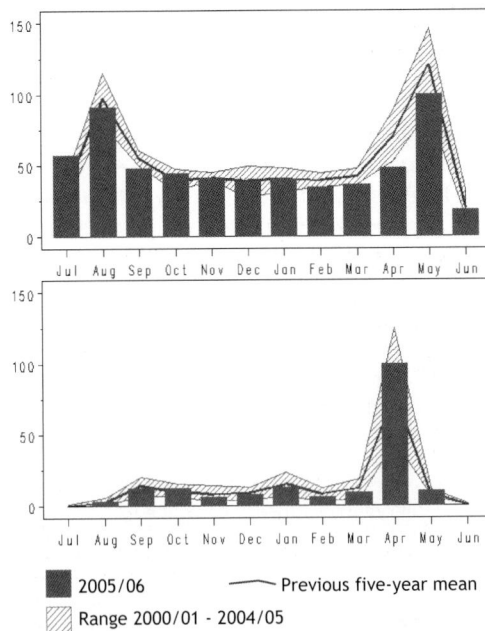

2005/06   Previous five-year mean
Range 2000/01 - 2004/05

Figure 48.b, Monthly indices for Sanderling for GB (above) & NI (below).

The underlying trend for Sanderling in Britain, which is now based on counts from November to March inclusive rather than December to February, shows that numbers have been fairly stable over the past 30 years. During the first part of the winter numbers were about average, however, from February to April numbers were lower than during the past five years. Numbers in Northern Ireland are very much lower, especially during the winter months, but with a notable movement through the province in April.

The Ribble Estuary remains the primary site for this species, although at the nearby Alt Estuary winter numbers have fallen and are at their lowest level for ten years. The same was also true for the North Norfolk Coast where peak winter numbers fell by over 40% since 2004/05. Peak winter numbers on the Wash were the highest ever recorded at the site being two and a half times the previous five-year mean. Peak numbers at Morecambe Bay were also the highest recorded at the site.

Passage numbers were in general similar to those in 2004/05. Internationally important numbers were again noted at four sites in which North Norfolk Coast featured in place of Thames Estuary compared to 2004/05. In total, nine sites held nationally important numbers during this passage, six of which also held nationally important numbers during the previous year.

	01/02	02/03	03/04	04/05	05/06	Mon	Mean
Sites of international importance in the UK							
Ribble Estuary	3,004	2,680	2,400	(1,453)	(2,155)	Dec	2,695
Sites of national importance in Great Britain							
Carmarthen Bay	1,600	1,770	833	769	800	Mar	1,154
Alt Estuary	1,556	1,431	913	815	624	Nov	1,068
North Norfolk Coast	1,319	1,150	601	889	506	Feb	893
Thames Estuary	552	875	385	562	457	Nov	566
The Wash	504	496	317	395	1,091	Nov	561
Dee Estuary (England & Wales)	550	286	(379)	274	(1,020)	Nov	502
Tiree					468 [48]	Feb	468 ▲
Ardivachar Point (South Uist)		398	460	400			419
Jersey Shore	391						391
Thanet Coast	434	444	342	418	307	Nov	389
Humber Estuary	358	440	370 [10]	(96)	(159)	Dec	389
Duddon Estuary	486	287	585	361	192	Nov	382
Solway Estuary	(218)	(266)	(370)	(302)	(165)	Dec	(370)
Morecambe Bay	275	240	306	225	652	Jan	340
Swansea Bay	356	410	200	234	467	Feb	333
Howmore Estuary SSSI Coast			312 [52]				312
Lindisfarne	321 [10]	283 [10]	221	388 [10]	294	Dec	301
Forth Estuary	274	389	269 [10]	181	256	Nov	274
South Ford		120	250	430			267
Tees Estuary	259	280	240	199	253	Nov	246
North Bay (South Uist)		67	235	340			214
Sites of all-Ireland importance in Northern Ireland							
Killough Harbour	76 [10]						76
Dundrum Bay	(0)	(30)	(0)	(48)	(5)	Jan	(48)
Sites no longer meeting table qualifying levels in Winter 2005/2006							
Lade Sands	236	140	118	350	120	Jan	193
Pegwell Bay	123	373 [10]	(115)	180	34	Feb	178
Other sites surpassing table qualifying levels in Winter 2005/2006 in Great Britain							
Tay Estuary	77	160	65	88	635	Dec	205
Ryde Pier to Puckpool Point	368	100	133	143	240	Nov	197

Sites surpassing international passage threshold in the UK in 2005/2006				
Ribble Estuary	3,491	May	Alt Estuary	2,317	Aug
The Wash	3,291	Jul	North Norfolk Coast	1,241	May

Sites surpassing national passage threshold in Great Britain in 2005/2006

Thames Estuary	1,072	Aug	Dee Estuary (England & Wales)	450	Oct
Morecambe Bay	925	Oct	Duddon Estuary	332	Oct
Humber Estuary	576	May	Lade Sands	330	Sep
Carmarthen Bay	567	Oct	Ryde Pier to Puckpool Point	305	Oct
Solway Estuary	524	Apr	Swansea Bay	300	Oct

Sites surpassing national passage threshold in Northern Ireland in 2005/2006

Bann Estuary	268	Apr

Little Stint
Calidris minuta

International threshold: 2,000
Great Britain threshold: ?[†]
All-Ireland threshold: ?[†]

GB max: 56 Sep
NI max: 0

Little Stints were recorded at 43 sites and in every month during the year. Most records were during autumn with birds being noted at 27 sites during September. Overall numbers during autumn 2005 were low, with the highest count from a single site being just seven at the North Norfolk Coast. Up to 13 birds were counted for WeBS during the winter, although the British average between November and March was seven. Previously we have presented a list of sites at which at least ten birds were counted during the passage months. However, in 2005/06 no site surpassed this threshold so we have presented the peak counts of five or more. There were no records from Northern Ireland during 2005/06.

Sites with 5 or more birds during passage periods in Great Britain in 2005/06[†]

North Norfolk Coast	7 Sep		North West Solent	5 Sep
Dee Estuary (Eng & Wal)	6 Jun		Severn Estuary	5 Sep
Humber Estuary	5 Aug			

[†] as no British or All-Ireland thresholds have been set a qualifying level of 5 has been chosen to select sites for presentation in this report

Temminck's Stint
Calidris temminckii

Scarce

GB max: 1 Sep
NI max: 0

The only autumn Temminck's Stint was one at Titchwell, North Norfolk Coast, in September. Similarly, just one was seen in spring at Old Moor, South Yorkshire, in May.

White-rumped Sandpiper
Calidris fuscicollis

Vagrant
Native Range: America

GB max: 3 Oct
NI max: 0

This species has been occurring more regularly in recent years, although still largely in late summer and autumn. All records this year were in Scotland, with two recorded at North Ronaldsay Lochs in September, and one at Loch Gruinart Floods and two at Loch Paible (North Uist) in October. The only previous records from Scotland have both been of singles at South Ford, Western Isles, in September 1999 and September 2000.

Baird's Sandpiper
Calidris bairdii

Vagrant
Native Range: America

GB max: 1 Sep
NI max: 0

A single Baird's Sandpiper was recorded at the North West Solent (Hurst to Lymington section) during September and October. This bird is the 13th recorded for WeBS since the first in 1974.

Pectoral Sandpiper
Calidris melanotos

Vagrant
Native Range: America, N Siberia, Australia

GB max: 3 Sep
NI max: 1 Aug

Pectoral Sandpipers were recorded at six sites, the peak month typically being September when singles were reported from the Severn Estuary, Nosterfield Gravel Pits and Staines Reservoirs. Other single birds were at the Bann Estuary in August, Loughs Neagh and Beg in November and WWT Martin Mere in May. This species was new for WeBS at four of these six sites, having previously being recorded at Nosterfield Gravel Pits in 2003/04 and Severn Estuary in both 1974/75 and 1989/90.

Curlew Sandpiper
Calidris ferruginea

International threshold:	10,000	
Great Britain threshold:	?[†]	
All-Ireland threshold:	?[†]	

GB max:	168	Aug
NI max:	22	Aug

The British peak of Curlew Sandpipers was almost half that of the previous year though more akin to that of 2003/04. Peak numbers in both Britain and Northern Ireland were recorded during August, one month earlier than usual, indicating that more adults than juveniles moved through. The Northern Irish maximum was the highest since 1998/99.

Throughout the UK a total of nine sites held passage numbers of ten or more birds, including one, Bann Estuary, in Northern Ireland. Particularly high counts were noted at North Norfolk Coast and the Swale Estuary, both of which were the highest at the respective sites for over five years.

Sites with 10 or more birds during passage periods in Great Britain in 2005/2006[†]

North Norfolk Coast	88	Aug	Breydon Wtr & Berney Marshes	17	Sep
Swale Estuary	87	Jul	Thames Estuary	12	Aug
Pegwell Bay	32	Jul	Humber Estuary	10	Aug
Exe Estuary	25	Sep	Mersey Estuary	10	Sep

Sites with 10 or more birds during passage periods in Northern Ireland in 2005/2006[†]

Bann Estuary	22	Aug

[†] *as no British or All-Ireland thresholds have been set a qualifying level of 10 has been chosen to select sites for presentation in this report*

Purple Sandpiper
Calidris maritima

International threshold:	750	
Great Britain threshold:	180[†]	
All-Ireland threshold:	10*	

GB max:	1,281	Jan
NI max:	63	Mar

50 is normally used as a minimum threshold

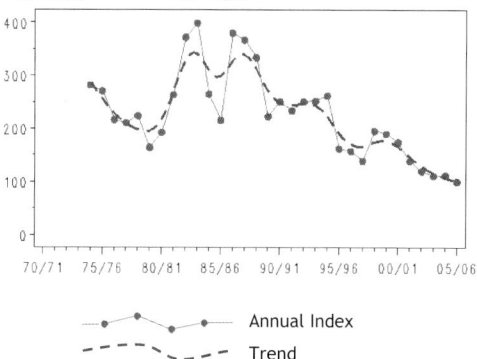

Figure 49.a, Annual indices & trend for Purple Sandpiper for GB.

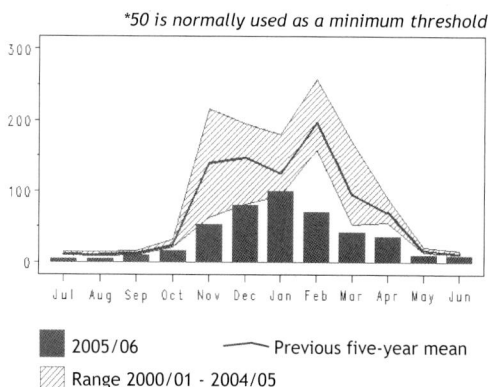

Figure 49.b, Monthly indices for Purple Sandpiper for GB.

The decline of Purple Sandpiper continued with the British annual index reaching its lowest ever level. This was also reflected in the monthly indices being well below average for the main winter period. The maximum count in Britain was slightly lower than for 2004/05 whilst the Northern Ireland maximum was the lowest counted since 2001/02. This species has, however, has always been under-represented by WeBS, and the results of the Non-estuarine Coastal Waterbird Survey (NEWS) in January 2007 may reveal a truer picture of its current status.

Following a supplementary count from Tiree, three sites now support nationally important numbers of Purple Sandpipers. The

count from the Island of Papa Westray was the highest recorded there so far by WeBS whilst again there were unfortunately no counts available for the Farne Islands. The Outer Ards Shoreline remains the key site in Northern Ireland though the peak count here was the lowest since 1997/98, whilst the count at Belfast Lough remained at an all time low.

As so few sites qualify for national importance, additional sites supporting a mean peak of over 100 birds are also listed. Of sites for which WeBS counts were received in 2005/06, the counts on the Forth Estuary and Island of Egilsay were higher than for 2004/05 though the Moray Coast and Dee Estuary both recorded reduced numbers. The area of

Seahouses to Budle Point, however, had its lowest total since 2002/03 and no longer has a mean peak over 100 birds.

	01/02	02/03	03/04	04/05	05/06	Mon	Mean
Sites of national importance in Great Britain							
Tiree					368 [48]	Feb	368 ▲
Island of Papa Westray	330	120	216	385	431	Feb	296
Farne Islands	194	(185)					194
Sites of all-Ireland importance in Northern Ireland							
Outer Ards Shoreline		122	83	84	60	Mar	87
Belfast Lough	16	15	17	6	5	Feb	12
Sites with mean peak counts of 100 or more birds in Great Britain[†]							
Forth Estuary	172	248	72	93	112	Feb	139
Ardivachar Point (South Uist)		120	110	144			125
Balranald Nature Reserve		190	180	0			123
Howmore Estuary SSSI Coast			120 [52]				120
Dee Estuary (Scotland)	71	92	81	185	157	Jan	117
Bornish & Ormiclate Machairs SSSI			112 [52]				112
East Unst	110 [9]						110
Island of Egilsay	4	141	195	81	130	Feb	110

[†] as few sites exceed the British threshold a qualifying level of 100 has been chosen to select sites

Dunlin
Calidris alpina

International threshold: 13,300
Great Britain threshold: 5,600
Great Britain passage threshold: 2,000
All-Ireland threshold: 1,250

GB max: 301,392 Dec
NI max: 9,967 Jan

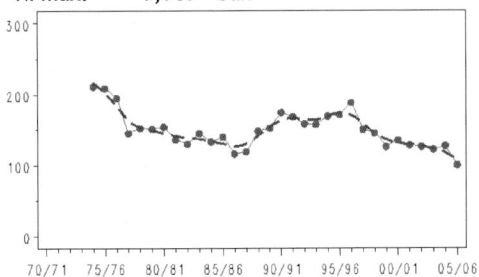

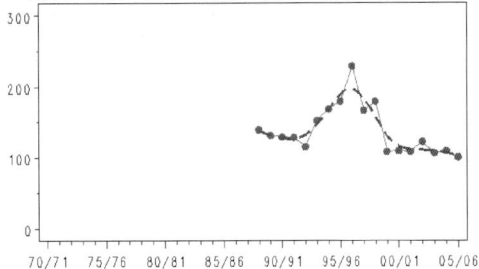

Annual Index
Trend

2005/06
Range 2000/01 - 2004/05
Previous five-year mean

Figure 50.a, Annual indices & trend for Dunlin for GB (above) & NI (below).

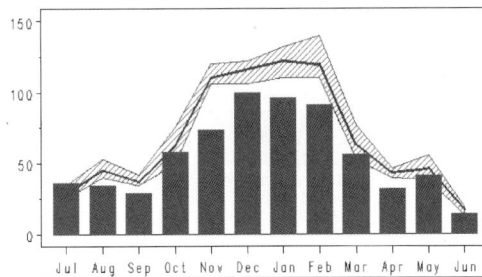

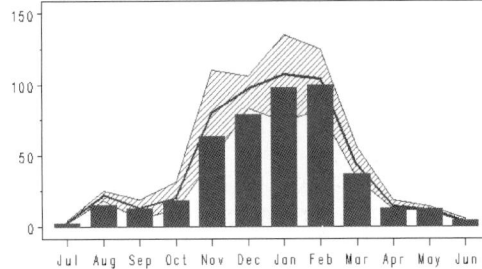

Figure 50.b, Monthly indices for Dunlin for GB (above) & NI (below).

The British maximum for Dunlin fell to its lowest level since 1970, a time when far fewer sites were being monitored. This decline in Dunlin numbers, and indeed the steady decrease over the past nine years, is clearly documented in the British annual index, which has fallen to its lowest level ever. Monthly indices show that numbers were lower than during the past five years through most of the winter. This decline is mirrored by a rise in the Netherlands, strongly suggesting that birds migrating here from the northeast, and possibly northwest, are short-stopping on the Wadden Sea, aided by amelioration of winter weather.

The overall decline was evident at many of the key sites with almost half of internationally or nationally important sites supporting their lowest total of the past five years. At the top of the table, the Thames Estuary deposed the

Mersey Estuary as the principal site for Dunlin, although the count at the Thames was lower than average. The peak core count for the Mersey Estuary of 26,970 was the lowest at the site for over 15 years and the five-year mean fell by 10% on that of the previous year. Low totals were experienced at key sites from either side of Britain from the Severn Estuary in the west to the Blackwater Estuary in the east. However, numbers at some sites, such as Chichester Harbour, remained similar to those of the past few years. Passage numbers on the Wash were slightly lower than average, although passage numbers on the Humber surpassed those for the previous six years.

Although the Northern Ireland maximum was comparable to that of recent years, the index reveals a decline not dissimilar to that of Britain. Birds here are likely to be from Greenland and Iceland breeding populations. However, numbers at Strangford Lough, the principal site for Dunlin in the province, were somewhat higher than in recent years although declines were noted at other key locations. The February count at the adjacent Upper River Quoile was the highest ever recorded at the site during WeBS being almost five times more than the previous peak.

	01/02	02/03	03/04	04/05	05/06	Mon	Mean
Sites of international importance in the UK							
Thames Estuary	48,104	54,205	(27,318)	40,838	37,352	Dec	45,125
Mersey Estuary	45,756	58,463	40,170	43,020	34,731 [10]	Nov	44,428
The Wash	31,069	42,794	31,624	39,041	19,605	Dec	32,827
Dee Estuary (England & Wales)	34,448 [10]	21,266	41,679	16,878	19,867	Feb	26,828
Humber Estuary	24,378	24,168	19,182 [10]	(14,733)	(15,378)	Dec	22,576
Severn Estuary	20,401	25,734	23,801	(16,069)	19,561	Feb	22,374
Langstone Harbour	17,500	17,320	24,286	28,239	22,356	Dec	21,940
Morecambe Bay	18,947	18,214	18,847	(17,848)	(27,110)	Feb	20,780
Ribble Estuary	11,141	11,423	24,445	24,024	13,866	Mar	16,980
Blackwater Estuary	15,004	18,806	13,958	(16,007)	12,189	Dec	15,193
Solway Estuary	12,861	12,850	17,576	(14,628)	(8,280)	Mar	14,479
Chichester Harbour	17,947 [10]	15,661	12,552	12,651	12,989	Jan	14,360
Sites of national importance in Great Britain							
Forth Estuary	13,296	12,143	7,840 [10]	9,132	6,422	Feb	9,767
Swale Estuary	11,280	14,761	5,034	9,181	7,830	Dec	9,617
Lindisfarne	9,085	(9,991)	(9,503)	5,885	(5,540)	Jan	8,616
Alt Estuary	8,438	6,885	12,743	8,540	5,184	Nov	8,358
Stour Estuary	11,211	3,315	9,268	7,235	7,019	Mar	7,610
Medway Estuary	5,872	6,901	8,086	9,373 [10]	7,367	Dec	7,520
Dengie Flats	15,720	7,710	2,700	3,040	8,200	Jan	7,474
Poole Harbour	(6,929)	(6,323)	(5,463)	(7,026)	(2,182)	Jan	(7,026)
Burry Inlet	6,654	4,955	10,150	6,318	6,965	Jan	7,008
Portsmouth Harbour	2,269	8,139 [10]	9,641	3,933	(9,228)	Feb	6,642 ▲
Duddon Estuary	5,415	3,942	7,680 [10]	6,970 [10]	8,741 [10]	Dec	6,550
Hamford Water	10,686 [10]	3,064	(3,476)	4,290	(2,851)	Nov	6,013
Breydon Wtr & Berney Marshes	6,280	5,273 [10]	4,100	4,387	8,072 [10]	Jan	5,622 ▲
Sites of all-Ireland importance in Northern Ireland							
Strangford Lough	3,352	4,408 [10]	4,967 [10]	4,934	7,669 [10]	Dec	5,066
Lough Foyle	2,804	4,209	4,212	1,688	3,334	Feb	3,249
Carlingford Lough	(2,090)	(2,872)	(2,339)	2,238	1,573	Dec	2,222
Sites no longer meeting table qualifying levels in Winter 2005/2006							
Colne Estuary	6,823	(350)	4,411	3,359	(5,323)	Dec	4,979
Belfast Lough	1,278	1,193	1,461 [10]	1,136 [10]	920	Jan	1,198
Other sites surpassing table qualifying levels in Winter 2005/2006 in Northern Ireland							
Upper Quoile River	0	120	0	85	1,400	Feb	321

Sites surpassing international passage threshold in the UK in 2005/2006				
The Wash	35,468	Jul	Humber Estuary	26,305	Jul
Ribble Estuary	29,305	May	Blackwater Estuary	15,178	Oct

Sites surpassing national passage threshold in Great Britain in 2005/2006				
Dengie Flats	13,018	Oct	Hamford Water	3,534	Oct
Thames Estuary	10,957	Oct	Solway Estuary	3,368	May
Morecambe Bay	8,822	Oct	North Norfolk Coast	3,333	Aug
Swale Estuary	7,728	Oct	Dee Estuary (England & Wales)	3,126	Aug
Stour Estuary	5,949	Oct	Severn Estuary	2,985	May
Lindisfarne	5,320	Sep	Mersey Estuary	2,750	Oct
Langstone Harbour	5,026	Oct	Chichester Harbour	2,473	Oct
Alt Estuary	4,958	May			

Buff-breasted Sandpiper

Tryngites subruficollis

Vagrant
Native Range: N America

GB max: 0
NI max: 2 Sep

Two Buff-breasted Sandpipers (the 15th & 16th recorded during WeBS counts) were present together at Black Brae and Donnybrewer, Lough Foyle, in September (by far the most regular month for this species in the UK). This is only the second time that more than one individual of this species has been recorded at one site for WeBS, the first being in October 1977 on the Thames Estuary.

Ruff

Philomachus pugnax

International threshold: 12,500
Great Britain threshold: 7*
All-Ireland threshold: +†

GB max: 713 Dec
NI max: 21 Sep

*50 is normally used as a minimum threshold

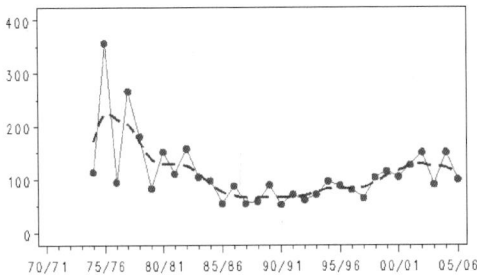

Annual Index
Trend

Figure 51.a, Annual indices & trend for Ruff for GB.

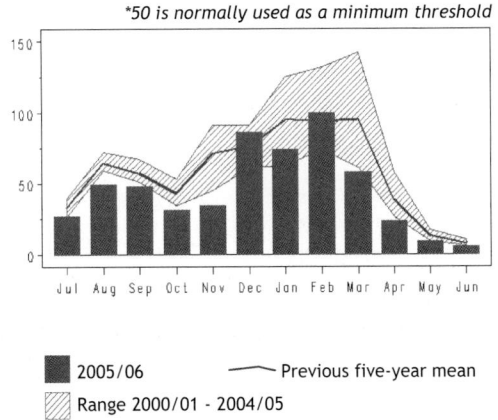

2005/06   Previous five-year mean
Range 2000/01 - 2004/05

Figure 51.b, Monthly indices for Ruff for GB.

Ruff have shown an increased tendency to winter in Britain since the late 1990s, although the numbers concerned remain relatively small. The current index suggests a slight decrease in wintering Ruff; however, this is likely to be due to lower numbers at just a couple of key sites. For example, peak wintering numbers at the Ouse Washes were somewhat lower than during the previous year. Numbers on the North Norfolk Coast peaked at 193 in September then fell to just two in November before rising again to three-figure counts in January and February. Low numbers during the first part of the WeBS year were a feature across Britain with national figures being lower than during the past five-years up until December.

Throughout Britain, a total of 30 sites held nationally important numbers of Ruff, five fewer than in 2004/05. During passage four sites held numbers exceeding 50 birds and a further 24 held seven birds or more. The highest single site count in Northern Ireland was of 11 at Lough Foyle in September, followed by seven at Loughs Neagh and Beg in the same month.

	01/02	02/03	03/04	04/05	05/06	Mon	Mean
Sites of national importance in Great Britain							
Ouse Washes	334	359	232 [12]	(431)	323	Feb	336
WWT Martin Mere	190	151	86	83	(50)	Mar	128
North Norfolk Coast	66	105	155	140	117	Jan	117
Lower Derwent Ings	101	179	99	73	50	Jan	100
Nene Washes	30	275	128	16	2	Nov	90
Breydon Wtr & Berney Marshes	155	55	43	86	59	Feb	80
Swale Estuary	46	95	41	128	37	Feb	69
Middle Yare Marshes	37	82	(17)	53	40	Dec	53
Crouch-Roach Estuary	(0)	(4)	(0)	42 [10]	(2)	Mar	42

	01/02	02/03	03/04	04/05	05/06	Mon	Mean
Blackwater Estuary	49	82	19	24	18	Dec	38
Ouse Fen & Pits (Hanson/RSPB)		(106)	42	2	2	Nov	38
Thames Estuary	34	35	43	28	38	Dec	36
Arun Valley	29	22	52	24	6	Mar	27
Ribble Estuary	5	76	21	5	15	Nov	24
Walland Marsh	32	0	7	16	55	Mar	22
Somerset Levels	(15)	29	33	10	12	Dec	21
Hamford Water	6 [10]	26	20	17	18	Dec	17
Humber Estuary	20 [12]	25	5 [10]	(7)	(6)	Feb	17
Sandbach Flashes		26	11	13	13	Jan	16
Holland Marshes	41	6	12	14	3	Dec	15
Abberton Reservoir	0	2	51	2	(6)	Nov	14
Dee Estuary (England & Wales)	(12)	5	10	29	10	Feb	14
Severn Estuary	3 [12]	21	(0)	13	16	Jan	13
Hardley Flood		33	7	0			13
Dungeness Gravel Pits	0	42	7	10	1	Mar	12
Stour Estuary	1	1	0	3	55	Dec	12 ▲
Cresswell Pond	(32)	1	12	0	6	Mar	10
Stodmarsh NNR & Collards Lagn	15	11	7	0	14	Jan	9
Hagnaby Lock Fen	6	19	9	11	1	Jan	9
Rutland Water	8	4	10	2	11	Nov	7
Sites no longer meeting table qualifying levels in Winter 2005/2006							
Minsmere	8	5	6	9	4	Dec	6
Tees Estuary	6	8	9	4	4	Feb	6
East Chevington Pools	14	(0)	7	4	0		6
Fairfield SSSI	9	12	0	0	0		4
Fen Drayton Gravel Pits	0	0	7	0	0		1
Other sites surpassing table qualifying levels in Winter 2005/2006 in Great Britain							
Pegwell Bay	7	10	1	3	11	Feb	6
North West Solent	4	(3)	3	9	7	Mar	6

Sites surpassing national passage threshold in Great Britain in 2005/2006

North Norfolk Coast	193	Sep	Hamford Water	15	Oct
Ouse Washes	157	Oct	The Wash	14	Aug
Humber Estuary	84	Sep	Barmston Ponds	12	Sep
Breydon Wtr & Berney Marshes	72	Jul	Dungeness Gravel Pits	12	Oct
Abberton Reservoir	36	Sep	Minsmere	12	Aug
WWT Martin Mere	34	Aug	Sandbach Flashes	12	Oct
Rutland Water	32	Aug	Blackwater Estuary	11	Aug
Swale Estuary	31	Sep	Holland Marshes	10	Sep
Tees Estuary	29	Aug	Hurworth Burn Reservoir	10	Aug
Thames Estuary	29	Oct	Forth Estuary	9	Aug
East Chevington Pools	28	Aug	Solway Estuary	8	Aug
Loch Leven	22	Sep	Dee Estuary (England & Wales)	7	Oct
Loch of Strathbeg	21	Sep	Filey Dams	7	Sep
Ribble Estuary	17	Apr	Ythan Estuary	7	Sep

[†] as no All-Ireland threshold has been set, a qualifying level of seven has been chosen to select sites for presentation in this report

Jack Snipe
Lymnocryptes minimus

International threshold:	?
Great Britain threshold:	?[†]
All-Ireland threshold:	250[†]

GB max:	135	Nov
NI max:	11	Nov

Despite Jack Snipe being secretive and often difficult to detect, the numbers recorded at key WeBS sites vary little with the same six sites featuring at the top of the table. The count of 14 at Chat Moss was the lowest at this site for some years, although the site remains a key area for this species. The highest single-site count was of 24 at the Lower Derwent Ings.

However, it is likely that all counts, including these higher ones, are best treated as minimum estimates given the nature of the species.

Whilst the British maximum was only three-quarters that of the previous year, the Northern Ireland maximum was the highest ever recorded. This was largely influenced by a count of ten at Dundrum Bay in November.

	01/02	02/03	03/04	04/05	05/06	Mon	Mean
Sites with mean peak counts of 5 or more birds in Great Britain[†]							
Doxey Marshes SSSI	64	(30)	16	61	18	Jan	40
Chat Moss	68 [24]	46 [24]	28 [24]	34 [24]	14 [24]	Nov	38
Dornoch Firth	25 [10]						25
Lower Derwent Ings	7	11	22	27	24	Nov	18
Chichester Harbour	16	39	7	6	18	Mar	17
Bickershaw Colliery Area	10 [24]	17 [24]	11 [24]	14 [24]	18 [24]	Nov	14
Fiddlers Ferry Pwr Stn Lagoons	32	6	0	0	16	Nov	11
Dee Estuary (England & Wales)	22 [10]	13	(9)	3	1	Jan	10
Severn Estuary	9	7 [12]	5	10	(19)	Feb	10
Humber Estuary	5	13	(5) [10]	(2)	(1)	Dec	9
Inner Moray and Inverness Firth	19	13	8	5	2	Feb	9
Kemerton Lake			9				9
Waulkmill Glen & Littleton Res	6	10	12	10	6	Jan	9
Langstone Harbour	0	13	0	12	9	Mar	7
River Eden - Ormside					7	Mar	7
Ardrossan-West Kilbride	6	8	2	6	6	Nov	6
Boat of Garten Pools				6			6
Morecambe Bay	10	5	3	6	4	Dec	6
Other sites surpassing table qualifying levels in Winter 2005/2006 in Great Britain[†]							
Kinsham Pool		3		1	8	Feb	4
Grouville Marsh	1	1	2	1	7	Feb	2
Island of Papa Westray	0	(0)	0	0	7	Dec	2
Bevercotes Colliery			4	2	6	Mar	4
Pugneys Country Park Lakes	0	3	0	3	6	Nov	2
Tyttenhanger Gravel Pits	0		5	3	6	Mar	4
Other sites surpassing table qualifying levels in Winter 2005/2006 in Northern Ireland[†]							
Dundrum Bay	0	1	1	1	10	Nov	3

[†] as few sites exceed the All-Ireland threshold and no British threshold has been set, a qualifying level of 5 has been chosen to select sites for presentation in this report

Snipe
Gallinago gallinago

International threshold: 20,000**
Great Britain threshold: ?[†]
All-Ireland threshold: ?[†]

GB max:	7,372	Dec
NI max:	274	Nov

*50 is normally used as a minimum threshold

Due to their secretive nature, Snipe are always under-recorded during WeBS counts despite being a widely distributed and locally common bird on many wetland sites in winter. The British maximum count was lower than in 2004/05, but was similar to the average for the last five years. The Northern Ireland maximum was higher than for 2004/05, but overall numbers recorded here are low and, consequently, show high variability.

The Somerset Levels remained at the top of the list of sites, although the highest single site count was recorded at the Lower Derwent Ings, with 1,182 in March. At eight further sites, annual peaks exceeding 200 birds were recorded, and peak counts at Camel Estuary, Minsmere, Island of Papa Westray and Duddon Estuary were the highest ever recorded at their respective sites.

	01/02	02/03	03/04	04/05	05/06	Mon	Mean
Sites with mean peak counts of 200 or more birds in Great Britain[†]							
Somerset Levels	854	972	308	1,513	713	Jan	872
Lower Derwent Ings	243	997	269	384	1,182	Mar	615
Severn Estuary	(217)	(240)	(519)	(349)	(337)	Feb	(519)
Doxey Marshes SSSI	544	(239)	390	716	310	Jan	490
Ouse Washes	1,685 [12]	126 [12]	233	302	87	Nov	487
North Norfolk Coast	1,169 [10]	92	121	77	155	Feb	323
Maer Lake	510	0	403	378	280	Dec	314
Middle Yare Marshes	(545)	257	124	(210)	(237)	Dec	275
Cleddau Estuary	189	283	311	144	233	Jan	232
Malltraeth RSPB	55	131	54	570	251	Jan	212
Sites with mean peak counts of 50 or more birds in Northern Ireland[†]							
Loughs Neagh and Beg	(16)	129	151	22	31	Feb	83
Belfast Lough	61	48	86 [10]	45 [10]	170	Nov	82
Strangford Lough	29	97 [10]	55	56	68 [10]	Jan	61
Ballysaggart Lough		51	53				52

	01/02	02/03	03/04	04/05	05/06	Mon	Mean
Other sites surpassing table qualifying levels in Winter 2005/2006 in Great Britain[†]							
Loch Saintear	144	34	220	46	400	Nov	169
Camel Estuary	(22)	36	65	(10)	(320)	Dec	140
Morecambe Bay	109	101	(147)	(265)	304	Nov	195
Minsmere	59	21	24	53	280	Dec	87
Island of Papa Westray	83	(41)	47	73	255	Jan	115
Duddon Estuary	27	13	20	96	226	Feb	76
Chichester Harbour	80	289	113	100	219	Jan	160
Southampton Water	107	166	212	184	210	Dec	176
Other sites surpassing table qualifying levels in Winter 2005/2006 in Northern Ireland[†]							
Dundrum Bay	40	40	21	(45)	(54)	Jan	40

[†] *as no British or All-Ireland thresholds have been set qualifying levels of 200 and 50 have been chosen to select sites, in Great Britain and Northern Ireland respectively, for presentation in this report*

Long-billed Dowitcher

Limnodromus scolopaceus

Vagrant

Native Range: America

GB max: 3   Oct
NI max: 0

Single Long-billed Dowitchers were at Drift Reservoir in October, Dee Estuary (England & Wales) in October and November and the Blackwater Estuary in November. In addition a single bird remained at Hayle Estuary from October to February. Long-billed Dowitcher was a new WeBS species for all of these sites except for the Dee Estuary (England & Wales); one had previously been recorded there in March 2003.

Woodcock

Scolopax rusticola

International threshold: 20,000**
Great Britain threshold: ?
All-Ireland threshold: ?

GB max: 65   Dec
NI max: 3   Feb

*50 is normally used as a minimum threshold

Due to its secretive habits and occurrence away from wetland habitats Woodcock remain poorly recorded during WeBS counts. The peak British total of just 65 in December represents just a tiny fraction of the total numbers present in the country during the winter. Birds were recorded at 162 sites during 2005/06, with the majority of records being of single birds. Peaks of five or more birds were noted at six sites: 16 at Grouville Marsh in January, eight at Hurleston Reservoir and Cors Caron in February and March respectively, seven at Hamford Water in December and January, seven at Killington Reservoir in December and six at Longueville Marsh in December and February.

Black-tailed Godwit

Limosa limosa

International threshold: 470
Great Britain threshold: 150
All-Ireland threshold: 90

GB max: 35,130   Sep
NI max: 1,158   Sep

Black-tailed Godwits have been increasing as a wintering species in Britain for the past 25 years and although there was no change in the annual index value for 2005/06 the underlying trend is still one of increase. Numbers were above average throughout the year, particularly during the late summer months. Numbers in Northern Ireland have doubled in the past five years and continued to rise in 2005/06, with the monthly indices showing a particularly high number present in March.

Following a re-evaluation of the international threshold for Icelandic Black-tailed Godwits, a total of 21 UK sites qualify as internationally important. By far the largest concentrations in the early winter are at the Wash and the Dee Estuary, whilst in March many birds traditionally move to the Ouse Washes. Peak counts at these three sites were a little lower than in 2004/05. Conversely, the winter peak at the Humber Estuary was by far the largest on record for this site, with almost all Humber birds in November counted at North Killingholme Haven Pits. Prior to the main winter period, a particularly high peak was recorded from the Wash, whilst as usual, much higher autumn numbers occurred on the Mersey Estuary than are found there during the winter months.

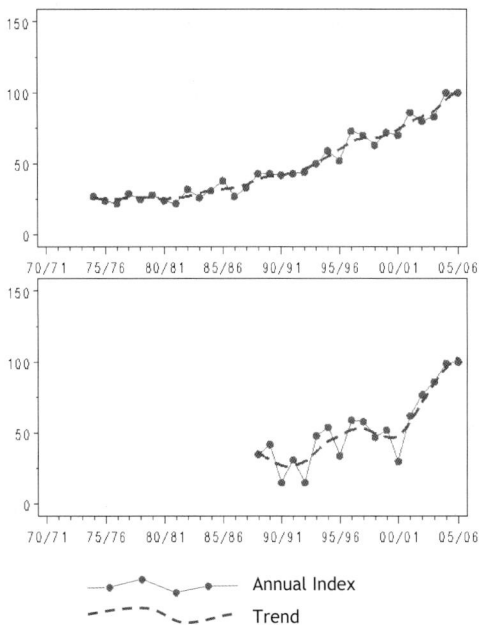

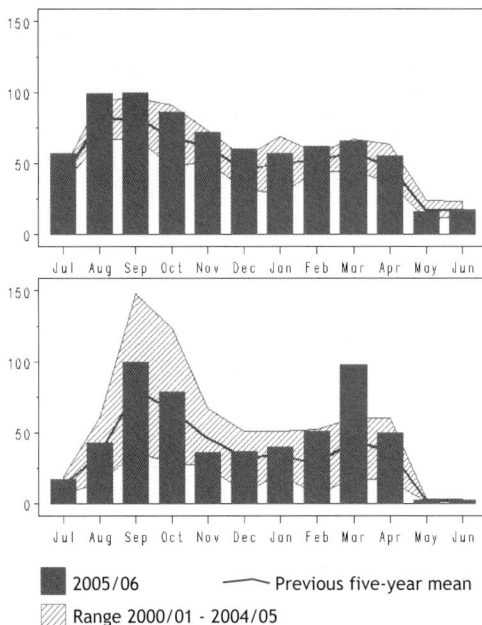

| | Annual Index |
| | Trend |

| | 2005/06 | | Previous five-year mean |
| | Range 2000/01 - 2004/05 |

Figure 52.a, Annual indices & trend for Black-tailed Godwit for GB (above) & NI (below).

Figure 52.b, Monthly indices for Black-tailed Godwit for GB (above) & NI (below).

	01/02	02/03	03/04	04/05	05/06	Mon	Mean
Sites of international importance in the UK							
The Wash	9,163	2,773	3,031	5,492	3,535	Nov	4,799
Dee Estuary (England & Wales)	4,624 [10]	3,955	4,493	5,362	5,115	Nov	4,710
Ouse Washes	3,273	3,468	3,137 [12]	3,424	3,206	Mar	3,302
Blackwater Estuary	(926)	(2,939)	(1,232)	2,356	(901)	Dec	2,648
Poole Harbour	(2,115)	(2,691)	(2,133)	1,732	(1,431)	Nov	2,168
Thames Estuary	1,967	1,584	1,380	1,931	1,944	Nov	1,761
Ribble Estuary	1,733	975	1,385	2,629	1,962	Mar	1,737
Humber Estuary	921	1,311	914 [10]	(629)	(2,848)	Nov	1,499
Swale Estuary	1,580 [10]	1,045	1,511	1,782	1,362	Jan	1,456
Breydon Wtr & Berney Marshes	1,607	1,142	1,277	1,566	1,675	Dec	1,453
Stour Estuary	2,037	981	1,316	1,024	1,173	Mar	1,306
Exe Estuary	737	890	(1,079)	1,054	1,090	Nov	970
Crouch-Roach Estuary	(260)	(162)	(261)	729 [10]	(259)	Feb	729
Chichester Harbour	552	715	1,050	545	672	Nov	707
Medway Estuary	(662)	(199)	(154)	(518)	(190)	Mar	(662)
River Avon: Ringwood/Christch	3	3,002	170	26	1	Nov	640
North Norfolk Coast	233	477	631	998	811	Feb	630
Pagham Harbour	252	826	541	664	340	Dec	525
Mersey Estuary	313	1,002	740	241	250	Dec	509
Belfast Lough	492	545	367 [10]	479	532	Mar	483 ▲
Morecambe Bay	(117)	143	(403)	655	626	Jan	475 ▲
Sites of national importance in Great Britain							
Hamford Water	366 [10]	490	414	314	625	Nov	442
North West Solent	452	(261)	373	(300)	474	Nov	433
Alde Complex	113	355	600	298	556	Nov	384
Newtown Estuary	231	510	(173)	(113)	374	Dec	372
Orwell Estuary	260 [10]	407 [10]	389 [10]	255 [10]	471	Mar	356
Southampton Water	(358)	196	(434)	291	489	Dec	354 ▼
Deben Estuary	260	304	258	305	575	Dec	340
Colne Estuary	344	(190)	253	472	142	Dec	303
Beaulieu Estuary	725	147	116	326	179	Dec	299
Carmarthen Bay	(8)	(29)	(331)	307	237	Mar	292
Forth Estuary	232	243	291	348	334	Jan	290
Burry Inlet	30	60	222	845	250	Nov	281
Portsmouth Harbour	(84)	246 [10]	78	340	443	Feb	277
Langstone Harbour	442	314	245 [10]	290	81	Mar	274
Blyth Estuary	244						244
Severn Estuary	141	193	200	(450)	215	Dec	240

	01/02	02/03	03/04	04/05	05/06	Mon	Mean
Eden Estuary	221 [10]	206	220	305	181	Nov	227
Sites of all-Ireland importance in Northern Ireland							
Strangford Lough	153	189 [10]	267	176	717 [10]	Nov	300
Sites no longer meeting table qualifying levels in Winter 2005/2006							
Meadow Lane Gravel Pits St Ives	0	(3)	0		0		1
Other sites surpassing table qualifying levels in Winter 2005/2006 in Great Britain							
Duddon Estuary	4	1	12	69 [10]	242	Mar	66
St Andrews Bay			0	(0)	236	Mar	118
Ingrebourne Valley	0	0	0	0	200	Dec	40
Tamar Complex	106	53 [10]	80	175	153	Dec	113
Other sites surpassing table qualifying levels in Winter 2005/2006 in Northern Ireland							
Loughs Neagh and Beg	0	0	0	39	178	Mar	43

Sites surpassing international passage threshold in the UK in 2005/2006

The Wash	8,205	Oct	Chichester Harbour	995	Oct
Dee Estuary (England & Wales)	5,379	Oct	Burry Inlet	994	Sep
Thames Estuary	5,221	Oct	Orwell Estuary	975	Sep
Humber Estuary	3,296	Sep	North Norfolk Coast	940	Aug
Ribble Estuary	2,921	Sep	Poole Harbour	847	Oct
Mersey Estuary	2,510	Aug	Exe Estuary	837	Oct
Stour Estuary	1,507	Apr	Morecambe Bay	747	Sep
Breydon Wtr & Berney Marshes	1,483	Oct	Langstone Harbour	665	Aug
Swale Estuary	1,389	Jul	Belfast Lough	642	Sep
Blackwater Estuary	1,243	Oct	Deben Estuary	506	Oct
Alde Complex	1,181	Oct	Portsmouth Harbour	494	Oct

Sites surpassing national passage threshold in Great Britain in 2005/2006

Severn Estuary	435	Oct	Southampton Water	210	Oct
Forth Estuary	380	Oct	Beaulieu Estuary	190	Apr
Pagham Harbour	338	Apr	Medway Estuary	175	Oct
St Andrews Bay	280	Apr	Colne Estuary	171	Sep
Crouch-Roach Estuary	265	Apr	Newtown Estuary	171	Apr
Sandbach Flashes	254	Jun	Yar Estuary	171	Jun
North West Solent	236	Jul	Nene Washes	156	Apr

Sites surpassing national passage threshold in Northern Ireland in 2005/2006

Strangford Lough	418	Oct
Lough Foyle	397	Oct

Bar-tailed Godwit
Limosa lapponica

International threshold:	1,200
Great Britain threshold:	620
All-Ireland threshold:	175

GB max: 32,255 Dec
NI max: 1,518 Feb

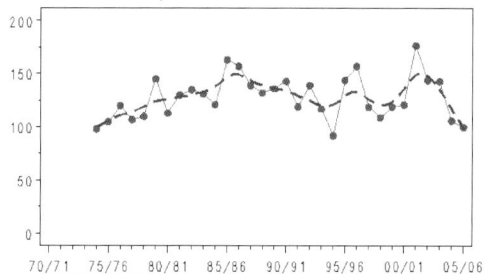

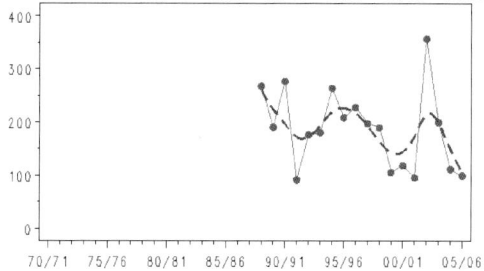

Annual Index

Trend

2005/06   Previous five-year mean

Range 2000/01 - 2004/05

Figure 53.a, Annual indices & trend for Bar-tailed Godwit for GB (above) & NI (below).

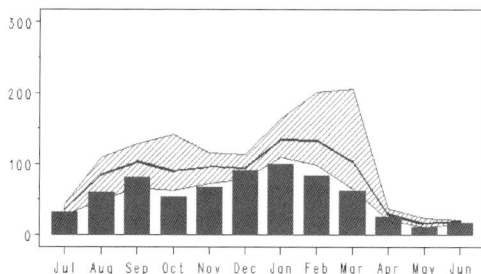

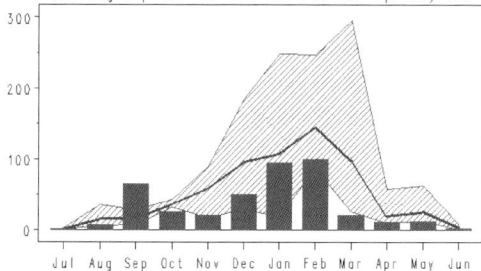

Figure 53.b, Monthly indices for Bar-tailed Godwit for GB (above) & NI (below).

Following the steep decline of 2004/05, numbers of Bar-tailed Godwit in both Great Britain and Northern Ireland showed further signs of reduction this winter. These indices are now based on the five months from November to March, rather than December to February. The counted British maximum was at its lowest ever level and, for the first time, was lower than that recorded for Black-tailed Godwit.

The decline in wintering Bar-tailed Godwits is evident at many of the key sites, in particular the Wash, where numbers were the lowest for over 15 years, whilst mean numbers at the Swale Estuary have fallen below the national importance threshold and an unusually low count was also recorded at the Cromarty Firth (possibly involving some local redistribution to the Dornoch Firth however). Few sites held peak winter numbers that were higher than their recent averages, the exceptions including the North Norfolk Coast and Chichester Harbour.

Passage numbers were, in general, similar to the previous few years, though only six British sites were of international, and none of national, importance compared to eight and four, respectively, in the previous year. Nevertheless, all but one of these sites held greater numbers than during the previous year. Numbers at the Wash and the Thames Estuary increased by approximately 50% and those at the Ribble Estuary and North Norfolk Coast doubled. In Northern Ireland, two sites held nationally important numbers during the passage period, with numbers at the key site, Strangford Lough, being only slightly higher than in 2004/05.

	01/02	02/03	03/04	04/05	05/06	Mon	Mean
Sites of international importance in the UK							
The Wash	23,751	18,374	16,280	11,268	8,594	Dec	15,653
Ribble Estuary	20,950	(3,111)	11,301	4,657	(3,510)	Mar	12,303
Alt Estuary	12,098	7,103	8,120	3,900	4,221	Jan	7,088
Thames Estuary	(6,460)	3,941	8,989	6,595	6,611	Feb	6,534
Morecambe Bay	(938)	5,718	4,424	1,752	(2,158)	Feb	3,965
Humber Estuary	3,669	2,688	4,291 [10]	(2,460)	(2,203)	Jan	3,549
Lindisfarne	5,237	(3,000)	(4,078)	2,900	1,787 [10]	Jan	3,501
Dee Estuary (England & Wales)	12,163 [10]	127	1,209	132	326	Nov	2,791
Dengie Flats	4,970	3,112	1,550	1,250	1,550	Jan	2,486
Cromarty Firth	1,044	2,212	3,439	2,311	651	Feb	1,931
Lough Foyle	1,328	4,108	1,019	630	1,133	Jan	1,644
Solway Estuary	(2,106)	1,761	1,572	1,050	(958)	Dec	1,622
Strangford Lough	1,949 [10]	1,079	2,019	1,422	1,378	Feb	1,569
North Norfolk Coast	1,678 [10]	1,555	1,271	1,203	1,743	Feb	1,490
Forth Estuary	964	1,793	1,750 [10]	1,599	940	Dec	1,409
Dornoch Firth	1,136 [10]	1,561	1,068	1,495	1,681	Dec	1,388 ▲
Tay Estuary	1,944	1,351	910	1,680	1,050	Jan	1,387
Sites of national importance in Great Britain							
Chichester Harbour	910	872	(910)	863	(1,200)	Mar	951
Inner Moray and Inverness Firth	995	997	830	901	770	Jan	899
South Ford		549	950	1,040			846
Hamford Water	1,002	485	803	(431)	(657)	Jan	763
Loch Bee SSSI Coast			713 [52]				713
Sites of all-Ireland importance in Northern Ireland							
Sites no longer meeting table qualifying levels in Winter 2005/2006							
Swale Estuary	595	606	462	922	481	Dec	613

Sites surpassing international passage threshold in the UK in 2005/2006

The Wash	9,849	Sep	Ribble Estuary	3,380	Apr
Thames Estuary	6,097	Sep	North Norfolk Coast	3,273	Sep
Alt Estuary	4,131	Aug	Lindisfarne	1,516	Oct

Sites surpassing national passage threshold in Great Britain in 2005/2006

Forth Estuary	1,188	Sep	Tay Estuary	1,050	Oct

Sites surpassing national passage threshold in Northern Ireland in 2005/2006

Strangford Lough	506	Sep
Lough Foyle	419	Sep

Whimbrel

Numenius phaeopus

International threshold: 6,800
Great Britain threshold: +[†]
All-Ireland threshold: +[†]

GB max: 1,205 Jul
NI max: 8 Aug

Whimbrel were recorded at 107 sites throughout Britain and five in Northern Ireland. Birds were noted in every month with, as usual, by far the largest numbers being in autumn and spring. The highest count during the winter months was of 24 on the Wash in November, but thereafter, most winter records were from southern England with just five records outside of this area (Carmarthen Bay, Cleddau Estuary and Lavan Sands in Wales and Carlingford Lough and Outer Ards Shoreline in Northern Ireland). The British maximum was slightly lower than during 2004/05, with peak counts at many sites being lower than in 2004/05. However, with the majority of birds being recorded during passage, monthly counts will always underestimate the total number of birds passing through during this period. Moreover, a single count date towards the middle of each month is unlikely to coincide with the peak of migration, particularly in the spring where the main pulse is typically in the last few days of April and first few days of May.

Sites with more than 50 birds during passgae periods in Great Britain in 2005/06[†]

The Wash	292	Jul	Langstone Harbour	96	Jul
Barnacre Res & Grizedale Lea	270 [53]	May	Lower Derwent Ings	95 [50]	May
Rye Harbour & Pett Level	222 [47]	May	Taw-Torridge Estuary	89	May
Brockholes Quarry	154 [54]	May	Chichester Harbour	78	Aug
North Norfolk Coast	129	Jul	Newtown Estuary	63	Apr
Burry Inlet	111	Jul	Southampton Water	63	Jul
Humber Estuary	107	Jul	Morecambe Bay	60	May
Severn Estuary	101	May	Breydon Wtr & Berney Marshes	59	May

[†] *as no British or All-Ireland thresholds have been set a qualifying level of 50 has been chosen to select sites for presentation in this report*

Curlew

Numenius arquata

International threshold: 8,500
Great Britain threshold: 1,500
All-Ireland threshold: 875

GB max: 74,713 Sep
NI max: 5,323 Jan

Following a steady increase in the 1980s and 1990s, the British annual index has seen a slight decrease since the turn of the century. Although not as dramatic a decline as seen for several other wader species, it will be of interest to see how this develops in the coming years. Allied to this, the maximum count has also been dropping annually, with the maximum for 2005/06 being the lowest peak since 1987/88. The index for Northern Ireland also shows a decline though the underlying trend has been fairly stable since the early 1990s. Typically, Curlew numbers peaked in Britain in August, remaining fairly constant through the winter before falling sharply in April as birds leave for their moorland breeding grounds. The Northern Ireland monthly index shows a more rapid drop between February and March than is seen for Britain, presumably as it involves a greater proportion of British breeders.

Following a re-evaluation of the international 1% threshold level, from 4200 up to 8500, only Morecambe Bay continues to support internationally important numbers of Curlew, both as a wintering site and during passage months. The Solway Estuary no longer qualifies due to this adjustment. Counts at many other key sites, such as Dee Estuary, The Wash, Thames Estuary, Forth Estuary and Severn Estuary were at their lowest for some years. Both Chichester Harbour and Medway Estuary now support nationally important numbers of Curlew, with the peak at the former site being the highest recorded there by WeBS. Peak counts from the top two sites in Northern Ireland were slightly below those recorded in 2004/05 but remained about average for the five-year period.

Figure 54.a, Annual indices & trend for Curlew for GB (above) & NI (below).

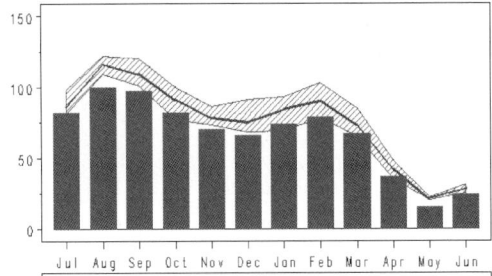

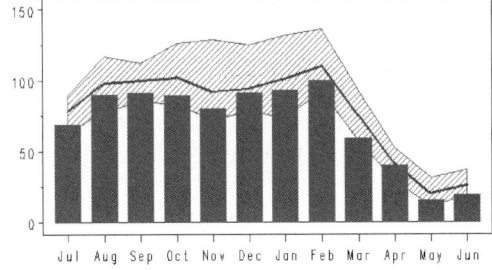

Figure 54.b, Monthly indices for Curlew for GB (above) & NI (below).

	01/02	02/03	03/04	04/05	05/06	Mon	Mean	
Sites of international importance in the UK								
Morecambe Bay	9,522	10,868	(10,866)	(7,338)	8,688	Jan	9,986	
Sites of national importance in Great Britain								
Solway Estuary	(4,311)	(3,701)	(4,561)	(3,328)	(2,971)	Mar	(4,561)	
Humber Estuary	4,277	3,941	3,530 [10]	3,751	(4,183)	Feb	3,936	
Dee Estuary (England & Wales)	4,305 [10]	3,270	4,978	3,668	2,566	Dec	3,757	
The Wash	4,339	4,774	4,036	2,937	2,520	Mar	3,721	
Thames Estuary	(2,354)	4,093	(2,651)	2,786	2,533	Dec	3,137	
Forth Estuary	(3,638)	3,229	(2,897)	2,669	2,403	Jan	2,967	
Severn Estuary	(2,164)	3,615 [10]	2,528	(2,545)	2,514	Dec	2,886	
Duddon Estuary	2,041	2,280	2,756	1,326	1,816	Nov	2,044	
Inner Moray and Inverness Firth	1,473	1,961	1,809	2,137	1,838	Jan	1,844	
North Norfolk Coast	2,302 [10]	1,430	1,539	1,523	1,734	Feb	1,706	
Lavan Sands	2,381	1,922 [10]	1,433	1,212	1,464	Jan	1,682	
Poole Harbour	1,577	1,605	1,427	(2,472)	1,013	Mar	1,619	
Lindisfarne	1,822 [10]	1,338 [10]	(1,072)	1,715 [10]	1,334	Jan	1,552	
Chichester Harbour	1,511	1,414	1,670	1,262	1,889	Mar	1,549	▲
Mersey Estuary	1,562	1,270	1,804	1,632	1,408	Nov	1,535	
Medway Estuary	1,639	(1,088)	(789)	(448)	1,367 [10]	Dec	1,503	▲
Sites of all-Ireland importance in Northern Ireland								
Lough Foyle	1,358	1,956	2,127	3,115	2,038	Feb	2,119	
Strangford Lough	1,676	1,189	1,342 [10]	1,594	1,523	Nov	1,465	
Sites no longer meeting table qualifying levels in Winter 2005/2006								
Inner Firth of Clyde	(1,294)	(1,455)	(1,485)	(1,133)	836	Feb	1,241	

Sites surpassing international passage threshold in the UK in 2005/2006

Morecambe Bay	9,515	Oct

Sites surpassing national passage threshold in Great Britain in 2005/2006

The Wash	5,140	Jul	Lavan Sands	1,955	Aug
Dee Estuary (England & Wales)	4,666	Sep	Blackwater Estuary	1,914	Aug
Humber Estuary	3,730	Aug	Langstone Harbour	1,811	Sep
Thames Estuary	3,611	Aug	Mersey Estuary	1,792	Oct
Forth Estuary	3,599	Sep	Duddon Estuary	1,752	Oct
Solway Estuary	3,216	Oct	Chichester Harbour	1,683	Oct
Burry Inlet	2,587	Sep	Lindisfarne	1,548	Aug
Severn Estuary	2,434	Aug	Montrose Basin	1,536	Aug
North Norfolk Coast	2,284	Aug			

Sites surpassing national passage threshold in Northern Ireland in 2005/2006

Lough Foyle	1,715	Sep
Strangford Lough	1,306	Oct

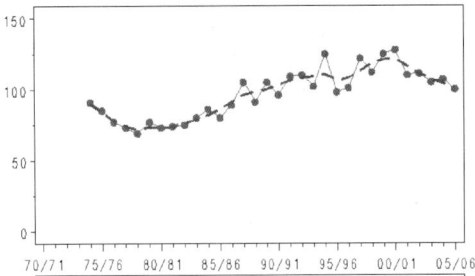

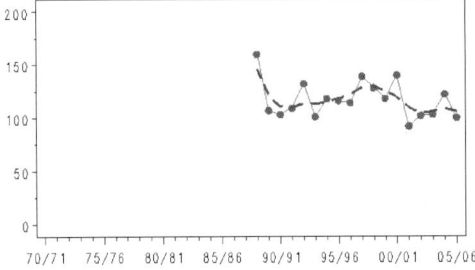

Legend:
- ●——● Annual Index
- - - - - - Trend
- ■ 2005/06
- ▨ Range 2000/01 - 2004/05
- ⌒ Previous five-year mean

Spotted Redshank
Tringa erythropus

International threshold: 900
Great Britain threshold: +[t]
All-Ireland threshold: +[t]

GB max: 164 Sep
NI max: 1 Aug

Typically, the national maximum for Spotted Redshank was recorded during autumn with the 2005/06 peak being slightly higher than that for the previous year. Birds were noted at a total of 52 sites outside of the winter period. The peak count of 39 at the Wash during August was the same as during August 2004. Numbers at key passage sites were similar to recent years, the exception being Humber Estuary where the number recorded has fallen recently.

Spotted Redshank were recorded at 43 sites between November and March. The highest count of 26 on the North Norfolk Coast during February was the highest winter total there to date, whilst notably high winter totals were also noted at the Wash and the Blackwater Estuary. There were just four records from Northern Ireland, all of single birds. These were at Dundrum Bay in August, March and April and Lough Foyle in October.

	01/02	02/03	03/04	04/05	05/06	Mon	Mean
Sites with mean peak counts of 10 or more birds in winter in Great Britain[t]							
Thames Estuary	(10)	26	3	(3)	(2)	Jan	15
North Norfolk Coast	6	6	11	16	26	Feb	13
Dee Estuary (England & Wales)	14	12	4	12	(8)	Nov	11
Other sites surpassing table qualifying levels in Winter 2005/2006 in Great Britain[t]							
Blackwater Estuary	5	8	4	9	18	Nov	9
The Wash	(2)	2	4	0	14	Nov	5

Sites with 10 or more birds during passage periods in Great Britain in 2005/2006[t]

The Wash	39	Aug	Blackwater Estuary	24	Sep
North Norfolk Coast	35	Aug	Minsmere	14	Jul
Abberton Reservoir	26	Sep	Humber Estuary	10	Oct

[t] *as no British or All-Ireland thresholds have been set a qualifying level of 10 has been chosen to select sites for presentation in this report*

Redshank
Tringa totanus

International threshold: 2,800
Great Britain threshold: 1,200
All-Ireland threshold: 245

GB max: 85,133 Oct
NI max: 10,668 Oct

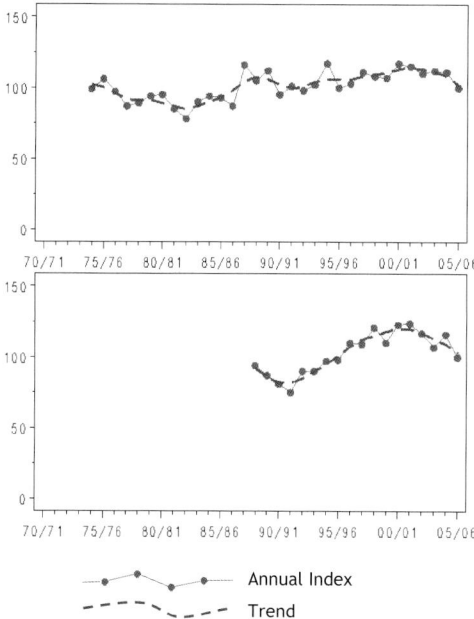

Annual Index
Trend

2005/06
Range 2000/01 - 2004/05
Previous five-year mean

Figure 55.a, Annual indices & trend for Redshank for GB (above) & NI (below).

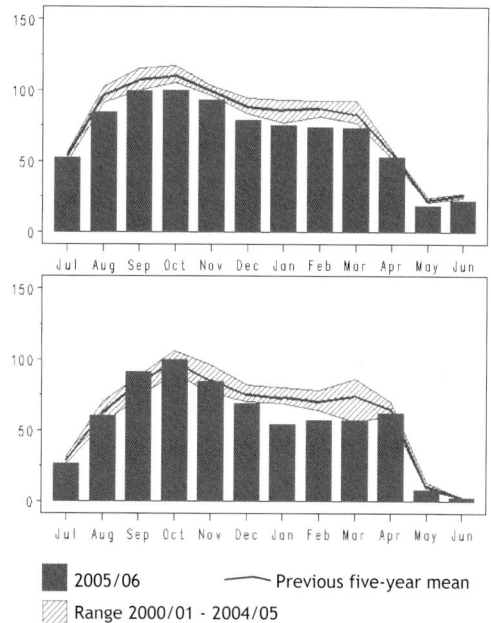

Figure 55.b, Monthly indices for Redshank for GB (above) & NI (below).

Whilst the underlying trend still appears relatively stable, the British maximum fell to its lowest peak since 1996/97, with the annual index also showing a slight drop. This was reflected in the monthly indices, which were substantially lower than average for the whole period August to March. The Northern Ireland index has also declined somewhat since 2004/05, continuing a slight downward trend.

Following a re-evaluation of the international 1% threshold level, upwards from 1300 to 2800, the number of UK sites supporting internationally important numbers has dropped sharply, from the 35 reported in last year's report to the 11 listed here. Peak counts on the majority of the top sites were down slightly on 2004/05, with the sustained decline at the Wash particularly of note. The peak on the Ribble Estuary, however, was the highest winter count there since 1973. During the passage months, the peak count at the Dee Estuary was the highest ever total there, and only exceeded by some historical Morecambe Bay totals from the 1970s.

	01/02	02/03	03/04	04/05	05/06	Mon	Mean
Sites of international importance in the UK							
Morecambe Bay	6,274	6,650	6,715	7,106	6,908	Nov	6,731
Dee Estuary (England & Wales)	8,579 [10]	5,847	5,736	5,812	(4,469)	Nov	6,494
Humber Estuary	4,526	4,787	8,229 [10]	(5,247)	4,682	Nov	5,556
Forth Estuary	4,204	4,194	4,587	5,501	4,066	Feb	4,510
Mersey Estuary	4,690	4,143	6,050	3,290	3,622	Nov	4,359
Thames Estuary	4,479	3,763	(4,383)	3,735	4,082	Dec	4,088
The Wash	4,501	3,619	3,410	2,902	2,172	Feb	3,321
Crouch-Roach Estuary	(1,220)	(592)	(496)	3,299 [10]	(556)	Nov	3,299
Strangford Lough	3,273 [10]	2,879 [10]	3,146 [10]	2,692	3,339	Nov	3,066
Ribble Estuary	1,877	3,882	(1,911)	2,211	4,078	Mar	3,012 ▲
Blackwater Estuary	(3,539)	2,849	(1,818)	3,034	2,158	Dec	2,895
Sites of national importance in Great Britain							
Inner Moray and Inverness Firth	2,714	2,942	2,317	2,846	1,531	Dec	2,470
Severn Estuary	2,616	2,439 [10]	(1,865)	2,516	1,930	Dec	2,375
Solway Estuary	1,668 [10]	(2,528)	(2,324)	(2,786)	(1,145)	Mar	2,327
Duddon Estuary	1,596	1,849	2,508	1,658	3,698	Nov	2,262
Hamford Water	2,575 [10]	2,334	1,892	1,699	1,695	Jan	2,039
Chichester Harbour	2,422	1,829	2,450	1,695	1,754	Nov	2,030
Inner Firth of Clyde	(2,433)	(1,589)	1,974	1,964	1,725	Dec	2,024
Orwell Estuary	2,279 [10]	1,825 [10]	1,939 [10]	1,799	1,813 [10]	Nov	1,931
Deben Estuary	1,999	2,017	1,869	1,707	1,930	Nov	1,904
Colne Estuary	1,871	(97)	(868)	(797)	(1,013)	Dec	1,871
Cromarty Firth	1,849	1,604	2,569	1,784	1,551	Dec	1,871
North Norfolk Coast	3,915 [10]	1,299	1,416	1,180	1,150	Dec	1,792
Montrose Basin	2,511	1,830	1,803	1,349	1,137	Nov	1,726
Alde Complex	2,071 [10]	1,456	1,430	1,957	1,608	Dec	1,704
Ythan Estuary	985	670	621	(914)	(5,274)	Nov	1,693 ▲
Tees Estuary	1,332	1,398	1,926	1,183	1,731	Nov	1,514
Lavan Sands	1,126	1,525 [10]	1,248	1,947	1,644	Nov	1,498
Blyth Estuary	1,481						1,481
Breydon Wtr & Berney Marshes	1,207	1,497 [10]	1,630 [10]	1,406	1,663 [10]	Dec	1,481
Swale Estuary	2,481	959	(1,352)	974	(1,288)	Nov	1,471
Lindisfarne	1,825	(1,371)	1,503	1,365	1,104	Nov	1,449
Stour Estuary	1,549	797	1,540	1,431	1,814	Nov	1,426
Medway Estuary	(1,537)	(972)	(814)	1,068 [10]	(1,405)	Nov	1,337
Sites of all-Ireland importance in Northern Ireland							
Belfast Lough	2,261	1,540	1,452	1,547 [10]	1,476	Nov	1,655
Lough Foyle	1,104	1,606	1,198	1,404	976	Nov	1,258
Carlingford Lough	1,525	1,211	1,027	1,324	1,197	Dec	1,257
Outer Ards Shoreline		1,351	1,228	1,121	1,307	Mar	1,252
Dundrum Bay	(696)	(530)	(942)	(256)	(424)	Dec	(942)
Larne Lough	363	427	356	462	737	Dec	469
Bann Estuary	260	324	240	290	304	Mar	284

Sites surpassing international passage threshold in the UK in 2005/2006

Dee Estuary (England & Wales)	12,367	Sep	Strangford Lough	4,099	Sep
Morecambe Bay	7,283	Oct	Humber Estuary	3,671	Aug
The Wash	6,052	Aug	Mersey Estuary	2,977	Oct
Forth Estuary	6,039	Sep	Ythan Estuary	2,886	Sep
Thames Estuary	4,811	Oct			

Sites surpassing national passage threshold in Great Britain in 2005/2006

Duddon Estuary	2,595	Oct	Ribble Estuary	1,799	Sep
Blackwater Estuary	2,472	Oct	Swale Estuary	1,727	Sep
Cromarty Firth	2,266	Oct	North Norfolk Coast	1,608	Oct
Montrose Basin	2,237	Sep	Chichester Harbour	1,596	Oct
Deben Estuary	2,037	Sep	Tees Estuary	1,544	Oct
Inner Firth of Clyde	1,984	Apr	Breydon Wtr & Berney Marshes	1,359	Sep
Inner Moray and Inverness Firth	1,910	Oct	Severn Estuary	1,280	Oct

Sites surpassing national passage threshold in Northern Ireland in 2005/2006

Belfast Lough	1,754	Oct	Outer Ards Shoreline	1,166	Oct
Carlingford Lough	1,554	Oct	Dundrum Bay	723	Aug
Lough Foyle	1,314	Oct	Bann Estuary	400	Sep

Greenshank
Tringa nebularia

International threshold:	2,300
Great Britain threshold:	6*[†]
All-Ireland threshold:	9*[†]

GB max: 1,360 Aug
NI max: 186 Oct

*50 is normally used as a minimum threshold

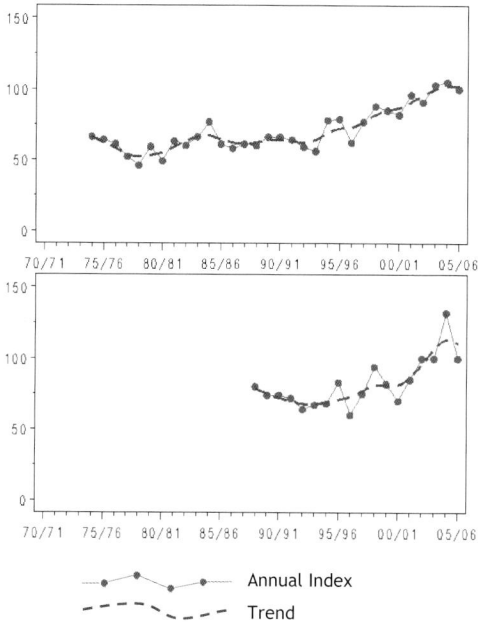

Figure 56.a, Annual indices & trend for Greenshank for GB (above) & NI (below).

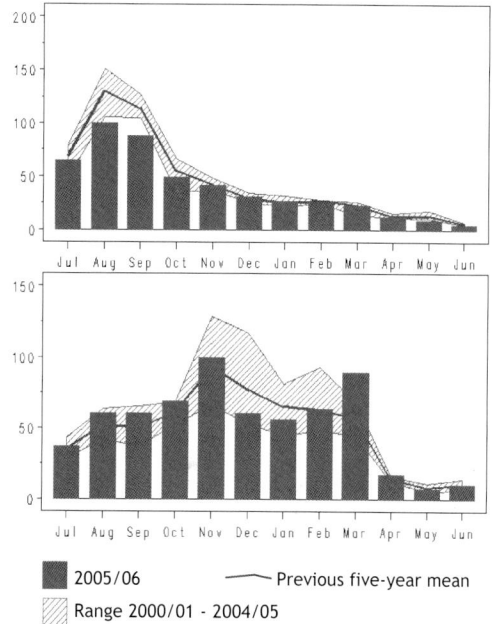

Figure 56.b, Monthly indices for Greenshank for GB (above) & NI (below).

As usual, peak numbers of Greenshank were recorded in August in Britain, although the August 2005 total was the lowest since 1993. The British annual index, which reflects counts from November to March, shows a steady increase in wintering Greenshank since the early 1990s. Although the current figure is slightly lower than the last two years it is well within the expected variation and as yet the trend shows no sign of slowing. The monthly indices emphasise the relatively low numbers on autumn passage but average numbers during the rest of the winter. The Northern Ireland maximum was the highest since 1978

and the index shows that this species has wintered in Northern Ireland in increasing numbers over the past few years. Although the index declined in 2005/06 this was largely due to low numbers during December and January with counts throughout the rest of the year being above average.

In total, 35 British sites held nationally important wintering numbers of Greenshank, one less that in 2004/05. However, the highest winter total from any site was in Northern Ireland with 84 at the regular key site of Strangford Lough, which, along with four other sites in the province, surpassed the all-

Ireland importance threshold. Passage numbers, as usual, far exceeded the number of wintering birds and numbers at four sites surpassed 100 birds, the highest total being at the Wash in August. As usual, the largest numbers during autumn passage were concentrated in the southeast corner of England.

	01/02	02/03	03/04	04/05	05/06	Mon	Mean
Sites of national importance in Great Britain							
Fal Complex	26	27	32	37	58	Nov	36
Chichester Harbour	44	35 [12]	42	20	41	Nov	36
Kingsbridge Estuary	26	41	(36)	45	27	Jan	35
Cleddau Estuary	34	28	20	(23)	42 [10]	Jan	31
Tamar Complex	30	31 [10]	26	42	23	Nov	30
Queens Valley Reservoir	20	19		22			20
Exe Estuary	14	18	14	22	20	Nov	18
Blackwater Estuary	(12)	27 [10]	12	12	(9)	Dec	17
Kentra Bay					17	Feb	17 ▲
Grouville Marsh	11	31	25	15	0		16
Southampton Water	15	13	19	(14)	(10)	Jan	16
Camel Estuary	(6)	(17)	17	15	12	Mar	15
Taw-Torridge Estuary	14	16	19	11	9	Nov	14
Carmarthen Bay	(1)	(0)	(2)	15	12	Jan	14
Jersey Shore	(13)						(13)
Inner Firth of Clyde	(12)	14	10	9	13	Nov	12
Morecambe Bay	7	6	15	12	13	Dec	11
Yealm Estuary	7	15	8	15	6	Dec	10
North West Solent	11	(5)	11	(13)	4	Dec	10
Foryd Bay	8	10	16	9	8	Dec	10
Tyninghame Estuary	7	11	9	11	8	Nov	9
Poole Harbour	6	11	(6)	7 [10]	(8)	Nov	8
Thames Estuary	8	(3)	6	6	13	Nov	8
The Wash	3	18	7	4	7	Nov	8
Solway Estuary	11	8	6	(6)	(0)		8
Lavan Sands	7	9	9	9	7	Dec	8
Hunterston Lagoon	3		9	10	8	Feb	8
Eden Estuary	(0)	6	9	7	9	Nov	8
Loch nan Capull (South Uist)			8				8
Medway Estuary	8	(1)	5	(0)	(1)	Dec	7 ▲
Forth Estuary	2	(6)	7 [10]	9	8	Nov	7
Broadford Bay	(8)	(7)	6	7	(1)	Nov	7
Brading Harbour	5	6	8	4	5	Nov	6
North Norfolk Coast	9 [10]	3	3	3	11	Nov	6 ▲
Rough Firth	7	2	6	7			6
Sites of all-Ireland importance in Northern Ireland							
Strangford Lough	56	72	61	117	84	Nov	78
Lough Foyle	20	22	27	37	18	Feb	25
Carlingford Lough	18	14	16	21	23	Nov	18
Dundrum Bay	18	15	15	11	(4)	Jan	15
Larne Lough	(15)	15	11	11	7	Nov	12
Sites no longer meeting table qualifying levels in Winter 2005/2006							
Dee Estuary (England & Wales)	6	3	9	5	4	Nov	5
Ceann a Bhaigh	4	8		5	4	Nov	5
Other sites surpassing table qualifying levels in Winter 2005/2006 in Great Britain							
Auchendores Reservoir	0	1	0	3	6	Jan	2
St Andrews Bay	0	0	0	(1)	6	Feb	2

Sites with 50 or more birds during passage periods in Great Britain in 2005/2006[†]

The Wash	258	Aug	Chichester Harbour	91	Aug
North Norfolk Coast	147	Aug	Blackwater Estuary	84	Sep
Thames Estuary	144	Aug	Stour Estuary	55	Sep
Hamford Water	104	Sep	Swale Estuary	55	Sep

Sites with 50 or more birds during passage periods in Northern Ireland in 2005/2006[†]

Strangford Lough	80	Oct
Lough Foyle	74	Oct

[†] *as no British or All-Ireland passage thresholds have been set a qualifying level of 50 has been chosen to select sites for presentation in this report*

Green Sandpiper
Tringa ochropus

International threshold: 17,000
Great Britain threshold: ?[†]
All-Ireland threshold: ?[†]

GB max: 309 Aug
NI max: 0

August was typically the peak month for Green Sandpiper numbers although the maximum count was well below average and at its lowest level since 2001/02. As is frequently the case, there were no records from Northern Ireland in 2005/06. Only four British sites held peaks of more than 15 birds during passage periods, all in early autumn. In the winter months of November to March, birds were noted from 135 sites, with 13 at Abberton Reservoir in December the highest count, although 12 at Beddington Sewage Farm and 11 in the Avon Valley were also noteworthy. This species does not show a pronounced spring passage in the UK, with numbers declining from March through April, to a low point in May, before increasing slightly in June as a result of the beginning of the autumn passage.

Sites with 15 or more birds during passage periods in Great Britain in 2005/2006[†]

| Abberton Reservoir | 27 | Sep | Blackwater Estuary | 17 | Aug |
| Rutland Water | 21 | Aug | Tophill Low Reservoirs | 16 | Aug |

[†] *as no British or All-Ireland thresholds have been set a qualifying level of 15 has been chosen to select sites for presentation in this report*

Wood Sandpiper
Tringa glareola

International threshold: 10,500[†]

GB max: 18 Aug
NI max: 0

Wood Sandpipers were noted at 24 sites during 2005/06, with all records being of passage birds and none recorded from December to April. The August peak, which was only a quarter of the record total in August 2004, was made up of counts from 13 sites, the highest being of four at Drift Reservoir. In the spring, following three at Dolydd Hafren in April, there were 11 birds in May including counts of two at Castle Lake and Stoney Beck Lake in May. The numbers recorded during WeBS counts are clearly only a small proportion of the total numbers passing through Britain on passage.

Sites with 3 or more birds during passage periods in the GB in 2005/2006[†]

| Drift Reservoir | 4 | Aug |
| Dolydd Hafren | 3 | Apr |

[†] *as no British or All-Ireland thresholds have been set a qualifying level of three has been chosen to select sites for presentation in this report*

Common Sandpiper
Actitis hypoleucos

International threshold: 17,500
Great Britain threshold: ?[†]
All-Ireland threshold: ?[†]

GB max: 643 Jul
NI max: 12 Aug

Passage numbers of Common Sandpipers were notably low during 2005/06, particularly during the late summer/early autumn period. WeBS annual reports normally display sites where more than 40 birds were noted during the year, but no sites reached this level in 2005/06 and so sites with a peak in excess of 20 birds have been shown instead. All of these sites were along the eastern coasts of England, with the exception of Morecambe Bay where the large numbers are largely a function of the huge size of the site.

Numbers declined sharply between September and October, and then more gradually to a fairly stable wintering population of about 35 birds between January and March. Most of the sites supporting Common Sandpipers during these months were in the southwest quarter of Britain; most held only one or two birds, with higher counts of up to six at several sites in March. Increasing numbers were noted in April and spring passage peaked in May, although some of the higher counts in May and June may well have involved breeding birds.

Sites with more than 20 birds during passage periods in Great Britain in 2005/06[†]

Pegwell Bay	39	Sep		Hornsea Mere	24	Aug
Rye Harbour & Pett Level	37	Jul		Blackwater Estuary	23	Jul
Morecambe Bay	32	Jul		Stour Estuary	23	Aug
North Norfolk Coast	25	Jul		The Wash	23	Jul

[†] *as no British or All-Ireland thresholds have been set a qualifying level of 20 has been chosen to select sites for presentation in this report*

Turnstone
Arenaria interpres

International threshold: 1,500
Great Britain threshold: 500
All-Ireland threshold: 225

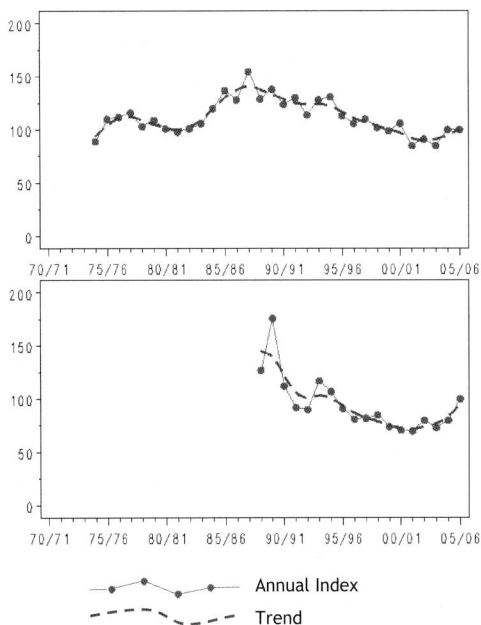

GB max: 14,389 Oct
NI max: 2,118 Oct

Annual Index
Trend

2005/06
Previous five-year mean
Range 2000/01 - 2004/05

Figure 57.a, Annual indices & trend for Turnstone for GB (above) & NI (below).

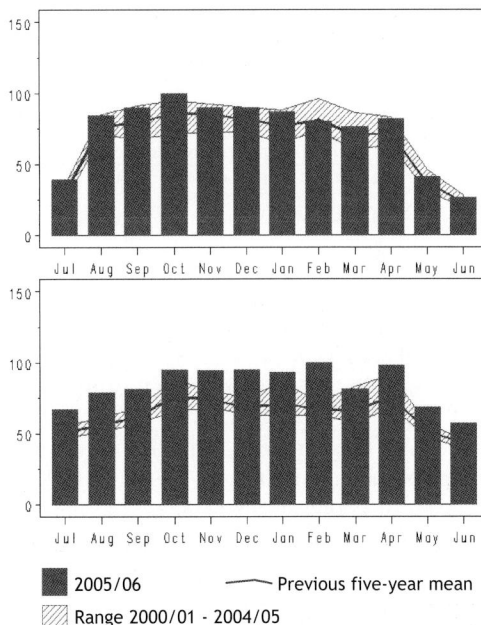

Figure 57.b, Monthly indices for Turnstone for GB (above) & NI (below).

Following a period of steady decline, the British annual index appears to be stabilising, with very little change since 2004/05; the counted British maximum was slightly higher. The Northern Ireland annual index rose substantially over the previous year, which was reflected in the monthly indices that showed peaks in all months that were higher than the five year average.

Nine British sites held numbers of national importance, with supplementary counts resulting in Tiree again attaining the top position following several years where there have been no counts from this important site. Counts from other nationally important sites were generally higher than previous years

although at sites where they were lower, they were still around average for the five-year period. In Northern Ireland, the Outer Ards Shoreline supported its highest peak for over 15 years and remains the key site in the province whilst peak counts at Belfast Lough, Carlingford Lough and Strangford Lough were all above average. Indeed, the Strangford Lough count was the highest there for a decade. On the North Norfolk Coast, passage numbers exceeded those seen in the winter, whilst away from nationally important wintering sites, the Burry Inlet, Farne Islands and Langstone Harbour all surpassed the national passage threshold level.

146

	01/02	02/03	03/04	04/05	05/06	Mon	Mean
Sites of national importance in Great Britain							
Tiree					1,191 [48]	Feb	1,191 ▲
Thanet Coast	964	694	1,192	1,130	949	Jan	986
Forth Estuary	610	940	701	619	847	Feb	743
Morecambe Bay	825	588	766	691	715	Dec	717
North Norfolk Coast	744	833	473	655	768	Dec	695
Thames Estuary	(879)	488	(465)	702	680	Dec	687
Humber Estuary	499	(529)	723 [10]	(570)	(183)	Mar	611
Stour Estuary	430	494	403	675	655	Nov	531
Solway Estuary	(308)	(300)	(259)	(509)	(277)	Jan	(509)
Sites of all-Ireland importance in Northern Ireland							
Outer Ards Shoreline		1,086	1,081	923	1,203	Jan	1,073
Belfast Lough	432	401	485	305	418	Dec	408
Carlingford Lough	140	157	181	624	354	Mar	291
Strangford Lough	227	206	225	235	435	Feb	266
Sites no longer meeting table qualifying levels in Winter 2005/2006							
Dee Estuary (England & Wales)	(405)	726	415	421	390	Dec	488
Other sites surpassing table qualifying levels in Winter 2005/2006 in Great Britain							
The Wash	(270)	579	354	488	515	Dec	484

Sites surpassing national passage threshold in Great Britain in 2005/2006

Morecambe Bay	1,269	Apr	Thanet Coast	776	Oct
The Wash	1,169	Aug	Langstone Harbour	742	Oct
North Norfolk Coast	928	Aug	Burry Inlet	672	Oct
Forth Estuary	799	Sep	Farne Islands	606	Aug

Sites surpassing national passage threshold in Northern Ireland in 2005/2006

Outer Ards Shoreline	1,001	Oct	Carlingford Lough	356	Sep
Belfast Lough	395	Oct	Strangford Lough	318	Oct

Wilson's Phalarope

Phalaropus tricolor

Vagrant
Native Range: America

GB max: 0
NI max: 1   Aug

A single Wilson's Phalarope was recorded at the Bann Estuary in Northern Ireland in August. This is only the third time that this species has been recorded during WeBS counts in Northern Ireland, the others being at Larne Lough in September 1996 and at Lough Foyle in October 2001.

Red-necked Phalarope

Phalaropus lobatus

Scarce

GB max: 4   Jun
NI max: 0

Three Red-necked Phalaropes were recorded in September, at the Blackwater Estuary, Staines Reservoirs and Toft Newton Reservoir. Four were also reported at a breeding site in the Western Isles in June, the highest single-site total recorded by WeBS.

Grey Phalarope

Phalaropus fulicarius

Scarce

GB max: 13   Nov
NI max: 0

There was a notable "wreck" of Grey Phalaropes in much of Britain during November, with singles recorded from Arun Valley, Belvide Reservoir, Blagdon Lake, Chew Valley Lake, Doddington Pool, Kenfig Pool, Kennet and Avon Canal (Seend to Devizes), Morecambe Bay, Portsmouth Harbour and Tees Estuary, whilst three were at the Ogmore Estuary. The only record outside of November was of one at Guernsey Shore in December.

Mediterranean Gull
Larus melanocephalus

International threshold: 6,600[†]

GB max: 326 Aug
NI max: 3 Mar

The British maximum was the highest ever recorded for WeBS, being over twice as high as the previous year's total. Birds were noted at 94 sites across the UK, the majority of records coming from southeast England and the Isle of Wight maintaining its importance for the species. The highest single-site total was of 148 at Brading Harbour in August, a new record for the site. Several other sites also supported new peak totals during 2005/06, including the Thames Estuary, Tamar Complex, Pagham Harbour, Foreland, Aberarth and Fishguard Harbour. In Northern Ireland, single birds were recorded at Lough Foyle and Belfast Lough, with two present at the Outer Ards Shoreline in March.

	01/02	02/03	03/04	04/05	05/06	Mon	Mean
Sites with mean peak counts of 5 or more birds in Great Britain[†]							
Folkestone: Copt Pnt/East Wear Bay				157 [42]			157
Brading Harbour	28	126	57	92	148	Aug	90
Newtown Estuary	65	80	(15)	(42)	57	Apr	67
Ryde Pier to Puckpool Point	8	45	9	47	45	Aug	31
Breydon Wtr & Berney Marshes					27 [42]	Jan	27
Thames Estuary	(13)	20	27	27	30	Aug	26
Tamar Complex	14	30	0	(26)	39	Sep	22
Pagham Harbour	0	16		2	60	Jan	20
Camel Estuary	(1)	8	25	26	18	Sep	19
Swansea Bay	20	16	19	12 [42]	28	Jul	19
Chichester Harbour	4 [12]	(16)	(14)	(8)	(22)	Apr	13
Foreland	3	1	4	4	50	Jan	12
Fleet and Wey	2	2	4	8	23	Jan	8
Morecambe Bay	(1)	(4)	(4)	4	12	Jul	8
North Norfolk Coast	(6)	(13)	8	5	6	Apr	8
Poole Harbour	(2)	3	(7)	(12)	8	Mar	8
Aberarth		0	0	6	22	Aug	7
Taw-Torridge Estuary	3	7	(3)	(5)	12	Jul	7
The Wash	1	2	8	15	7	Apr	7
Ribble Estuary	9	7	8	(2)	0		6
Other sites surpassing table qualifying levels in WeBS-Year 2005/2006 in Great Britain[†]							
Fishguard Harbour	2	2	2	2	14	Sep	4
Beaulieu Estuary			0	0	9	Mar	3
Barton Pits	0	1	1	0	8	May	2
Lade Sands	1		3	0	6	Jul	3
Orwell Estuary	1	2 [10]	3 [10]	2 [10]	6 [10]	Jan	3
Tyne Estuary	1	1	0	1	5 [42]	Jan	2

[†] as no British or All-Ireland thresholds have been set a qualifying level of 5 has been chosen to select sites for presentation in this report

Laughing Gull
Larus atricilla

Vagrant
Native Range: America

GB max: 4 Nov
NI max: 0

There was a record influx of Laughing Gulls into the southwest in autumn 2005, which resulted in WeBS records at eight sites between November and March. All reports were of single birds and only at the Taw-Torridge Estuary and Traeth Bach did birds remain for more than one month. Reports at Welsh sites were from Kenfig Pools in November and Traeth Bach December to February. All other records were from English sites and included Taw-Torridge Estuary in October, January, February and March, Portsmouth Harbour and Swanpool (Falmouth) in November and Chichester Harbour, Fal Complex and Shibdon Pond in December.

Franklin's Gull
Larus pipixcan

Vagrant
Native Range: America

GB max: 1   Nov
NI max: 0

Prior to 2005/06 only one Franklin's Gull had been recorded during WeBS, in December 1994. During the current WeBS year two records were received. In November, one was recorded at Ruan Lanihorne in the Fal Complex, while in June, the first ever Guernsey record was reported from two adjacent stretches of the Guernsey Shore.

Little Gull
Larus minutus

International threshold: 1,230
Great Britain threshold: ?[†]
All-Ireland threshold: ?[†]

GB max: 340   Apr
NI max: 2   Mar

Following the record peak of 2004/05 the British maximum returned to more usual levels. Much of this was due to the lower numbers recorded at Hornsea Mere, which in 2005/06 were the lowest since 1999/2000. Numbers at most sites were unremarkable, although the Low Tide total from the Alt Estuary surpassed all previous WeBS counts at this site. In recent years, large numbers of Little Gulls have been present in the southern North Sea in autumn and the numbers recorded on WeBS counts depend greatly on the weather conditions around the dates of the counts.

	01/02	02/03	03/04	04/05	05/06	Mon	Mean
Sites of international importance in the UK							
Hornsea Mere	3,150 [11]	1,350 [11]	(940) [12]	7,000	160	Sep	2,915
Sites with mean peak counts of 5 or more birds in Great Britain[†]							
Alt Estuary	212 [11]	218	432	201 [11]	530 [11]	Apr	319
Forth Estuary	22	41	75 [28]	321	(0)		115
Tophill Low Reservoirs	0	10	110 [12]	75	(0)		49
Tay Estuary	(22)	50	36	28	26	Jul	35
Monikie Reservoirs		60	0	38	(0)		33
North Norfolk Coast	70	9	38	8	32	Jun	31
Minsmere	2	(15)	73	1	8	Aug	21
Morecambe Bay	31	1	36	7	3	Nov	16
East Chevington Pools	4	29	12	18	3	Jun	13
Lindisfarne	0	(0)	26	(0)	(0)		13
Alde Complex	0	0	0	0	49	Mar	10
Thames Estuary	(3)	17	(2)	3	(1)	Aug	10
Moray Firth				8 [1]	9 [1]	Nov	9
Outer Tay & St Andrews Bay			13 [28]	4 [28]			9
Dengie Flats	0		8	0	(22)	Nov	8
Humber Estuary	(0)	(2)	(12)	(0)	3	Jun	8
St Andrews Bay	8			7			8
Tring Reservoirs	0	1	40	0	0		8
Rescobie Loch	0	(11)					6

[†] *as no British or All-Ireland thresholds have been set a qualifying level of 5 has been chosen to select sites for presentation in this report*

Sabine's Gull
Larus sabini

Scarce

GB max: 1   Nov
NI max: 0

A single Sabine's Gull was recorded in Gerrans Bay during November. This was the 16th record of this species for WeBS and the first for three years.

Black-headed Gull
Larus ridibundus

International threshold: 20,000**
Great Britain threshold: 19,000[†]
All-Ireland threshold: ?[†]

GB max: 183,960 Jan
NI max: 16,573 Jan

The British maximum has fallen below 200,000 for the first time since gulls were first regularly counted for WeBS back in 1993/94. Although analysis of gull data is not easy, given the fact that recording them is optional for WeBS, the current figure is over 10% lower than in the previous year and almost 30% lower than the ten-year average. The Northern Ireland maximum was over 2,000 higher than the previous year and the highest since 1999/2000.

In total, four sites held mean numbers exceeding 20,000 birds, although at no individual site were counts above this number submitted during 2005/06. In recent years, counts of gulls at key sites have been largely influenced by the inclusion of Winter Gull Roost data, which have highlighted the importance of many sites for roosting birds, and further submission of gull roost counts for any sites would be most welcome. The highest single-site total during 2005/06 was at Morecambe Bay where over 16,000 birds were noted during July, when not all sections were covered. Peak totals from Tophill Low Reservoirs were particularly low and have fallen from 25,000 to just over 3,000 in the last five years. The key Northern Ireland site remains Belfast Lough where almost 10,000 birds were counted; this was the highest total ever recorded there by WeBS.

	01/02	02/03	03/04	04/05	05/06	Mon	Mean
Sites of international importance in the UK							
Thames Estuary	(22,911)	(30,275)	43,601 [42]	40,048	(14,737)	Dec	41,825
Bewl Water	33,000 [42]	57,000 [42]	31,000 [42]	69,000 [42]	440	Mar	38,088
Chew Valley Lake			29,800 [42]				29,800
Humber Estuary	(2,217)	(363)	21,450 [42]	(1,028)	(2,298)	Jul	21,450
Sites no longer meeting table qualifying levels in WeBS-Year 2005/2006[†]							
Lower Derwent Ings	4,500	25,300	28,000		11,000	Jan	17,200
Sites with mean peak counts of 10,000 or more birds in Great Britain[†]							
Poole Harbour	(7,386)	(12,461)	17,707 [42]	(11,811)	(5,720)	Mar	17,707
Lower Derwent Ings	4,500	25,300	28,000		11,000	Jan	17,200
Queen Mary Reservoir			16,836 [42]				16,836
The Wash	(16,136)	(15,999)	17,582 [42]	11,093	15,595	Aug	15,281
Church Wilne Reservoir			15,000 [42]				15,000
Southampton Water	(1,788)	(826)	14,822 [42]	(2,280)	(645)	Aug	14,822
Grafham Water			14,470 [42]				14,470
Morecambe Bay	7,795	(17,772)	12,574	16,757	(16,581)	Jul	14,296
Ribble Estuary	(24,460)	(821)	7,419 [42]	9,750 [42]	(10,228)	Jun	12,964
Hamilton Low & Strathclyde Parks			12,600 [42]				12,600
Southfield Reservoir			12,000 [42]				12,000
Exe Estuary				11,577 [11]			11,577
Tophill Low Reservoirs	25,000	11,900	8,900	8,385	3,170	Dec	11,471
Eyebrook Reservoir			11,300 [42]				11,300
Pitsford Reservoir	10,000 [11]	12,000 [11]	10,000 [42]				10,667
Inner Moray and Inverness Firth		7,452 [10]	12,760 [42]	(210)	(76)	Nov	10,106
Severn Estuary	(5,725)	9,209 [10]	13,139 [42]	9,656 [42]	8,278 [42]	Feb	10,071
Sites with mean peak counts of 1,000 or more birds in Northern Ireland[†]							
Belfast Lough	8,986 [12]	5,503 [10]	7,095 [10]	7,515 [10]	9,936 [10]	Feb	7,807
Outer Ards Shoreline		4,945	5,113	2,419	4,566	Jan	4,261
Loughs Neagh and Beg	(2,787)	(4,036)	(1,593)	(2,267)	(3,472)	Mar	(4,036)
Strangford Lough	3,503 [10]	3,518 [10]	3,388	3,111	4,011 [10]	Jan	3,506
Lough Foyle	1,627	2,780	1,300 [42]	1,057	2,565	Sep	1,866
Larne Lough	2,060	733	831	1,396	591	Sep	1,122
Other sites surpassing table qualifying levels in WeBS-Year 2005/2006 in Great Britain[†]							
Doddington Pool	0	600	150	11,000 [42]	12,000	Nov	4,750
Longnewton Reservoir	7,100	8,400	5,800	9,500	10,400	Sep	8,240
Lackford GPs	8,000 [11]	8,000	12,105 [42]	605	10,000	Jan	7,742

[†] *as few sites exceed the British threshold and no All-Ireland threshold has been set qualifying levels of 10,000 and 1,000 have been chosen to select sites, in Great Britain and Northern Ireland respectively, for presentation in this report*

Ring-billed Gull

Larus delawarensis

Vagrant
Native Range: N America

GB max: 3 Nov
NI max: 1 Mar

Ring-billed Gulls were reported from a total of 11 sites; eight in England and one each in Scotland, Wales and Northern Ireland. All records were of single birds, although there were multiple site records from the Alt Estuary in August and November, Thames Estuary in September, October, December, February and March and Fal Complex in November and March. The only record from Northern Ireland was from Belfast Lough in March.

Common Gull

Larus canus

International threshold: 20,000**
Great Britain threshold: 9,000[†]
All-Ireland threshold: ?[†]

GB max: 47,442 Mar
NI max: 5,362 Sep

The British maximum was almost 20% less than during the previous year and 25% less than in 2003/04. The Northern Ireland maximum was average compared to those of the past few years. Figures for key sites in the table below have, during the past few years, been augmented by the inclusion of totals from the Winter Gull Roost Survey. The importance of sites for roosting birds is highlighted by the figures recorded at Bewl Water, which has fallen from 75,000 during the Winter Gull Roost Survey to just 80 during standard Core Counts. As seen for Black-headed Gull, numbers at Tophill Low Reservoirs have fallen and are currently a fifth of those five years ago. Mean numbers there, as well as at Haweswater Reservoir, have fallen below the international threshold, although the sites remain of national importance. Nevertheless, Haweswater Reservoir held the highest single site-total during 2005/06, whilst peak numbers at Rutland Water were also worthy of note. The key site in Northern Ireland was Lough Foyle where peak numbers were almost twice those of 2004/05.

	01/02	02/03	03/04	04/05	05/06	Mon	Mean
Sites of international importance in the UK							
Bewl Water	63,000 [42]	42,000 [42]	75,000 [42]	75,000 [42]	80	Mar	51,016
Sites of national importance in Great Britain							
Haweswater Reservoir	16,566 [11]	13,674 [11]	27,986 [42]	22,000 [11]	12,535 [11]	Mar	18,552 ▼
Tophill Low Reservoirs	33,000	23,100	16,530	6,500	6,340	Dec	17,094 ▼
Hallington Reservoir	4	24,000 [11]	25,000 [42]				16,335
Eyebrook Reservoir			16,100 [42]				16,100
Humber Estuary	(366)	(2,077)	29,000 [42]	2,005	(120)	Dec	15,503
Blyth Estuary	(1,337)		12,000 [42]				12,000 ▲
West Water Reservoir			10,050 [42]				10,050
Ribble Estuary	(8,653)	(146)	(6,036)	(9,817)	(253)	Dec	(9,817)
Chew Valley Lake	(0)		18,200 [42]		0		9,100
Sites no longer meeting table qualifying levels in WeBS-Year 2005/2006							
Derwent Reservoir	6,500	11,800 [11]	(6,500) [42]	1,714	3,500	Dec	6,003
Ullswater	(0)		11,470 [42]	(0)	0		5,735
Colt Crag Reservoir		8,200 [11]	4,700 [42]				6,450
Sites with mean peak counts of 3,000 or more birds in Great Britain[†]							
Rye Harbour and Pett Level			8,600 [42]				8,600
Solway Estuary	(1,398)	7,193	9,564 [42]	(2,275)	(285)	Aug	8,379
Rutland Water	50 [11]	100	12,080 [42]	14,500	10,000	Mar	7,346
Lower Derwent Ings	5,500	14,200	3,720 [42]		5,500	Jan	7,230
Southwold Sole Bay			5,000 [42]				5,000
North Norfolk Coast	(1,420)	(1,283)	5,600 [42]	(2,163)	4,342	Jan	4,971
Tees Estuary	8,130	2,970	4,033	6,193 [42]	2,103	Mar	4,686
Moray Firth	5,961 [1]	5,037 [1]	5,208 [1]	809 [1]			4,254
Eccup Reservoir	9,000	5,000	579 [42]	1,200			3,945
Hule Moss	2,200 [12]	6,300 [12]	5,600 [12]	3,550	1,850	Nov	3,900

	01/02	02/03	03/04	04/05	05/06	Mon	Mean
Loch of Lintrathen	1,450		0	10,000			3,817
Loch of Skene	570	433	17,284 [42]	361	370	Dec	3,804
Lindisfarne	(2,920)	(370)	(3,644)	(580)	(252)	Sep	(3,644)
Forth Estuary	(1,658)	(1,356)	6,321 [42]	2,500 [42]	2,100 [42]	Dec	3,640
Morecambe Bay	3,632	3,194	4,358	3,633	2,802	Dec	3,524
Dee Estuary (England & Wales)	(1,519)	4,182	5,311	692	(309)	Jul	3,395
Wigtown Bay	(4,277)	1,427	3,251	7,269	675	Dec	3,380
Pitsford Reservoir	3,000 [11]	4,000 [11]	3,000 [42]				3,333
Hamilton Low Parks and Strathclyde Park			3,200 [42]				3,200
Chichester Harbour	2,062	4,142	3,389	3,778	2,379	Feb	3,150
Severn Estuary	1,500 [12]	746 [10]	3,714 [42]	4,259 [42]	5,110 [42]	Feb	3,066
Sites with mean peak counts of 1,000 or more birds in Northern Ireland[†]							
Lough Foyle	3,300	4,606	(5,930)	2,322	4,354	Sep	4,102
Belfast Lough	2,103 [12]	2,718	2,644 [10]	1,937 [10]	2,156	Feb	2,312
Outer Ards Shoreline		772	2,543	1,171	1,328	Mar	1,454
Other sites surpassing table qualifying levels in WeBS-Year 2005/2006 in Great Britain[†]							
Tyne Estuary	240	489	4,450 [42]	95	5,000 [42]	Dec	2,055
Burry Inlet	1,513	2,315	4,239	942	3,953	Aug	2,592
Thames Estuary	3,135	2,041	2,319 [42]	(3,669)	3,768	Aug	2,986
Lade Sands	0		2,000	2,000	3,000	Jan	1,750

[†] *as few sites exceed the British threshold and no All-Ireland threshold has been set qualifying levels of 3,000 and 1,000 have been chosen to select sites, in Great Britain and Northern Ireland respectively, for presentation in this report*

Lesser Black-backed Gull
Larus fuscus

International threshold: 5,500
Great Britain threshold: 500
All-Ireland threshold: ?[†]

GB max: 29,065 Jun
NI max: 1,839 Sep

The Great Britain maximum count was the lowest since 2001/02, whereas the Northern Ireland maximum was three times higher than that for 2004/05 and was the second highest in the province recorded for WeBS. The key site for this species is traditionally Morecambe Bay, notably around the breeding colony at South Walney, although the peak count here was much lower than in previous years being 30% below the five-year peak mean. Six sites (Lakenheath Fen, Ribble Estuary, Ouse Washes, Dee Flood Meadows, Heaton Park Reservoir and Heathfield Gravel Pits) attained nationally important status with four further sites surpassing table-qualifying levels for the year. In Northern Ireland, increased coverage at Loughs Neagh and Beg confirmed its status as the key site and a high peak at Belfast Lough was over double the five year peak mean for the site. As mentioned for other gull species, assessment of the relative importance of different sites for this species would be much enhanced by further submission of roost counts from key sites.

	01/02	02/03	03/04	04/05	05/06	Mon	Mean
Sites of international importance in the UK							
Morecambe Bay	31,620	36,461	31,479	33,004	21,932	Jun	30,899
Theale Gravel Pits	(0)	(3)	20,000 [42]	1,152 [42]	(74)	Dec	10,576
Chew Valley Lake	(0)		7,015 [42]		(0)		7,015
Queen Mary Reservoir			6,656 [42]				6,656
Cotswold Water Park (West)	(687)	(25)	5,800 [42]	(44)	(141)	Dec	5,800
Sites of national importance in Great Britain							
Severn Estuary	945	(3,072)	(8,073)	(10,036)	4,696 [42]	Feb	5,364
River Avon: Fordingbridg/Ringwd	3,478	2,309	6,550 [42]	3,500	5,100	Oct	4,187
Ribble Estuary	(91)	(244)	(106)	(113)	(3,011)	Jun	(3,011) ▲
Great Pool Westwood Park	1,350	2,000	3,800 [42]	2,500	2,500	Jan	2,430
Longnewton Reservoir	970	2,680	1,890	2,930	3,310	Sep	2,356
Rutland Water	2,000 [11]	5,000	2,500	200	1,200	Apr	2,180
Hule Moss	3,090 [12]	2,100 [12]	250 [12]	2,400	2,500	Sep	2,068
Thames Estuary	1,560	1,507	1,898 [42]	2,966	(775)	Sep	1,983
Alt Estuary	1,619	4,341	(945)	556	614	Aug	1,783
Alde Complex	767 [10]	4,474	388 [42]	1,833	1,162	Mar	1,725

	01/02	02/03	03/04	04/05	05/06	Mon	Mean	
Frampton Pools	1,500	(250)					1,500	
Belvide Reservoir			3,000 [42]	0			1,500	
Lower Windrush Valley GPs	3,166	871	484	1,343	1,071	Jan	1,387	
Blithfield Reservoir	0			2,620 [42]			1,310	
Roadford Reservoir	52	70	6,031 [42]	110	71	Apr	1,267	
Llangorse Lake	1,170 [12]	1,110 [11]	1,140 [12]	2,660 [11]	28	Oct	1,222	
Bartley Reservoir			1,200 [42]				1,200	
Pitsford Reservoir	550 [11]	1,000 [11]	2,000 [12]				1,183	
Hurleston Reservoir	65	700	1,500 [42]	3,500 [42]	84	Oct	1,170	
Chelmarsh Reservoir	500	(34)	3,500 [42]	83	56	Jan	1,035	
Solway Estuary	(243)	(673)	(971)	(154)	(253)	Aug	(971)	
The Wash	(582)	855	898	1,039	(1,075)	Jul	967	
Portworthy Mica Dam	(2,000)	419	700	960 [42]	469	Sep	910	
Cleddau Estuary	825	659	723	1,537	552	Oct	859	
Lakenheath Fen				0	1,500	Jan	750	▲
Wellington Gravel Pits		(1,400)	750	100			750	
Haweswater Reservoir	231 [11]	400 [11]	1,450 [12]				694	
Ouse Washes	37 [12]	104	760 [42]	256	2,305	Jan	692	▲
Hayle Estuary	(340)	130	940	980	(552)	Feb	683	
Llys-y-fran Reservoir	6	2,000	90	650	600	Jan	669	
Inner Firth of Clyde	(557)	544	705	509	769	Aug	632	
Burghfield Gravel Pits			618 [42]				618	
Dee Flood Meadows	(52)	(0)	(4)	(0)	600	Nov	600	▲
Heaton Park Reservoir	350 [12]	920 [11]	200 [42]	870 [42]			585	▲
Heathfield Gravel Pits	260	300	(1,000)				520	▲
Hollowell Reservoir			500 [42]				500	
Sites no longer meeting table qualifying levels in WeBS-Year 2005/2006								
Blyth Estuary	(93)		200 [42]				200	
Sites with mean peak counts of 500 or more birds in Northern Ireland[†]								
Loughs Neagh and Beg	(228)	1,218	1,115	(434)	997	Sep	1,110	
Other sites surpassing table qualifying levels in WeBS-Year 2005/2006 in Great Britain[†]								
Camel Estuary	452	117	769	142	606	Jan	417	
Fernworthy Reservoir	4	61	(139)	663	548	Nov	319	
Langford Lowfields Gravel Pits	6		30	112	540	Nov	172	
Heritage Park Loch	28	75	46	28	500	Jul	135	
Other sites surpassing table qualifying levels in WeBS-Year 2005/2006 in Northern Ireland[†]								
Belfast Lough	79	279	310	246	792	Sep	341	

[†] as no All-Ireland threshold has been set a qualifying level of 500 has been chosen to select sites for presentation in this report

Yellow-legged Gull / Caspian Gull

Larus michahellis / Larus (argentatus) cachinnans

International threshold: 7,000[†]

GB max:	178	Aug
NI max:	1	Nov

Yellow-legged Gulls were recorded from 34 sites, all of which were in England apart from one at Belfast Lough in November which was the first WeBS record in Northern Ireland. The monthly maximum was slightly higher than the previous year. The highest site total was again recorded at Southampton Water, whilst maximum figures from the Thames Estuary, Poole Harbour and River Avon – Fordingbridge to Ringwood were all much lower than in 2004/05.

Caspian Gulls were recorded from seven sites, with a notable cluster of records in the Nene Valley where birds were noted at Earls Barton, Ditchford and Clifford Hill Gravel Pits. Other records came from Minsmere, Weirwood Reservoir, Welbeck Estate and the Thames Estuary. All records were of single birds and all occurred between October and March, in contrast to the late-summer peak shown by Yellow-legged Gull. The taxonomic subcommittee of the British Ornithologists Union has recently recommended that Caspian Gull should be treated as a separate species, Larus michahellis, and, following adoption by the BOU Records Committee, this recommendation will be followed by WeBS in future years.

Southampton Water	157	Aug	Great Pool Westwood Park	6	Jan, Feb
Thames Estuary	16	Oct	Pagham Harbour	5	Jul
Poole Harbour	9	Oct	River Avon: Fordingbridg/Ringwd	5	Sep, Dec, Jan

[†] as no sites exceed the international threshold and no British or Northern Ireland thresholds have been set, a qualifying level of 5 has been chosen to select sites for presentation in this report

Herring Gull

Larus argentatus

International threshold: 5,900
Great Britain threshold: 4,500[†]
All-Ireland threshold: ?[†]

GB max: 47,346 Feb
NI max: 10,802 Jan

The British maximum was almost half that of the previous year though was more in line with figures from years preceding 2004/05. The Northern Ireland total was the highest ever for the province and was over 25% higher than in the previous year. However, as the recording of gulls remains optional during WeBS these totals are largely dependent on the level of coverage of this group.

Following a re-evaluation of the international 1% threshold level downwards from 13,000 to 5,900, a total of eight UK sites

now qualify as internationally important for Herring Gull, although assessment of the true importance of sites would be enhanced by submission of further roost counts where possible. The highest single-site total was recorded during Low Tide Counts at Belfast Lough, whereas the highest WeBS total was at Morecambe Bay, albeit the lowest ever recorded at the site. Mean numbers at the North Norfolk Coast have fallen below the national importance threshold; the count there was the lowest for almost ten years.

	01/02	02/03	03/04	04/05	05/06	Mon	Mean	
Sites of international importance in the UK								
Ribble Estuary	(9,767)	(209)	14,859 [42]	(31,090)	(2,060)	Jun	22,975	
Inner Moray and Inverness Firth		27,956 [10]	(2,341)	(2,003)	3,000	Nov	15,478	
Forth Estuary	(1,868)	(1,925)	7,376 [42]	(15,434)	(1,780)	Oct	11,405	
Morecambe Bay	12,170	14,373	10,551	8,311	7,488	May	10,579	
Belfast Lough	9,157	7,046	7,536 [10]	7,903 [10]	10,296 [10]	Feb	8,388	
Queen Mary Reservoir			8,279 [42]				8,279	
Moray Firth	9,564 [1]	10,335 [1]	6,468 [1]	2,349 [1]			7,179	
The Wash	(7,603)	7,640	10,703 [42]	3,258	3,527	Apr	6,546	
Sites of national importance in Great Britain								
Rye Harbour and Pett Level			5,850 [42]				5,850	
Hastings to Bexhill			5,700 [42]				5,700	
Isle of May					5,220 [42]	Jan	5,220	▲
Llandegfedd Reservoir				4,710 [42]			4,710	
Hamilton Low & Strathclyde Parks			4,600 [42]				4,600	▲
Thames Estuary	2,867	3,330	(4,349)	8,504	3,680	Aug	4,595	
Sites with mean peak counts of 2,500 or more birds in Great Britain[†]								
North Norfolk Coast	5,062	3,964	(3,047)	5,307	2,340	Mar	4,168	
Alt Estuary	1,440	3,153	3,825 [42]	7,155	(2,150)	Feb	3,893	
Chew Valley Lake			3,400 [42]				3,400	
Roughrigg Reservoir		47	1,121	15,144 [42]	416	210	Dec	3,388
Dee Estuary (England & Wales)	(778)	3,602	4,052 [42]	4,244	1,210	Jun	3,277	
Severn Estuary	(345)	2,981 [10]	3,500 [42]	(3,164)	(2,666)	May	3,241	
Troon Meikle Craigs			3,174 [42]				3,174	
Dungeness Gravel Pits		(400)	3,000 [42]	(500)			3,000	
Caldey Island			2,800 [42]				2,800	
Solway Estuary	(2,719)	3,281	2,189 [42]	(1,051)	(480)	Aug	2,735	
Guernsey Shore	1,972	2,127	(2,759)	(3,744)	(2,362)	Oct	2,593	
Sites with mean peak counts of 1,000 or more birds in Northern Ireland[†]								
Outer Ards Shoreline		1,001	(1,351)	1,179	1,304	Mar	1,209	
Other sites surpassing table qualifying levels in WeBS-Year 2005/2006 in Great Britain[†]								
Loch of Strathbeg		(51)	360	775	220	4,342	Dec	1,424
Burry Inlet	2,106	2,834	1,904	1,089	3,007	Sep	2,188	
Other sites surpassing table qualifying levels in WeBS-Year 2005/2006 in Northern Ireland[†]								
Lough Foyle		93	63	182	151	1,480	Jan	394

[†] as few sites exceed the British threshold and no All-Ireland threshold has been set qualifying levels of 2,500 and 1,000 have been chosen to select sites, in Great Britain and Northern Ireland respectively, for presentation in this report

Iceland Gull
Larus glaucoides

International threshold: 2,000

GB max: 9   Feb
NI max: 3   Jan

Iceland Gulls were recorded at 21 sites throughout the UK, although most reports were from Scotland and England. The highest total was of four at Loch Eriboll in February and there were reports of two and three at Belfast Lough in January and March respectively. All other records were of single birds with none reported in more than one month.

Glaucous Gull
Larus hyperboreus

International threshold: 10,000

GB max: 4   Feb
NI max: 0

Glaucous Gulls were recorded from nine sites, all but two of which were in Scotland. In England, singles were reported at the Otter Estuary in January and February and Hurworth Burn Reservoir in February. In Scotland, singles were present at Uyea Sound in October, North Ronaldsay Lochs in November, Burra Firth in January and February, Strathy Bay in February, and Loch of Skaill and Ugie Estuary in March.

Great Black-backed Gull
Larus marinus

International threshold: 4,400
Great Britain threshold: 400
All-Ireland threshold: ?[†]

GB max: 9,740   Oct
NI max: 1,918   Jan

Following a record low British maximum count in 2004/05, numbers were back to more usual levels, although still somewhat below the peaks of previous years. The Northern Ireland maximum increased to a new record high however. Despite a re-evaluation of the international 1% threshold downwards from 4,700 to 4,400, no UK site reaches this level. Twenty-five British sites surpassed national qualifying levels, one less than in 2004/05. The Wash remains the key site for this species whilst the counters at the Thames Estuary recorded their highest count to date. Conversely, counts at Tophill Low Reservoirs were the lowest ever recorded for WeBS. The peak count from Belfast Lough was the highest for over six years, as was the peak from Lough Foyle.

	01/02	02/03	03/04	04/05	05/06	Mon	Mean
Sites of national importance in Great Britain							
The Wash	4,515	1,959	4,628	(1,480)	1,773	Sep	3,219
Humber Estuary	(83)	(113)	2,200 [42]	(226)	(66)	Aug	2,200
Thames Estuary	(412)	1,236 [10]	857 [42]	1,648	1,972	Oct	1,428
Tees Estuary	(1,038)	702	1,523	1,657	(366)	Jan	1,294
Lower Derwent Ings	2,200	777	1,041 [42]		500	Jan	1,130
Lynemouth Ash Lagoons			1,074				1,074
Grafham Water			1,050 [42]				1,050
Dungeness Gravel Pits	(0)		1,000 [42]	(0)			1,000
Coquet Island			980 [42]				980
Ogston Reservoir			900 [42]				900
Pegwell Bay	1,000	1,305 [10]	305	610	1,190	Oct	882
Tophill Low Reservoirs	900	3,030	223 [42]	120	17	Aug	858
Durham Coast	(16)	(21)	(41)	(684)	776	Oct	776
Inner Moray and Inverness Firth		1,432 [10]	70	(93)	(4)	Nov	751
Moray Firth	884 [1]	1,001 [1]	674 [1]	336 [1]			724
North Norfolk Coast	748	617	1,051	327	471	Oct	643
Loch of Strathbeg	(129)	569	(606)	191	795	Dec	540
Hastings to Bexhill			520 [42]				520

	01/02	02/03	03/04	04/05	05/06	Mon	Mean
Eyebrook Reservoir			500 [42]				500
Don Mouth to Ythan Mouth	(67)	(55)	(200)	495	(347)	Aug	495
Guernsey Shore	(205)	353	560	404	(477)	Jan	449
Hanningfield Reservoir	(0)	1,098 [42]	437 [42]	0	140	Nov	419
Southfield Reservoir			408 [42]				408
Hoveringham & Bleasby GPs	0		1,600 [42]	2	0		401
Romney Sands			400 [42]				400
Sites no longer meeting table qualifying levels in WeBS-Year 2005/2006							
Morecambe Bay	331	353	(322)	(296)	(294)	Jul	342
Forth Estuary	(108)	(211)	286 [10]	(239)	(84)	Oct	286
Sites with mean peak counts of 500 or more birds in Northern Ireland[†]							
Belfast Lough	458	397	436 [10]	1,008	1,281	Jan	716
Other sites surpassing table qualifying levels in WeBS-Year 2005/2006 in Great Britain[†]							
Fleet and Wey	576	87	200 [42]	142	873	Nov	376
Cambois to Newbiggin	75	108	156	162	(600)	Oct	220
Camel Estuary	247	233	278	117	492	Jan	273
Lade Sands	0		200	250	400	Jan	213
Other sites surpassing table qualifying levels in WeBS-Year 2005/2006 in Northern Ireland[†]							
Lough Foyle	397	(221)	340	199	503	Dec	360

[†] as no All-Ireland threshold has been set a qualifying level of 500 has been chosen to select sites for presentation in this report

Kittiwake

Rissa tridactyla

International threshold: 20,000**
Great Britain threshold: ?[†]
All-Ireland threshold: ?[†]

GB max: 2,404 Jul
NI max: 25 Sep

The British maximum was over twice that of the preceding year, and was recorded in July rather than September. Half of this July total was recorded at Loch of Strathbeg, which lies close to known concentrations of breeding birds. Scotland remains the key region for this species with 11 out of the 16 sites holding mean numbers exceeding 200. Four of the remaining five are in northeast England. The Northern Ireland maximum was the highest for three years and all of the birds involved that month were recorded at Belfast Lough, whilst other counts were of 11 at Larne Lough in November and two at Loughs Neagh and Beg in March.

	01/02	02/03	03/04	04/05	05/06	Mon	Mean
Sites with mean peak counts of 200 or more birds in Great Britain[†]							
Loch of Strathbeg	0	940	6,300	152	1,130	Jul	1,704
Nigg Bay to Cove Bay				846			846
Lunan Bay	0	400	3,400	100	250	May	830
Arran	1,700	185	290	340	701	Sep	643
Tay Estuary	266	1,100	133	690	740	Aug	586
Loch Linnhe: Camas Shallachain	500						500
The Wash		(481) [30]					(481)
Farne Islands	920			0			460
Tweed Estuary	340	470	860	114	340	Jul	425
Tees Estuary	20	30	1,492	(56)	61	Apr	401
Loch a` Phuill (Tiree)	406	1,128	276	104	54	Jun	394
Forth Estuary	(274)	(453)	(426)	170	(276)	Sep	320
Durham Coast	(0)	(0)	(0)	(279)	(250)	Jul	(279)
Beadnell to Seahouses	0	160	350	140	512	Jun	232
Don Mouth to Ythan Mouth	(18)	0	153	534	(165)	Aug	229
Solway Estuary	(574)	(300)	(47)	0	(150)	Apr	214
Other sites surpassing table qualifying levels in WeBS-Year 2005/2006 in Great Britain[†]							
Anstruther Harbour	4	4	20	14	550	Jun	118

[†] as no British or All-Ireland thresholds have been set a qualifying level of 200 has been chosen to select sites for presentation in this report

Little Tern
Sternula albifrons

International threshold: 490
Great Britain threshold: ?[†]
All-Ireland threshold: ?[†]

GB max: 962 Jul
NI max: 0

The count of Little Terns at the Dee Estuary (England & Wales) was the second highest ever recorded at the site, the highest being of 700 in August 1995. This count was also the highest during 2005/06 and pushes the site mean above that of the North Norfolk Coast, traditionally the key site for this species. Numbers at several sites in the southern half of England were low, noticeably at Chichester Harbour where the peak was the lowest for over six years.

	2001	2002	2003	2004	2005	Mon	Mean
Sites with mean peak counts of 50 or more birds in Great Britain[†]							
Dee Estuary (England & Wales)	(0)	242	(256)	300	411	Jul	318
North Norfolk Coast	(265)	(280)	(405)	233	246	Jul	286
The Wash	(103)	(36)	68	(108)	182	Jul	125
Thames Estuary	(1)	(100)	(28)	33	74	Aug	69
Durham Coast		(6)		(0)	67	Jun	67
Chichester Harbour	200 [12]	42	28	36	14	May	64
Duddon Estuary		28	42	84	92	Jun	62

[†] *as no British or All-Ireland thresholds have been set a qualifying level of 50 has been chosen to select sites for presentation in this report*

Gull-billed Tern
Gelochelidon nilotica

Vagrant
Native Range: S & E Europe, America, Asia

GB max: 1 May
NI max: 0

A single Gull-billed Tern was reported at the Taw-Torridge Estuary (Caen to Whitehouse section) during May. This was the fourth record for WeBS, the last being at Drift Reservoir in September 2002.

Black Tern
Chlidonias niger

International threshold: 7,500
Great Britain threshold: ?
All-Ireland threshold: ?

GB max: 27 Aug
NI max: 2 Sep

Black Terns were recorded at 16 sites in 2005, well below the 40 sites in 2004. There were no spring records and the first birds of the year were not noted until August when the national maximum reached 27; this was less that a third of the 2004 figure. Up to nine birds were present at Doddington Pool in August, while counts of five and four were recorded at Dengie Flats and the Thames Estuary, respectively, during the same month. Birds were reported throughout September and October and the latest record for the year was of a single bird at Wigtown Bay in November.

White-winged Black Tern
Chlidonias leucopterus

Vagrant
Native Range: E Europe, Asia

GB max: 1 Sep
NI max: 0

Whilst there have been ten previous records of White-winged Black Tern during WeBS counts, the one recorded at River Clyde, Lamington, in September was the first time this species had been counted in Scotland.

Sandwich Tern

Sterna sandvicensis

International threshold: 1,700
Great Britain threshold: ?[†]
All-Ireland threshold: ?[†]

GB max: 8,628 Jul
NI max: 312 Sep

The British maximum fell by a quarter compared to the previous year, whereas the Northern Ireland figure rose slightly. As the recording of terns is optional for WeBS, these values can vary largely due to coverage as well as changes in numbers at the key locations. Sandwich Terns were recorded in every month in 2005 and from 117 sites across the UK. The highest count from any single site was of 3,228 on the North Norfolk Coast in July, although this was the lowest peak total for this, the principal UK site for this species, for over five years. Low counts were also noted at Dundrum Bay. Numbers at Cemlyn Bay & Lagoon have remained over 2,000 for the third year running helping this site maintain its internationally important status.

	2001	2002	2003	2004	2005	Mon	Mean
Sites of international importance in the UK							
North Norfolk Coast	(3,365)	(4,600)	4,170	5,533	3,228	Jul	4,383
Forth Estuary	(994)	(2,317)	2,802	(1,526)	(1,243)	Jul	2,802
Cemlyn Bay and Lagoon	0		2,455	2,700	2,000	Jun	1,789
Sites with mean peak counts of 200 or more birds in Great Britain[†]							
Dee Estuary (England & Wales)	(11)	1,632	716	759	829	Jul	984
Tees Estuary	35	974	2,601	(333)	221	Jun	958
Duddon Estuary	(0)	704	955	1,144	604	Jun	852
Solway Estuary	(235)	(206)	(548)	(282)	(209)	Aug	(548)
Ythan Estuary		930	150				540
Morecambe Bay	(0)	220	531	500	(110)	May	417
Humber Estuary	(124)	(396)	(303)	(324)	(325)	Jul	(396)
The Wash	512	150	223	208	307	Aug	280
Tay Estuary	167	461	310	96	126	May	232
Sites with mean peak counts of 200 or more birds in Northern Ireland[†]							
Dundrum Bay	296	722	264	173	133	Aug	318
Belfast Lough	409	357	136	99	255	Sep	251
Other sites surpassing table qualifying levels in Summer 2005 in Great Britain[†]							
Farne Islands	72	120		(0)	295	Apr	162
Don Mouth to Ythan Mouth	0	30	3	158	(254)	Sep	89
Lavan Sands	(7)	25	170	250	235	Aug	170

[†] as no British or All-Ireland thresholds have been set a qualifying level of 200 has been chosen to select sites for presentation in this report

Common Tern

Sterna hirundo

International threshold: 1,900
Great Britain threshold: ?[†]
All-Ireland threshold: ?[†]

GB max: 5,186 Jul
NI max: 136 May

The British peak was comparable with that of previous year while the maximum in Northern Ireland was the highest recorded in the province. However, the national totals depend on the level of coverage and as the recording of terns is optional these figures will depend on which sites terns were counted at. Furthermore, the majority of sites in Northern Ireland are usually only counted during the core winter period and further counts in the summer months would undoubtedly lead to higher peaks.

Numbers at the Alt Estuary, as well as being the highest ever recorded at the site, were the highest single site total during 2005/06. In general, numbers at most other sites were around average with several of the lower than average counts being due to incomplete coverage.

	2001	2002	2003	2004	2005	Mon	Mean
Sites with mean peak counts of 200 or more birds in Great Britain[†]							
Alt Estuary	800 [11]	900 [11]	1,664	1,135	2,010	Aug	1,302
Tees Estuary	(12)	696	1,678	1,251	(521)	Jun	1,208
Thames Estuary	(190)	(158)	(224)	(553)	(219)	Aug	(553)
North Norfolk Coast	(213)	(321)	419	476	450	Jun	448
Dee Estuary (England & Wales)	(3)	422	(384)	(180)	(109)	Jul	422
Forth Estuary	(40)	(691)	193	(183)	(287)	Jul	390
Chichester Harbour	500 [12]	130 [12]					315
Southampton Water	(300)	(50)	(7)	(63)	(62)	Aug	(300)
Humber Estuary	(6)	(291)	280	(160)	(61)	Aug	286
The Wash	(435)	(102)	122	199	129	Aug	221
Ythan Estuary		18	415				217
Other sites surpassing table qualifying levels in Summer 2005 in Great Britain[†]							
Loch of Strathbeg	0	108	199	151	449	Jul	181
Don Estuary	0	4	7	39	260	Jul	62

[†] as no British or All-Ireland thresholds have been set a qualifying level of 200 has been chosen to select sites for presentation in this report

Roseate Tern
Sterna dougallii

Scarce

GB max: 3 Jul
NI max: 0

Roseate Terns were reported from just five sites during 2005, the lowest number for several years with counts from one of the key areas, St Mary's Island, unavailable. The earliest record was of two at the Forth Estuary in May, with birds in June being one at East Chevington Pools and two at the Duddon Estuary, followed by two in July at Alnmouth and one at the Alt Estuary in the same month.

Arctic Tern
Sterna paradisaea

International threshold:	?
Great Britain threshold:	?[†]
All-Ireland threshold:	?[†]

GB max: 2,365 Jul
NI max: 0

Reflecting the northerly breeding distribution of Arctic Tern, 75% of the key sites for this species were in Scotland with the remainder in northern England. On average, the Farne Islands remains the key site but in recent years terns have not been counted during the summer months at this site. The July count of 2,100 at Loch of Strathbeg was the highest-ever count at a single site. The previous highest count, also at this site, was in August 1995.

The high count at Loch of Strathbeg greatly influenced this year's British maximum, which as always is greatly dependent on the number of sites at which terns are recorded. New site peaks were reached at Loch of Beith, Loch An Duin (Aird Point) (Lewis), Cemlyn Bay & Lagoon, North Norfolk Coast, Beadnell to Seahouses and Loch Paible (North Uist).

	2001	2002	2003	2004	2005	Mon	Mean
Sites with mean peak counts of 50 or more birds in Great Britain[†]							
Farne Islands	(600)	(0)		(0)	(0)		(600)
Loch of Strathbeg	0	35	(68)	40	2,100	Jul	544
Ythan Estuary		106	860				483
Loch of Beith		150 [12]	31		1,000	Aug	394
Forth Estuary	2	(1,214)	197	(186)	7	Aug	321
Loch An Duin (Aird Point) (Lewis)					300	Jun	300
Loch a` Phuill (Tiree)	473	477	150	120	58	Jun	256
Tay Estuary	32	660	290	0	10	Jul	198
Morecambe Bay		94	(178)	(59)	(16)	May	136

	2001	2002	2003	2004	2005	Mon	Mean
Eden Estuary	(53)	125	320	4	0		112
The Houb (Whalsay)	0	120	82	300	3	Aug	101
Cambois to Newbiggin	5	246	0				84
St Andrews Bay	44	29	(0)	192	70	Aug	84
Braewick Loch	70	170	50	30	8	Jun	66
Inner Loch Indaal		51	76				64
Lindisfarne	(0)		(1)	3	(120)	Jun	62
Other sites surpassing table qualifying levels in Summer 2005 in Great Britain[†]							
Cemlyn Bay and Lagoon	0		23	17	80	Jun	30
North Norfolk Coast	(0)	(2)	12	(16)	(65)	May	31
Balta Sound	7	9	0	55	55	May	25
Beadnell to Seahouses	0	0	1	0	53	Jul	11
Loch Bhasapoll (Tiree)	17	22	36	9	50	Jun	27
Loch Paible (North Uist)		10	36	20	50	Jun	29

[†] *as no British or All-Ireland thresholds have been set a qualifying level of 50 has been chosen to select sites for presentation in this report*

Kingfisher
Alcedo atthis

International threshold:	?
Great Britain threshold:	?[†]
All-Ireland threshold:	?[†]

GB max:	485 Oct
NI max:	7 Dec

Following a slight drop in 2004/05, the Great Britain peak rose again to its highest ever level. Twelve sites held five-year mean peaks of seven or more birds, two less than in the previous year, and gravel pit complexes again tended to produce the highest counts. The top site remained the Somerset Levels, though the peak count was slightly lower than for 2004/05. As ever, however, WeBS methodology is not ideal for surveying the fortunes of this widespread but often elusive species.

	01/02	02/03	03/04	04/05	05/06	Mon	Mean
Sites with mean peak counts of 7 or more birds in Great Britain[†]							
Somerset Levels	16	(14)	(12)	20	(18)	Oct	18
Wraysbury Gravel Pits	14	19	12	18	16	Sep	16
Ditchford Gravel Pits	7		13	12	13	Sep	11
Eversley Cross & Yateley GPs	11	8	10	6	8	Sep	9
Pitsford Reservoir	11	9	11	(3)	6	Sep	9
Lee Valley Gravel Pits	12	4	10	6	9	Sep	8
Southampton Water	5	(5)	(6)	9	(11)	Oct	8
Avon Valley: Salisbury/Fordingbridge	(3)	(4)	(6)	(5)	(7)	Nov	(7)
Chichester Gravel Pits	3	4	7	9	11	Feb	7
Colne Valley Gravel Pits	(3)	(4)	(4)	5	9	Sep	7
North Norfolk Coast	4 [10]	7	7	6	8	10	Oct	7
Thames Estuary	3	9	(7)	7	10	Oct	7
Other sites surpassing table qualifying levels in WeBS-Year 2005/2006 in Great Britain[†]							
Taw-Torridge Estuary	4	5	(5)	(5)	9	Sep	6
Chilham and Chartham Gravel Pits	1	2	3	3	9	Sep	4
Orwell Estuary	(2) [10]	(4)	10 [5]	(5)	7	Dec	6
Wraysbury Pond	1		0	2	7	Dec	3

[†] *as no British or All-Ireland thresholds have been set a qualifying level of 7 has been chosen to select sites for presentation in this report*

PRINCIPAL SITES

Table 6 below lists the principal sites for non-breeding waterbirds in the UK as monitored by WeBS. All sites supporting more than 10,000 waterbirds are listed, as are all sites supporting internationally important numbers of one or more waterbird species. Naturalised species (*e.g.* Canada Goose and Ruddy Duck) and non-native species presumed to have escaped from captive collections have been excluded from the totals, as have gulls and terns since the recording of these species is optional (see *Analysis*). Table 7 lists other sites holding internationally important numbers of waterbirds, which are not routinely monitored by standard WeBS surveys but rather by the Icelandic Goose Census and aerial surveys.

A total of 234 sites are listed in tables 6 and 7. Of these 215 supported one or more species in internationally important numbers and 83 held a five-year mean peak of 10,000 or more birds. Typically there are few changes to the top twenty sites listed in the principal sites table, with the order of the top ten rarely changing. The Wash remains as the key waterbird site with regard to numbers and in 2005/06 held the highest numbers of the preceding five years. Numbers on the North Norfolk Coast continue to rise and the five-year mean surpassed that of Morecambe Bay to become the third most important site in the UK in terms of numbers of waterbirds.

Following a low year in 2004/05, numbers on the Humber Estuary were back up to around average on the previous year, which saw the Humber Estuary climb back into the top five sites at the expense of the Thames. Total numbers on the Solway Estuary were over 40% lower than for 2004/05 and counts on both the Mersey and Forth Estuaries continue to fall and are now their lowest for over ten years. Conversely, numbers on the Ouse Washes rose by over 30% to their highest total during the past five years. Following recent declines in the Severn Estuary, numbers rose to well above the five-year average. The recent decline in the numbers of diving ducks at Loughs Neagh and Beg that have contributed to the fall in numbers at the site seem not to be as severe as first thought. The site remained internationally important for Scaup and total numbers at the site were average for the five-year period.

Five-year averages of sites holding 10,000 or more waterbirds were relatively similar compared to the previous year, with 75 of the 83 sites undergoing changes of less than 10%. The greatest increases were experienced at Abberton Reservoir (52%) and Ouse Washes (31%). The greatest decreases were seen at Solway Estuary (43%) and the Mersey Estuary (12%).

Table 6. Total number of waterbirds at principal sites in the UK, 2001/2002 to 2005/06 (includes data from all available sources) and species occurring in internationally important numbers at each. (Species codes for those listed are provided in Table 8.)

Site	01/02	02/03	03/04	04/05	05/06	Average	Int.Imp.species
The Wash	332,655	343,467	338,382	369,516	398,233	356,451	PG DB SU PT OC RP GP GV L. KN SS DN BW BA CU RK
Ribble Estuary	210,308	255,012	252,374	242,666	220,693	236,211	BS WS PG SU WN T. PT OC RP GV L. KN SS DN BW BA RK
North Norfolk Coast	185,106	212,442	284,590	235,276	246,946	232,872	PG DB WN PT RP KN BW BA
Morecambe Bay	211,422	250,769	249,321	204,094	201,456	223,412	PG SU PT OC KN DN BW BA CU RK
Humber Estuary	159,720	174,931	217,799	163,062	171,097	177,322	PG DB SU RP GP GV L. KN DN BW BA RK
Thames Estuary	169,963	197,181	160,183	172,439	176,503	175,254	DB T. SV OC AV RP GV KN DN BW BA RK
Dee Estuary (England and Wales)	195,534	127,015	171,901	115,285	130,282	148,003	SU PT OC KN DN BW BA RK
Solway Estuary	124,042	153,365	145,051	140,102	81,921	128,896	WS PG YS SU PT SP OC RP KN DN BA RK
Somerset Levels	114,998	102,800	85,154	99,759	87,810	98,104	MS WN GA T. PT SV L.
Mersey Estuary	102,668	108,764	97,787	85,571	75,333	94,025	SU T. DN BW RK
Forth Estuary	88,248	109,461	91,991	84,780	79,243	90,745	PG JI SU KN BA RK
Ouse Washes	76,122	66,419	85,715	89,847	130,682	89,757	MS BS WS WN GA T. PT SV BW

Site	01/02	02/03	03/04	04/05	05/06	Average	Int.Imp.species
Breydon Wtr / Berney Mrshs	78,429	64,807	75,827	110,697	106,431	87,238	BS PG WN SV AV GP L. BW
Swale Estuary	88,292	86,102	86,967	73,830	82,846	83,607	WN T. PT RP GP BW
Strangford Lough	80,423	79,525	88,519	78,451	83,271	82,038	MS QN SU GP KN BA RK
Loch of Strathbeg	68,135	49,492	79,327	81,645	86,258	72,971	WS PG YS
Blackwater Estuary	66,476	81,687	64,534	78,257	70,054	72,202	DB GP GV KN DN BW RK
Severn Estuary	60,735	68,658	64,675	64,055	80,250	67,675	MS BS SU PT DN
Inner Moray/Inverness Firth	59,617	59,711	80,226	65,878	68,475	66,781	PG JI
Alt Estuary	90,066	63,502	72,793	53,073	41,838	64,254	GV KN SS BA
Lindisfarne	63,760	64,120	57,578	53,365	59,790	59,723	PG YS QS WN BA
Loughs Neagh and Beg	63,479	53,401	59,026	56,259	58,255	58,084	MS WS PO SP
Stour Estuary	59,755	50,206	48,587	46,677	48,214	50,688	RP GV KN BW
Montrose Basin	63,766	37,034	35,478	50,208	57,045	48,706	PG
Burry Inlet	41,384	43,830	52,852	49,289	45,360	46,543	PT OC BW
Chichester Harbour	51,494	44,826	43,732	43,375	47,646	46,215	DB DN BW
Dengie Flats	55,929	39,325	23,895	45,747	58,302	44,640	GV KN BA
Carmarthen Bay	31,358	36,917	47,067	55,748	45,084	43,235	CX
Hamford Water	44,225	40,135	37,991	39,939	43,465	41,151	DB RP GV
Langstone Harbour	33,738	37,457	43,775	45,649	41,477	40,419	DB DN BW
Loch Leven	38,128	39,589	37,329	33,773	40,355	37,835	MS PG T.
Dornoch Firth	41,744	39,106	37,819	36,329	33,712	37,742	JI WN BA
WWT Martin Mere	37,482	39,106	30,884	45,320	30,938	36,746	WS PG
Cromarty Firth	36,642	26,278	41,307	37,970	36,785	35,796	JI BA
Lough Foyle	29,448	34,154	37,291	33,076	38,651	34,524	WS QN BA
Lower Derwent Ings	26,151	33,262	32,630	34,614	38,396	33,011	
Alde Complex	29,086	29,647	22,905	31,838	34,369	29,569	AV BW
Abberton Reservoir	20,604	20,677	31,382	24,133	50,286	29,416	MS SV PO
Duddon Estuary	25,067	22,185	32,592	29,175	35,351	28,874	PT
Nene Washes	19,343	52,857	20,912	29,263	20,701	28,615	BS PT
Medway Estuary	29,796	27,291	25,863	27,025	30,871	28,169	PT AV BW
Rutland Water	22,151	26,193	28,239	26,196	31,192	26,794	MS GA SV
WWT Caerlaverock (Inland)	19,217	34,020	25,174	26,988		26,350	WS YS
Crouch-Roach Estuary	21,024	21,377	18,219	34,182	29,553	24,871	DB RK
Tees Estuary	22,198	25,355	30,085	21,040	23,568	24,449	
Poole Harbour	24,561	25,955	24,856	26,360	17,672	23,881	AV BW
Orwell Estuary	21,444	24,794	25,228	19,983	25,192	23,328	BW
Cleddau Estuary	18,585	17,879	20,398	27,752	31,431	23,209	
Pegwell Bay	21,256	28,801	25,509	18,222	20,188	22,795	
Inner Firth of Clyde	22,904	23,199	23,739	19,916	23,242	22,600	
Tay Estuary	25,699	20,723	21,302	19,903	24,545	22,434	PG BA
Loch Spynie	15,467	19,523	15,748	30,734	27,247	21,744	PG JI
Exe Estuary	24,754	20,590	22,895	20,125	19,139	21,501	BW
Lavan Sands	17,281	21,637	21,062	22,205	19,284	20,258	
Belfast Lough	18,818	18,789	19,417	20,077	19,528	19,326	BW
Deben Estuary	16,969	17,080	17,873	19,050	19,081	18,011	
Wigtown Bay	18,134	14,868	22,295	19,461	15,144	17,980	PG YS
Colne Estuary	25,665	4,187	19,342	18,294	19,102	17,318	DB
Mersehead RSPB Reserve	19,864	24,166	16,820	24,490	124	17,093	YS PT
Loch of Skene	17,136	11,004	13,695	18,834	24,732	17,080	PG JI
Fleet and Wey	19,678	14,551	16,290	17,494	17,347	17,072	MS
Pagham Harbour	14,287	13,223	14,566	20,528	20,774	16,676	DB PT BW
Carsebreck & Rhynd Lochs	19,947	15,252	16,530	12,262	17,028	16,204	PG
Southampton Water	16,867	16,684	15,441	15,039	12,560	15,318	
Eden Estuary	15,169	15,124	15,496	14,216	12,191	14,439	
Taw-Torridge Estuary	14,195	11,847	10,315	16,422	17,243	14,004	
Middle Yare Marshes	13,852	10,682	9,739	17,519	18,133	13,985	
Blyth Estuary	13,713	.	.	.	.	13,713	BW
North West Solent	12,307	10,071	15,139	16,224	13,707	13,490	DB
Portsmouth Harbour	8,125	14,911	16,420	9,663	17,232	13,270	DB
Walland Marsh	21,016	5,109	5,951	21,164	12,904	13,229	
Ythan Estuary	13,569	10,757	8,049	13,599	16,182	12,431	
Rye Harbour and Pett Level	11,458	10,854	16,921	9,717	11,389	12,068	
Dungeness Gravel Pits	11,073	13,872	11,673	11,003	12,626	12,049	SV
R. Nith: Keltonbnk-Nunholm	12,215	11,697	14,855	8,972	.	11,935	PG YS
Loch of Lintrathen	10,476	9,243	16,418	11,070	10,330	11,507	PG
Dyfi Estuary	9,391	11,490	11,524	12,418	12,016	11,368	
Arun Valley	10,719	15,688	11,913	8,724	9,379	11,285	
Cotswold Water Park (West)	10,032	9,103	12,475	10,176	12,105	10,778	
Pitsford Reservoir	9,022	8,852	12,523	10,266	12,197	10,572	
Carlingford Lough	9,763	10,764	10,516	11,302	10,458	10,561	QN
Beaulieu Estuary	11,774	11,148	6,140	10,793	10,776	10,126	

Site	01/02	02/03	03/04	04/05	05/06	Average	Int.Imp.species
Chew Valley Lake	8,775	8,369	10,109	9,285	13,497	10,007	SV
Hule Moss	9,726	7,109	15,860	8,994	6,992	9,736	PG
Loch of Harray	9,129	10,010	12,330	8,501	7,566	9,507	MS JI
Outer Ards Shoreline	210	12,694	12,749	9,552	9,893	9,020	QN
Lee Valley Gravel Pits	9,592	9,316	8,996	8,394	8,053	8,870	GA
Upper Lough Erne	7,717	8,777	9,239	9,369	9,151	8,851	MS WS
Cameron Reservoir	18,249	4,495	11,228	5,088	2,847	8,381	PG
R. Avon: R'wood-Christch	4,838	24,594	4,766	2,794	2,795	7,957	BW
Loch Fleet Complex	11,677	8,275	6,486	5,177	7,923	7,908	JI
Dundrum Bay	9,257	6,859	6,959	7,490	8,421	7,797	QN
R. Avon: Fordingbr-R'wood	7,420	10,022	6,781	6,100	6,198	7,304	GA
R Clyde: Carstairs-Thankert	14,501	4,683	6,623	4,510	5,785	7,220	PG
Tring Reservoirs	6,838	6,420	6,912	8,405	6,831	7,081	MS
Hornsea Mere	5,720	8,905	7,341	6,912	6,340	7,044	MS
Loch Eye	5,210	1,926	4,479	8,352	15,006	6,995	WS JI
Orchardton and Auchencairn Bays	.	7,813	11,246	4,170	4,371	6,900	YS
Slains Lochs (Meikle and Sand and Cotehill)	2,832	360	576	17,311	12,615	6,739	PG
Loch of Stenness	9,129	7,098	5,636	5,072	6,475	6,682	JI
Loch of Boardhouse	5,096	5,257	7,111	5,904	7,159	6,105	JI
Hickling Broad	5,303	6,109	7,070	4,980	6,382	5,969	BS
Milldam & Balfour Mains Pools	5,037	7,755	6,217	3,918	4,841	5,554	JI
Holburn Moss	1,776	6,378	10,501	5,397	3,000	5,410	PG
Loch Gruinart Floods	.	3,541	5,021	4,299	4,145	4,252	JH
Severn Hams	6,080	2,320	4,245	4,639	3,065	4,070	PT
Kilconquhar Loch	3,284	3,942	5,871	2,985	3,734	3,963	JI
Tweed Estuary	3,807	3,516	4,181	3,527	3,523	3,711	MS
Lake of Menteith	732	4,958	4,639	6,462	732	3,505	PG
Dee Flood Meadows	4,603	3,887	2,859	1,550	4,109	3,402	PT
Loch a` Phuill (Tiree)	2,632	2,643	3,077	2,800	5,428	3,316	JH
Island of Egilsay	1,463	2,570	3,548	5,401	3,480	3,292	JI
R. Tay: Haughs of Kercock	1,804	1,306	808	6,027	6,257	3,240	JI
Loch Watten	5,356	3,024	2,199	3,100	2,448	3,225	JI
Loch of Hundland	3,123	4,343	3,100	2,562	2,821	3,190	JI
Loch of Skaill	2,372	3,363	3,053	2,652	4,312	3,150	JI
St Benet`s Levels	3,216	2,262	3,762	.	.	3,080	BS
Loch Scarmclate	2,025	2,870	3,198	2,700	3,868	2,932	JI
Loch Bee (South Uist)	1,416	3,048	2,674	4,038	3,428	2,921	JH
Lower Lough Erne	.	2,931	2,484	3,341	2,877	2,908	MS
Loch Lomond	1,718	4,224	2,968	3,194	2,185	2,858	NW
Loch of Swannay	2,880	2,975	2,660	2,910	2,779	2,841	JI
Loch Ken	2,893	4,220	4,037	2,333	618	2,820	JI
Loch of Wester	5,776	.	.	1,129	1,162	2,689	JI
Loch Hempriggs	2,785			3,891	1,242	2,639	JI
Balranald Nature Reserve	3,484	2,815	2,167	2,894	1,722	2,616	JH
Isle of Coll	1,638	2,296	2,027	3,072	4,000	2,607	NW JH YN
Killough Harbour	1,815	2,732	1,158	2,736	3,977	2,484	QN
Lochs Davan and Kinord	5,567	3,124	1,335	913	1,034	2,395	JI
Loch Paible (North Uist)	1,169	2,253	2,609	2,698	1,687	2,083	JH
Upper Quoile River	2,758	897	1,239	977	4,394	2,053	MS
Lower Teviot Valley	1,134	2,781	.	1,621	.	1,845	JI
Loch Garten	2,804	1,000	1,133	2,417	1,715	1,814	JI
Teviot Haughs	.	.	1,127	.	2,393	1,760	JI
Melbost Sands, Broad Bay & Tong Saltings (Lewis)	1,729	.	394	1,808	1,527	1,365	JH
Loch Bhasapoll (Tiree)	2,163	777	1,135	1,173	1,574	1,364	JH
Balnakiel Farm	.	.	.	.	1,198	1,198	YN
Loch Mor: Baleshare (North Uist)	686	789	1,353	1,112	770	942	JH
Loch Gorm	.	.	.	187	1,629	908	NW JH YN
Loch Riaghain (Tiree)	736	647	604	684	494	633	JH
Loch An Eilein (Tiree)	366	569	492	643	861	586	JH
Moine Mhor & Add Estuary	221	586	1,072	344	471	539	JH
Balnakeil Bay	.	.	925	16	52	331	YN
Branahuie Saltings	324	.	.	.	.	324	JH
Kentra Moss and Lower Loch Shiel	207	216	277	364	306	274	JH

Table 7. Other sites in the UK holding internationally important numbers of waterbirds in 2005/06, which are not routinely monitored by standard WeBS surveys. (Species codes for those listed are provided in Table 8.)

Site	Int.Imp.species	Site	Int.Imp.species
Aberlady Bay Roost	PG	Scolt Head Roost	PG
Banks Marsh West	PG	Snettisham Roost	PG
Beauly Firth Roost	JI	Wells-next-the-Sea	PG
Benbecula	JH	North Sutherland	YN
Berney Marshes	PG	North Uist	JH YN
Bridge of Crathies	WS	Baleshare and Carinish (Grimsay)	JH
Bute	JI JH	Balranald Clettraval and Tigharry	JH
Caithness Lochs	JI	Berneray	JH YN
Colonsay/Oronsay	YN	Boreray and Lingay	JH
Dingwall Bay	JI	Malaclate To Grenitote	JH
Dupplin Lochs	PG	Oronsay	JH
East Mainland	JI	Paible	JH
East Mains Flood	JI	Trumisgarry Clachan and Newton	JH
Easterton - Fort George	PG JI	Norton Marsh	PG
Fala Flow	PG	Orkney	JI YN
Findhorn Bay Roost	PG	South Walls (Hoy)	YN
Hule Moss (West)	PG	Pilling to Cockerham	PG
Inner Cromarty Firth - Conor Islands	JI	Read`s Island Flats	PG
Inner Cromarty Firth - Urquhart	JI	Rhunahaorine	NW
Inner Firth of Tay	PG JI	Rossie Bog	PG
Island of Eday	JI	Sanday	JI
Island of Islay	NW YN	Skinflats Roost	PG
Islands of Shapinsay	JI	Solway Firth	YS
Isle of Oronsay	YN	Sound of Harris (NW) (Harris)	YN
Isle of Colonsay	NW	South Mainland	JI
Isle of Lismore	NW	South Uist	JH
Isle of South Ronaldsay	JI	Askernish To Smerclate	JH
Keills Peninsula and Isle of Danna	NW	Bornish To Askernish	JH
Liverpool Bay	CX	Drimore To Howmore	JH
Loans of Tullich	WS	Howbeg To Bornish	JH
Loch Eye and Cromarty Firth	WS PG JI	Lochdar, Gerinish and Drimsdale	JH
Loch Fleet	JI	Lochdar Gerinish To Drimore	JH
Loch Tullybelton	PG	Southwest Lancashire	PG
Lune Estuary	PG	Stranraer Lochs	NW
Machrihanish	NW JH	Strathearn (West)	PG JI
Martin Mere and Ribble Estuary	BS WS	Stronsay (Whole Island)	JI
Meikle Loch - Slains	PG	Tay Estuary - Tentsmuir Point	PG
Munlochy Bay Roost	JI	Tay and Isla Valley	PG JI
Nigg Bay	JI	Tayinloan	JH
North Norfolk Coast & The Wash	PG	Tiree	NW JH YN
Heigham Holmes	PG	West Mainland	JI
Holkham Bay Roost	PG	West Water Reservoir	PG
Holkham and Wells	PG	Wyre Estuary	PG
Holme and Thornham	PG	Ythan Estuary and Slains Lochs	PG
Horsey Mere	PG		

Table 8. Species codes for species listed in tables 6, 7 and 9.

AV	Avocet	PG	Pink-footed Goose
BA	Bar-tailed Godwit	PO	Pochard
BS	Bewick's Swan	PT	Pintail
BW	Black-tailed Godwit	QN	Light-bellied Brent Goose (Nearctic population)
CA	Cormorant	QS	Light-bellied Brent Goose (Svalbard population)
CU	Curlew	RK	Redshank
DB	Dark-bellied Brent Goose	RP	Ringed Plover
DN	Dunlin	SP	Scaup
GA	Gadwall	SS	Sanderling
GN	Goldeneye	SU	Shelduck
GP	Golden Plover	SV	Shoveler
GV	Grey Plover	SZ	Slavonian Grebe
JH	Greylag Goose (Northwest Scotland population)	T.	Teal
JI	Greylag Goose (Icelandic population)	TT	Turnstone
KN	Knot	TU	Tufted Duck
L.	Lapwing	WN	Wigeon
MS	Mute Swan	WS	Whooper Swan
ND	Great Northern Diver	YN	Barnacle Goose (Nearctic population)
NW	Greenland White-fronted Goose	YS	Barnacle Goose (Svalbard population)
OC	Oystercatcher		

WeBS Low Tide Counts

AIMS

Estuarine sites in the UK provide the most important habitat for non-breeding waterbirds, acting as wintering grounds for many migrants but also as stopover feeding locations for other waterbirds passing along the East Atlantic Flyway. Core Counts on estuaries tend to quantify birds present at high tide roosts. Although important, knowledge of roost sites provides only part of the picture, and does not elucidate the use that waterbirds make of a site for feeding.

The WeBS Low Tide Counts scheme has flourished since its inception in the winter of 1992/93, with most of the major estuaries covered. The scheme aims principally to monitor, assess and regularly update information on the relative importance of inter-tidal feeding areas of UK estuaries for wintering waterbirds and thus to complement the information gathered by WeBS Core Counts.

The data gathered contribute greatly to the conservation of waterbirds by providing supporting information for the establishment and management of UK Ramsar sites and Special Protection Areas (SPAs), other site designations and whole estuary conservation plans. In addition, WeBS Low Tide Counts enhance our knowledge of the low water distribution of waterbirds and provide data that highlight regional variations in habitat use, whilst also informing protection of the important foraging areas identified. WeBS Low Tide Counts provide valuable information needed to gauge the potential effects on waterbirds of a variety of human activities which affect the extent or value of inter-tidal habitats, such as proposals for dock developments, recreational activities, tidal power barrages, marinas and housing schemes. Designing mitigation or compensation for such activities can be assisted using data collected under the scheme. Furthermore, the effects on bird distributions of climate change and sea level rise can be assessed.

METHODS

The scheme provides information on the numbers of waterbirds feeding on subdivisions of the inter-tidal habitat within estuaries. Given the extra work that Low Tide Counts entail, often by the same counters that carry out the Core Counts, WeBS aims to cover most individual estuaries about once every six years, although on some sites more frequent counts are made. Co-ordinated counts of waterbirds are made by volunteers each month between November and February on pre-established subdivisions of the inter-tidal habitat in the period two hours either side of low tide.

DATA PRESENTATION

Tabulated Statistics

Tables 9. and 10. present three statistics for 18 of the more numerous waterbird species present on 18 estuaries covered during the 2005/06 winter: the peak number of a species over the whole site counted in any one month (with checks for count synchronicity made from assessing proximity of count dates and consultation with Local Organisers); an estimate of the mean number present over the winter for the whole site (obtained by summing the mean counts of each species for each count section) and the mean density over the site (in birds per hectare), which is the mean number divided by the total area surveyed (in hectares). The area value used for these calculations is the sum of the inter-tidal and non-tidal components of each count section but omits the sub-tidal areas (*i.e.* those parts of the count section which are under water on a mean low tide).

Dot Density Maps

WeBS Low Tide Count data are presented as dot density maps, with subdivision of count sections into basic habitat elements. The reason for such a subdivision is to ensure species are plotted on appropriate habitat areas and to improve the accuracy of density estimates. Each section for which a count has been made is divided into a maximum of three different habitat components:

Inter-tidal: Areas that lie between mean high water and mean low water.

Sub-tidal: Areas that lie below mean low water. In more 'open-coast'-type situations, a sub-tidal zone reaching 500 m out from the inter-tidal sections has been created arbitrarily, to indicate the approximate extent of visibility offshore from land-based counts.

Non-tidal: Areas that lie above mean high water (usually saltmarsh although some grazing marshes are also covered).

The mean count for the sector is then divided amongst a varying number of the different components, dependent on the usual habitat preferences of the species involved. For example, Dunlin dots are plotted exclusively on inter-tidal sections whereas Wigeon dots are spread across inter-tidal, sub-tidal and non-tidal areas (in proportion to the relative areas of these three components).

Currently, throughout all WeBS Low Tide Count analyses, mean low tide and mean high tide are taken from the most recent Ordnance Survey 1:25000 maps (in Scotland, the lines on the OS maps are mean low water springs and mean high water springs instead). It is recognised, unfortunately, that these maps represent the current real shape of the mudflats, water channels and saltmarshes to varying degrees of accuracy. However, in the interests of uniformity across the UK, the Ordnance Survey outlines are adhered to throughout the analyses.

The maps display the average number of birds in each count section as dots spread randomly across habitat components of count sections, thus providing an indication of both numbers and density. **It is important to note that individual dots do not represent the precise position of individual birds; dots have been assigned to habitat components proportionally and are then randomly placed within those areas. No information about the distribution of birds at a finer scale than the count sector level should be inferred from the dot density maps.** For all maps in the present report, one dot is equivalent to one bird, except where stated. The size of individual dots has no relevance other than for clarity.

As most estuaries have now been covered more than once at low tide, density maps show the relative distributions of species in the winter of 2005/06 compared to an earlier winter of survey. It is hoped that comparative dot density distributions will lead to an easier and fuller appreciation of low tide estuarine waterbird distribution, and changes therein. The following colour conventions apply to density maps: red dots = 2005/06 winter; blue dots = earlier winter; pale blue = water; yellow = inter-tidal habitat (*e.g.* mudflat, sandflat); pale green = non-tidal habitat (*e.g.* saltmarsh, reedbed); grey or brown = not covered in one survey winter; dark blue = sector never covered. More detailed information concerning analysis and presentation of WeBS Low Tide Counts can be obtained from the National Organiser (WeBS Low Tide Counts), or from the publication *Estuarine Waterbirds at Low Tide* (Musgrove *et al.* 2003)

ESTUARY ACCOUNTS

The main estuaries counted at low tide in the winter of 2005/06 are discussed, including comprehensive coverage of Morecambe Bay utilising aerial survey methods. WeBS Low Tide Counts were carried out on 24 different sites, with estuary accounts encompassing 18 of these. Other counts, usually on limited numbers of sectors, were made in the winter of 2005/06 on Adur Estuary, Burry Inlet, Duddon Estuary, Langstone Harbour, Medway Estuary and Swansea Bay. These sites are not included in the estuary accounts, but data can be obtained from the WeBS Low Tide Count National Organiser upon request.

For the main site accounts, data were collected during the period November to February. Assessment of national and international importance is based on five-year peak mean counts from the main species accounts in this volume of *Wildfowl & Wader Counts*. Figure 58. shows the location of the sites discussed, and a site description is presented for each estuary. Distribution maps are presented for selected species, which are those of international or national importance, or are known to be undergoing site-level changes, where possible. General bird distribution is described for the winter of 2005/06, focusing on species held in important numbers at the site in question.

Table 9. Sites with Estuary Accounts and important bird numbers held. Numbers in parentheses refer to the location in figure 58. For species codes see table 8.

	International Importance	National Importance
Auchencairn Bay *(16)*	YS	*None*
Belfast Lough *(18)*	BW	SU, SP, E., GN, RM, RH, BV, GG, OC, RP, PS, RK, TT
Blackwater Estuary *(5)*	DB, GP, GV, KN, DN, BW, RK	SU, WN, T., PT, SZ, CA, AV, RU, GK
Breydon Water *(3)*	PG, WN, T., SV, GP, BW, RK	BS, EW, PT, AV, RU
Chichester Harbour *(8)*	DB, DN, BW	SU, RM, LG, SZ, GP, GV, BA, CU, RK, GK
Cleddau Estuary *(12)*	*None*	WN, T., GP, GK
Lindisfarne *(2)*	PG, JI, YS, QS, WN, BA	WS, SU, PT, E., RX, SZ, GP, GV, KN, SS, DN, CU, RK
Mersey Estuary *(13)*	SU, T., DN, BW, RK	WS, CU
Montrose Basin *(1)*	PG	WS, SU, WN, E., KN, RK
Orwell Estuary *(4)*	*None*	DB, GA, PT, AV, KN, BW, RK
Morecambe Bay *(14)* *including Piel Channel Flats, Kent, Leven, Lune & Wyre Estuaries*	PG, SU, PT, OC, KN, DN, BW, BA, CU, RK	WS, WN, T., SV, E., GN, RM, RH, GG, CA, RP, GP, GV, SS, GK, TT
Stour Estuary *(4)*	BW	DB, SU, PT, AV, GV, DN, RU, RK, TT
Strangford Lough *(17)*	MS, WS, QN, SU, GP, KN, BA, RK	WN, T., MA, PT, SV, E., GN, RM, RH, BV, GG, CO, OC, RP, GV, L., DN, BW, CU, GK

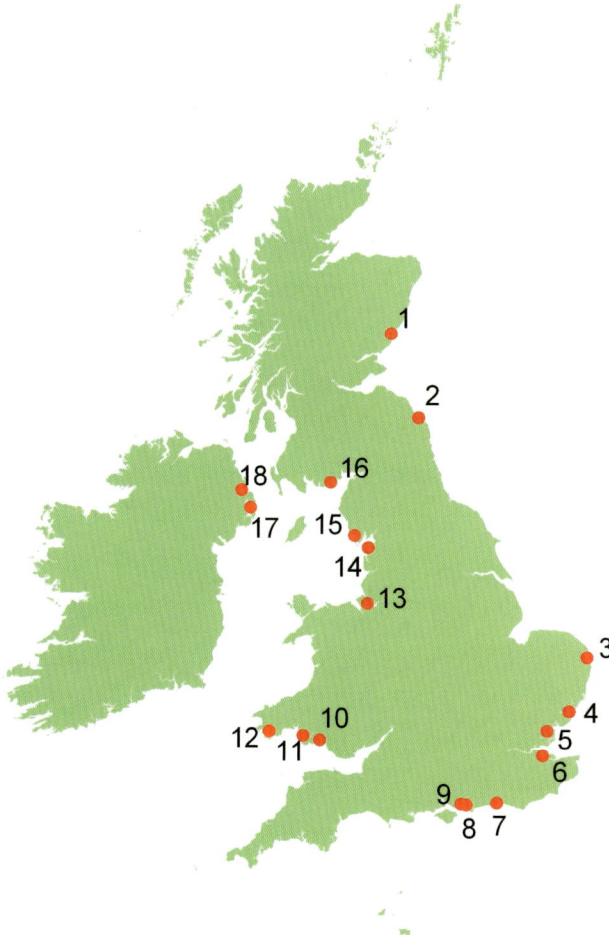

Figure 58. Map showing estuaries covered at low tide in the winter of 2005/06. 1: Montrose Basin; 2: Lindisfarne; 3: Breydon Water; 4: Stour & Orwell Estuaries; 5: Blackwater Estuary; 6: Medway Estuary; 7: Adur Estuary; 8: Chichester Harbour; 9: Langstone Harbour; 10: Swansea Bay; 11: Burry Inlet; 12: Cleddau Estuary; 13: Mersey Estuary; 14: Morecambe Bay (comprising aerial survey plus South Walney & Piel Channel Flats, Kent, Leven, Lune and Wyre Estuaries); 15: Duddon Estuary; 16: Auchencairn Bay; 17: Strangford Lough; 18: Belfast Lough.

Table 10. Peak and mean counts, and mean density (birds per ha) of 18 waterbird species across 18 estuaries covered by the 2005/06 WeBS Low Tide Counts. Orwell Estuary displayed by Stour Estuary. "+" indicates non-zero densities of <0.01 birds per ha.

Species	Adur Estuary Peak No.	Mean No.	Mean Dns.	Auchencairn Bay Peak No.	Mean No.	Mean Dns.	Belfast Lough Peak No.	Mean No.	Mean Dns.
Brent Goose	2	1	0.01	0	0	0	91	49	0.11
Shelduck	0	0	0	75	49	0.12	347	242	0.53
Wigeon	6	3	0.04	1,000	522	1.26	262	185	0.41
Teal	111	95	1.28	2	1	+	573	431	0.95
Mallard	19	10	0.14	17	10	0.02	248	236	0.52
Pintail	0	0	0	0	0	0	0	0	0
Oystercatcher	6	4	0.05	530	408	0.99	4,546	3,935	8.63
Ringed Plover	73	44	0.59	11	24	0.03	121	168	0.27
Golden Plover	0	0	0	101	33	0.08	177	59	0.13
Grey Plover	11	7	0.09	0	0	0	0	0	0
Lapwing	1,047	633	8.55	552	138	0.33	1,710	912	2
Knot	0	0	0	0	0	0	109	44	0.1
Dunlin	256	172	2.32	2	1	+	908	583	1.28
Black-tailed Godwit	1	0	+	0	0	0	503	371	0.81
Bar-tailed Godwit	0	0	0	0	0	0	123	89	0.2
Curlew	0	0	0	98	55	0.13	494	469	1.03
Redshank	43	35	0.47	56	41	0.1	1,529	1,365	2.99
Turnstone	42	20	0.27	0	0	0	286	253	0.55

Species	Blackwater Estuary Peak No.	Mean No.	Mean Dns.	Breydon Water Peak No.	Mean No.	Mean Dns.	Chichester Harbour Peak No.	Mean No.	Mean Dns.
Brent Goose	3,759	2,569	1.09	0	0	0	6,510	3,812	1.52
Shelduck	1,740	1,824	0.77	283	155	0.39	567	406	0.16
Wigeon	1,976	2,323	0.99	12,245	7,187	17.88	1,665	1,246	0.5
Teal	2,664	3,332	1.41	1,090	534	1.33	669	544	0.22
Mallard	79	80	0.03	278	179	0.45	64	51	0.02
Pintail	294	297	0.13	90	55	0.14	147	69	0.03
Oystercatcher	351	529	0.22	27	19	0.05	772	665	0.26
Ringed Plover	91	75	0.04	13	20	0.03	64	81	0.03
Golden Plover	8,783	7,390	3.14	14,300	11,251	27.99	1,910	1,253	0.5
Grey Plover	687	943	0.4	45	26	0.06	1,227	581	0.23
Lapwing	2,779	2,406	1.02	17,175	9,168	22.81	2,067	1,663	0.66
Knot	4,199	3,907	1.66	430	185	0.46	762	523	0.21
Dunlin	10,764	16,764	7.11	8,072	5,412	13.46	11,265	10,224	4.07
Black-tailed Godwit	624	720	0.31	1,298	1,055	2.62	472	434	0.17
Bar-tailed Godwit	163	142	0.06	0	0	0	463	328	0.13
Curlew	662	681	0.29	564	334	0.83	746	507	0.2
Redshank	2,131	2,627	1.11	1,663	1,400	3.48	751	664	0.26
Turnstone	103	168	0.07	5	2	+	117	82	0.03

Species	Cleddau Estuary Peak No.	Mean No.	Mean Dns.	Kent Estuary Peak No.	Mean No.	Mean Dns.	Leven Estuary Peak No.	Mean No.	Mean Dns.
Brent Goose	7	4	+	0	0	0	0	0	0
Shelduck	721	607	0.46	215	329	0.19	191	84	0.06
Wigeon	2,724	2,239	1.69	157	414	0.14	537	234	0.17
Teal	1,782	1,510	1.14	86	243	0.08	77	42	0.03
Mallard	186	177	0.13	240	506	0.22	247	153	0.11
Pintail	1	0	+	497	1,080	0.45	24	8	0.01
Oystercatcher	442	300	0.23	401	451	0.36	322	222	0.17
Ringed Plover	42	37	0.03	11	22	0.01	3	11	+
Golden Plover	2,251	1,150	0.87	5	20	+	66	37	0.03
Grey Plover	12	11	0.01	0	0	0	3	1	+
Lapwing	4,531	2,997	2.27	19	40	0.02	348	148	0.11
Knot	1	0	+	30	80	0.03	173	58	0.04
Dunlin	2,736	2,409	1.82	1,524	4,422	1.38	308	221	0.16
Black-tailed Godwit	1	0	+	0	0	0	3	1	+
Bar-tailed Godwit	3	1	+	0	0	0	2	1	+
Curlew	1,092	986	0.75	440	628	0.4	243	118	0.09
Redshank	617	553	0.42	252	630	0.23	398	284	0.21
Turnstone	48	30	0.02	0	0	0	2	1	+

Table 10. continued

Species	Lindisfarne			Lune Estuary			Mersey Estuary		
	Peak No.	Mean No.	Mean Dns.	Peak No.	Mean No.	Mean Dns.	Peak No.	Mean No.	Mean Dns.
Brent Goose	2,203	2,231	0.76	0	0	0	0	0	0
Shelduck	1,180	1,315	0.45	158	125	0.12	4,044	2,558	0.75
Wigeon	1,708	1,620	0.55	1,499	1,063	1.03	337	172	0.05
Teal	1,311	1,007	0.34	99	72	0.07	9,200	3,297	0.97
Mallard	494	450	0.15	368	221	0.21	273	231	0.07
Pintail	271	173	0.06	0	0	0	200	112	0.03
Oystercatcher	858	836	0.28	302	293	0.28	360	335	0.1
Ringed Plover	53	58	0.02	1	5	+	54	72	0.02
Golden Plover	1,255	880	0.3	2,600	1,280	1.23	1,500	1,290	0.38
Grey Plover	393	358	0.12	1	0	+	597	196	0.06
Lapwing	2,913	2,126	0.72	5,805	5,325	5.14	10,098	7,904	2.32
Knot	1,010	854	0.29	14,006	3,582	3.45	40	28	0.01
Dunlin	1,847	1,745	0.59	48	21	0.02	34,731	23,535	6.92
Black-tailed Godwit	4	4	+	0	0	0	312	267	0.08
Bar-tailed Godwit	1,787	1,223	0.41	738	288	0.28	2	2	+
Curlew	1,520	1,206	0.41	268	157	0.15	931	761	0.22
Redshank	853	692	0.23	568	334	0.32	2,283	1,549	0.46
Turnstone	61	62	0.02	4	2	+	414	167	0.05

Species	Montrose Basin			S. Walney & Piel Channel Flats			Orwell Estuary		
	Peak No.	Mean No.	Mean Dns.	Peak No.	Mean No.	Mean Dns.	Peak No.	Mean No.	Mean Dns.
Brent Goose	22	5	0.01	29	27	0.01	1,357	920	0.75
Shelduck	1,239	900	1.12	836	833	0.37	659	450	0.37
Wigeon	4,849	3,551	4.43	260	337	0.15	1,642	1,523	1.24
Teal	263	109	0.14	6	2	+	949	578	0.47
Mallard	241	175	0.22	43	40	0.02	503	395	0.32
Pintail	112	74	0.09	4	1	+	308	249	0.2
Oystercatcher	1,140	882	1.1	5,117	5,146	2.31	1,490	1,232	1
Ringed Plover	0	0	0	44	44	0.02	162	330	0.13
Golden Plover	82	26	0.03	829	829	0.37	1,003	614	0.5
Grey Plover	0	0	0	8	8	+	268	223	0.18
Lapwing	464	219	0.27	1,249	1,369	0.61	2,438	1,512	1.23
Knot	942	537	0.67	557	293	0.13	3,569	1,691	1.38
Dunlin	20	10	0.01	1,470	721	0.32	3,468	2,878	2.35
Black-tailed Godwit	89	47	0.06	0	0	0	634	567	0.46
Bar-tailed Godwit	15	9	0.01	0	0	0	13	7	0.01
Curlew	353	196	0.24	1,185	1,102	0.49	782	708	0.58
Redshank	671	436	0.54	1,123	1,154	0.52	1,813	1,590	1.3
Turnstone	15	5	0.01	103	103	0.05	205	178	0.15

Species	Stour Estuary			Stangford Lough			Wyre Estuary		
	Peak No.	Mean No.	Mean Dns.	Peak No.	Mean No.	Mean Dns.	Peak No.	Mean No.	Mean Dns.
Brent Goose	1,308	824	0.51	5,105	3,691	0.85	0	0	0
Shelduck	1,592	1,173	0.72	4,451	3,678	0.85	321	217	0.32
Wigeon	4,530	3,558	2.19	1,467	827	0.19	304	206	0.3
Teal	1,394	926	0.57	1,497	1,272	0.29	855	293	0.43
Mallard	303	259	0.16	339	265	0.06	408	254	0.37
Pintail	259	231	0.14	643	634	0.15	15	5	0.01
Oystercatcher	1,402	1,167	0.72	5,298	6,162	1.42	1,282	786	1.16
Ringed Plover	239	354	0.15	252	261	0.06	4	11	0.01
Golden Plover	911	644	0.4	7,489	4,827	1.12	855	337	0.5
Grey Plover	1,856	1,610	0.99	249	92	0.02	0	0	0
Lapwing	6,378	3,899	2.4	5,855	4,035	0.93	2,772	2,348	3.46
Knot	7,762	7,001	4.3	8,014	5,614	1.3	230	127	0.19
Dunlin	13,678	12,470	7.66	7,669	6,484	1.5	663	555	0.82
Black-tailed Godwit	1,123	771	0.47	717	400	0.09	221	85	0.13
Bar-tailed Godwit	66	46	0.03	652	646	0.15	7	4	0.01
Curlew	835	764	0.47	1,250	1,283	0.3	363	268	0.39
Redshank	1,940	1,789	1.1	2,679	2,705	0.63	849	490	0.72
Turnstone	481	443	0.27	219	184	0.04	21	11	0.02

AUCHENCAIRN BAY

Site description

Located on the north shore of the Solway Firth, Auchencairn Bay is a small sheltered bay with a rocky shoreline, narrow shingle beach and a mixture of sediment types, with mud and sand intermixed with poorly sorted boulders and pebbles. The site is designated as a SSSI but lies just outside the Solway Firth SPA. On the western shore of Auchencairn Bay there is some minor development, with a small road that runs along the shoreline; the area is otherwise largely undeveloped and very scenic. The major human activity potentially affecting the waterbirds on this site is disturbance from shell fishing. Auchencairn Bay was counted at low water in 2005/06 for the first time under the WeBS Low Tide Count scheme.

General bird distribution 2005/06

Area covered 414 ha; Mean total birds 1,293; Mean bird density 3.1 birds per ha.

Sixteen different species were recorded at Auchencairn Bay at low water, many of which were present in low numbers. Figure 59. shows how two species used the bay. Wigeon was the most abundant wildfowl species (1.2 birds per ha) and favoured the outer bay and subtidal areas south and east of Torr Point towards Hestan Island. Conversely, Lapwing were recorded exclusively on the more sheltered inner bay west of Torr Point towards Duncraig.

Both Redshank and Ringed Plover favoured typically the more sheltered inner bay and saltmarsh areas. Oystercatcher was the most abundant wader species recorded (1.0 birds per ha) and was generally fairly widespread, though the species was more numerous towards the outer bay. Both Shelduck and Curlew were widely distributed throughout the site in moderate numbers, with Shelduck averaging 48 birds throughout the winter and Curlew averaging 54 birds.

Other species recorded at low densities included Snipe, Dunlin, Great Crested Grebe and Grey Heron; in November a party of 40 Pink-footed Geese was present in the outer bay area.

Comparative bird distribution

Auchencairn Bay was covered for the first time under the WeBS Low Tide Scheme in 2005/06; it is therefore not possible to make comparisons with distributions from other years.

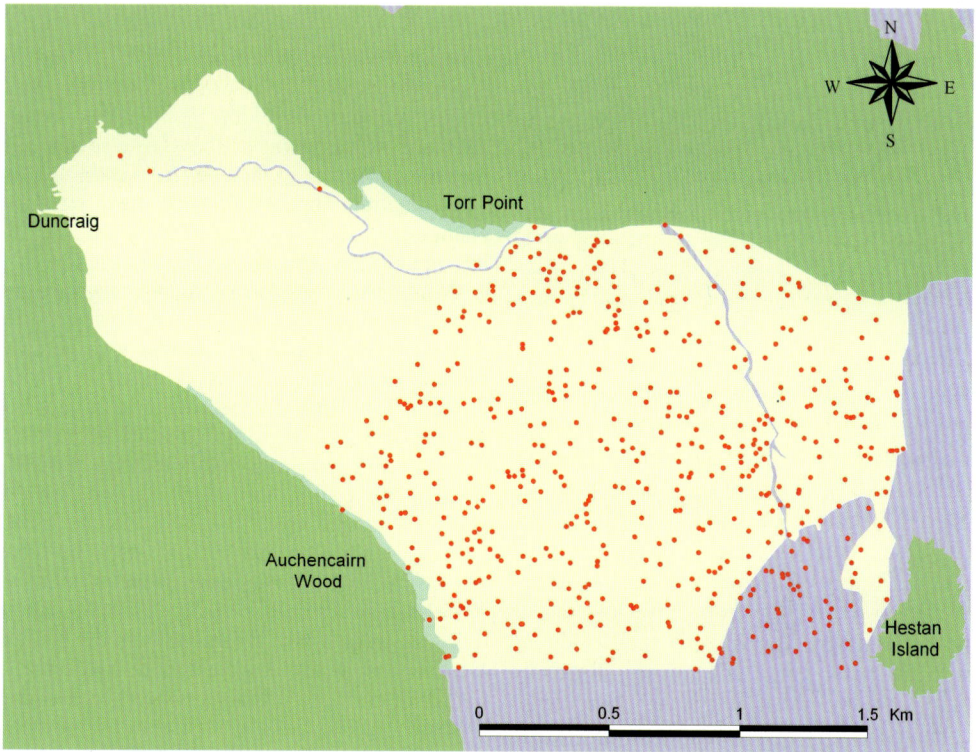

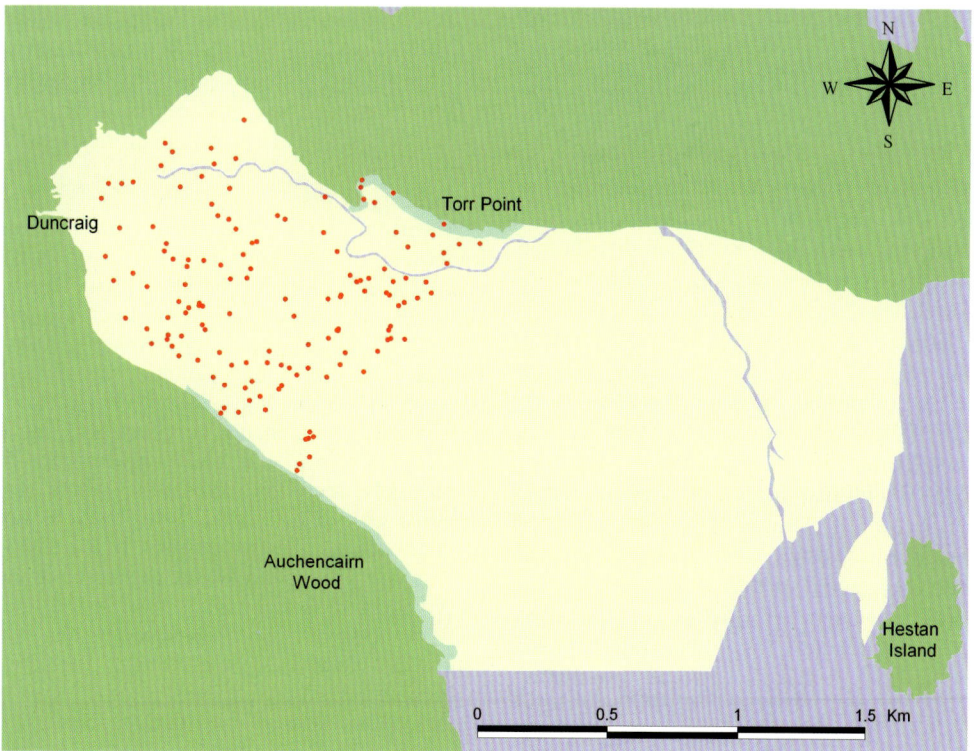

Figure 59. Low tide distribution of Wigeon (above) and Lapwing (below) at Auchencairn Bay for the winter of 2005/06. Yellow = intertidal; pale blue = subtidal; pale green = nontidal.

BELFAST LOUGH

Site description

Belfast Lough is a large sea lough in the northeast of Ireland, with the city of Belfast at its head. The area surveyed comprised the coast from Carrickfergus on the north shore around to the eastern end of Bangor on the south shore. Much of the site is afforded SPA and Ramsar status, with a further proposed SPA over open water. The outer parts of the lough's shore are generally rocky with some sandy bays, although more extensive areas of intertidal mud are found towards Belfast. Industrial land claim has reduced the area of the mudflats over the last 150 years, and Belfast has become the main port in Northern Ireland for heavy cargo. More recently, all of the area, including the important Belfast Harbour Pools, has been given a degree of protection. Extensive areas of the lough support commercial shellfisheries. There are problems of refuse disposal, pollution and general disturbance, but notably bait diggers on the north shore can pose potentially high levels of disturbance.

General bird distribution 2005/06

Area covered 456 ha; Mean total birds 12,448; Mean bird density 27.3 birds per ha.

In 2005/06, one species was found in internationally important numbers according to WeBS Core Count data – Black-tailed Godwit. This species had a fairly restricted distribution at low tide, underlining the importance of preferred feeding sites. Belfast Harbour Pools was notable in supporting very high mean densities of 17.5 birds per ha; Victoria Park Lake and intertidal areas of the south west corner of the lough were the other main centres of distribution, with a scattering north of Macedon Point marking the edge of the range. Other species of national importance included Eider, Great Crested Grebe and Scaup, all densely distributed on the west of the lough, generally south of Carrickfergus. Exposed areas of Whitehouse Lake were important for Shelduck and Redshank, as was the Belfast area in general; Victoria Park Lake held Ringed Plover and Redshank. Oystercatcher, also nationally important at the site, was densely and widely distributed.

Comparative bird distribution

As reported in *WITUK 2004/05* (Banks *et al.* 2006), declines of Goldeneye at Belfast Lough SPA are well known, as they are elsewhere in Northern Ireland (Maclean & Austin 2006). The situation appears to show few signs of changing at Belfast Lough and so the species has again been selected for discussion here.

Although Low Tide Counts may not be ideal for some wildfowl, the narrow inlet formed by the lough means that it is perhaps unlikely that the species is beyond visible range at low water. Accepting this, Figure 60. demonstrates the distribution of Goldeneye in 2005/06 and ten years previously in 1995/96. Across the lough, the mean density of the species changed from 0.2 birds per ha (itself half the density on 1994/95) to just 0.03 birds per ha in 2005/06, reflecting a mean count of 55 birds. At the finer scale, 1995/96 saw a dense aggregation of these ducks in the Belfast Harbour area (unlike the more widespread distribution of the previous winter), peaking at a density of 7.0 birds per ha and with a mean of 4.44 birds per ha. In 2005/06, the mean density had dropped to 0.22. A small congregation was recorded just to the north in the later winter, but not in comparable densities. Around the rest of the lough, the species was still widespread, albeit much more thinly than in 1995/96, but absent in the Carrickfergus area.

Turnstone is a species of national importance at Belfast Lough that has shown decline over a similar period to Goldeneye, being issued with a High Alert over the ten-year period (Maclean & Austin 2006). However, Low Tide Counts reveal that between 1995/96 and 2005/06, Turnstone density did not change appreciably, mean figures of 0.62 and 0.56 birds per ha respectively recorded across the whole site (Figure 60). Changes in distribution are evident between the two winters. In the earlier winter, many more birds were registered in the Holywood and Bangor areas than in the later winter; however, in the later winter the southwest corner of the lough, including Whitehouse Lake, were much more densely occupied. It is possible this represents a response to human activity or change in habitat use, although Turnstone are typically less susceptible to disturbance than other less adaptable waders.

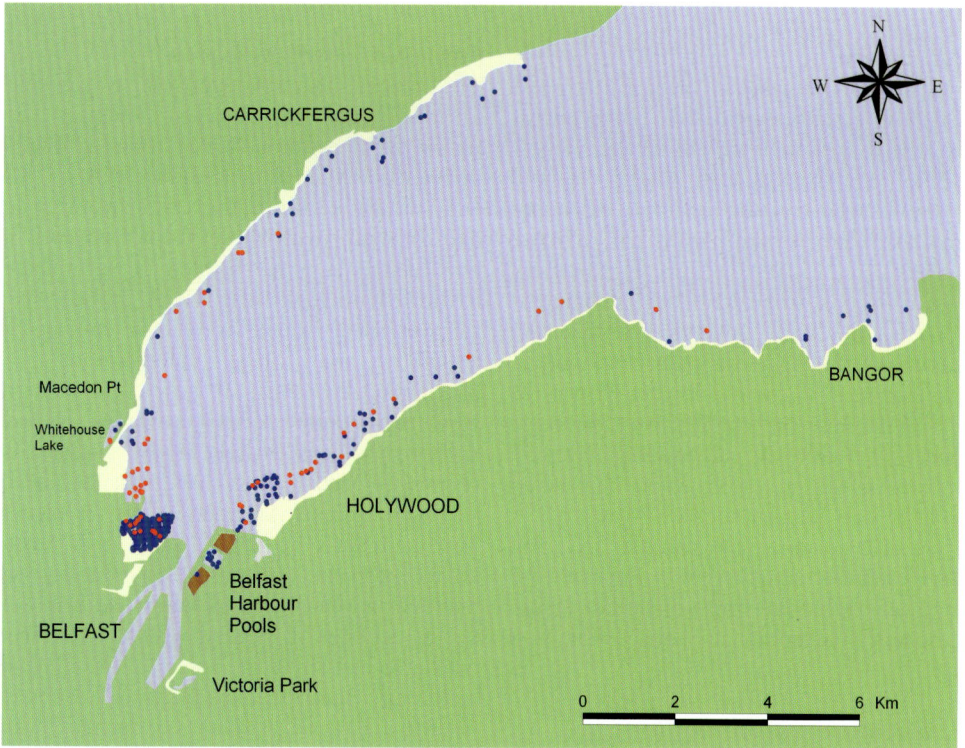

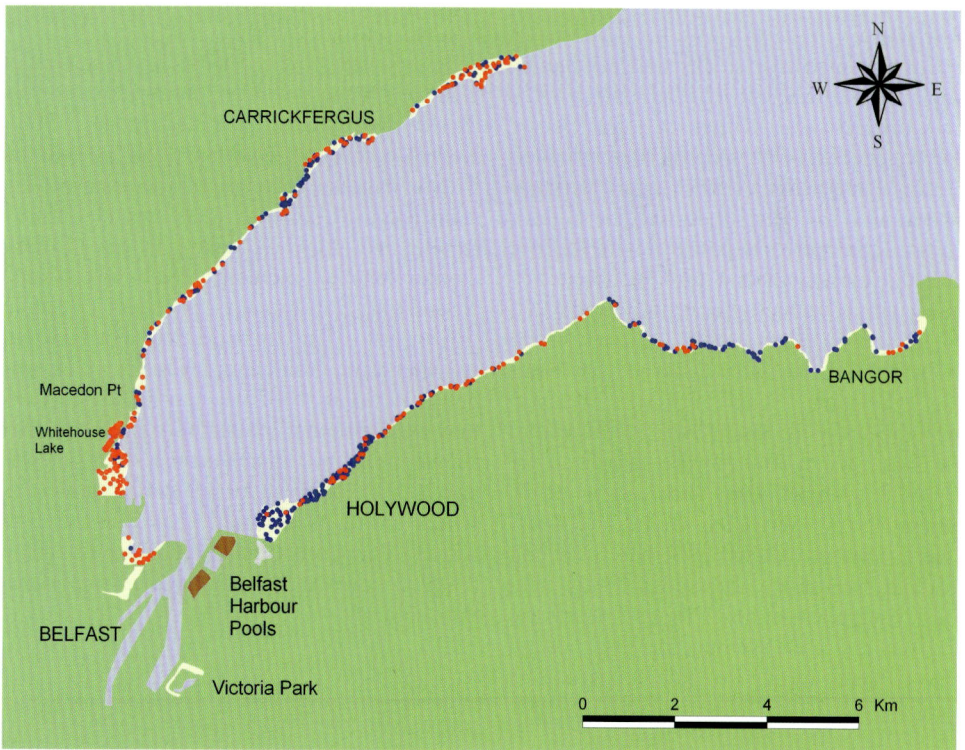

Figure 60. Low tide distribution of Goldeneye (above) and Turnstone (below) in Belfast Lough for the winters of 1995/96 (blue) and 2005/06 (red). Yellow = intertidal; pale green = non-tidal; blue = subtidal. Grey areas not covered in earlier winter; brown in later winter.

BLACKWATER ESTUARY

Site description

The Blackwater Estuary is located on the coast of Essex in eastern England. It is the largest estuary in Essex and is one of the largest estuarine complexes in East Anglia. Its mud flats are fringed by saltmarsh on the upper shores, with shingle, shell banks and offshore islands a feature of the tidal flats. The diversity of estuarine habitats results in the site being of importance for a wide range of over-wintering waterbirds, including geese, ducks and waders. The importance of the Blackwater Estuary as both a wintering and staging post for large numbers of wildfowl and wading birds is underlined by its protective legislation, being designated as a Ramsar site, a SPA and SSSI. The site also contains both national and local nature reserves. Erosion of saltmarsh is a major issue on this site whilst disturbance from bait diggers, watersports and shell fishing also occurs.

General bird distribution 2005/06

Area covered 2,379ha; Mean total birds 46,746; Mean bird density 19.6 birds per ha.
Many of the marshes, creeks and sandbanks around the perimeter of the Blackwater Estuary offer suitable habitat for some species at low water. Forty-two species were recorded using the estuary at low tide, many in large numbers, with this site supporting the highest mean total birds of sites covered with 46,746. By far the most abundant bird recorded was Dunlin, with a winter mean of over 10,000 birds, although mean counts of over 1,000 birds were recorded for Dark-bellied Brent Goose, Wigeon, Teal, Golden Plover, Lapwing, Redshank and Knot. Additionally, single figure mean counts were calculated for Greenshank, Spotted Redshank, Ruff, Common Scoter, Slavonian Grebe and Great Northern Diver. Collier's Reach, Lawling Creek and Tollesbury Fleet consistently held the largest numbers of birds, though it is worth noting that the area surrounding Osea Island was not covered by the low tide counts in this winter.

Comparative bird distribution

The change in distributions of two species Shelduck and Black-tailed Godwit, are investigated between 1994/95 and 2005/06; both are classified interests of the SPA.

It should be noted that comparisons were only possible with those sectors covered in both winters as some areas on the north shore, in particular Goldhanger Creek and the area around Osea Island, were not covered in 2005/06 and thus interpretation of distributional changes is subject to caution. However, across sectors covered in both winters, Shelduck declined by 7.3 birds per ha (Figure 61.), indicating changes at the site in step with Medium Alerts issued for the species over a similar time period (Maclean & Austin 2006). Despite increases in density of Shelduck around Collier's Reach and Bradwell Creek, declines of larger magnitude were seen on the creeks along the south of Osea Island, including Southey, Lawling and Southey Creeks. Likewise, large-scale declines were apparent on the outer estuary at Salcott Channel and Tollesbury.

In contrast to Shelduck, Black-tailed Godwit mean winter numbers on the Blackwater Estuary have undergone a 568% increase between 1994/95 and 2005/06, in line with a national increase of the species. The main areas of increase were around Collier's Reach and Lawling and Southey Creeks, though smaller increases were also visible around Old Hall Marshes and Salcott Channel (Figure 61.). This increase does not take into consideration Goldhanger Creek which was not covered in 2005/06, and so the increase may be even more dramatic. Avocets too have shown a massive increase between 1994/95 and 2005/06, with no birds recorded in the first winter and a mean count of 212 in the latter winter. The Core Count figure for Avocet in 2005/06 now exceeds the national importance threshold.

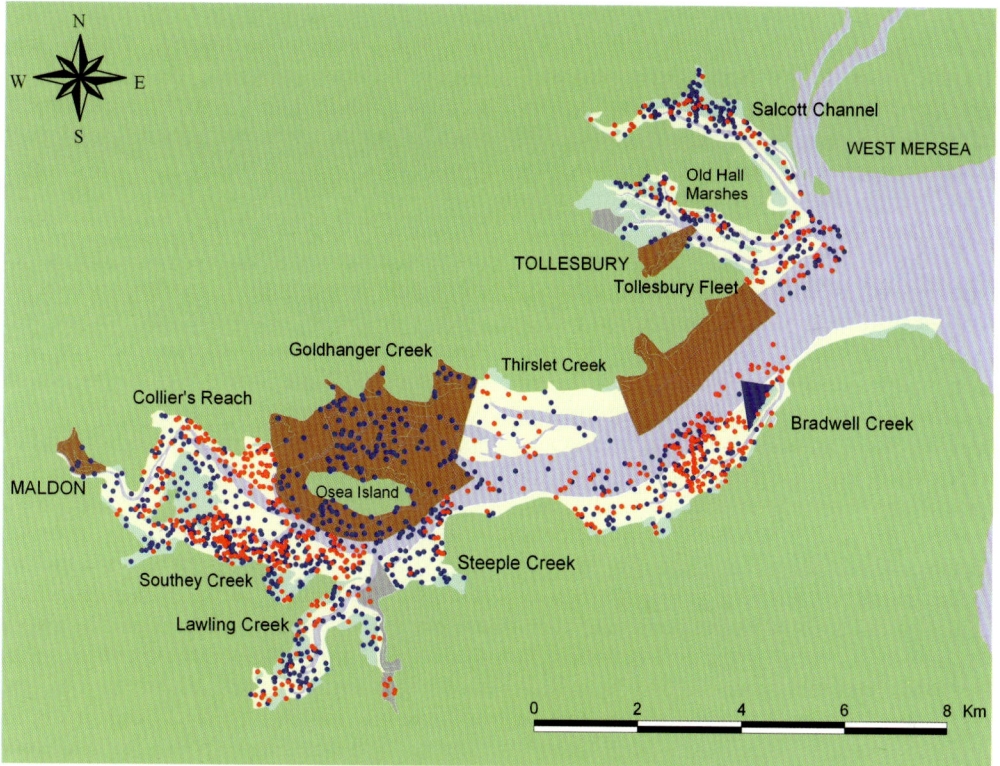

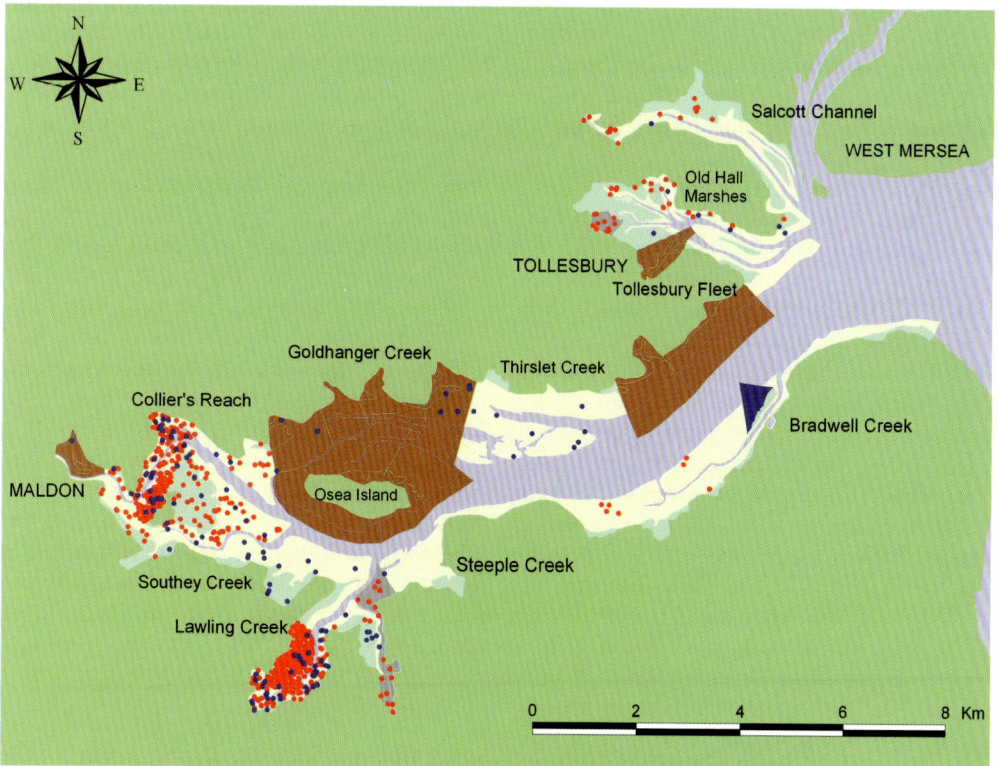

Figure 61. Low tide distribution of Shelduck (above: 1 dot = 3 birds) and Black-tailed Godwit (below) in the Blackwater Estuary for the winters of 1994/95 (blue dots) and 2005/06 (red). Yellow = intertidal; pale blue = subtidal; pale green = intertidal. Grey areas not covered in earlier winter; brown in later winter.

BREYDON WATER

Site description

Breydon Water is a bar-built estuary separated from the North Sea by the spit of land on which Great Yarmouth sits. The estuary forms the lower reaches of the Yare and Waveney rivers, which drain much of central East Anglia. The rivers are tidal for many miles inland but only the estuary area from the confluence of the rivers is considered here. At high tide, Breydon Water forms a large lake but as the tide recedes, the only water that remains forms a narrow channel, well marked by buoys for the numerous leisure cruisers. There are small areas of saltmarsh, principally at the eastern end. To the north of the estuary stretches the huge expanse of the Halvergate Levels, Breydon Marshes and Berney Marshes. These form an extensive area of grazing marsh that has been subject to varying degrees of drainage in recent years. The main high tide roosts occur at the RSPB reserve at Berney Marshes (only accessible by boat, train or a very long walk) and in the eastern saltmarsh. The site is classified as a SPA and is judged in favourable condition. The main conservation issues in the area involve boating, shooting and grazing marsh management. The river channel leading out through Great Yarmouth to the sea is highly industrialized.

General bird distribution 2005/06

Area covered 402 ha; Mean total birds 37,178; Mean bird density 92.5 birds per ha.

Breydon Water supports a large number of birds with 25 different species recorded, many in large numbers, typically holding the highest mean densities of birds of all sites included in Low Tide Counts; indeed, Breydon Water once again repeated the trend with an overall density of 92.5 birds per ha. Golden Plover were present in the highest mean numbers (11,250 birds) and densities (28 birds per ha on average across the winter), and were distributed densely from Acle Marshes to Breydon Junction. Lapwing (22.8 birds per ha), Wigeon (14.97 birds per ha) and Dunlin (13.7 birds per ha) were also present in high densities as large numbers of these species occupy a restricted feeding zone at low tide. Black-tailed Godwit averaged over 1,000 birds

and were widely distributed, though they favoured Berney Marshes, the Pump House area and Burgh Flats. Avocet, increasing at this site as elsewhere in England, congregated at Berney and Acle Marshes.

Comparative bird distribution

Regular counts at Breydon Water at low tide are received by WeBS, the earliest survey made in 1998/99. This site was issued with Alerts for only two species, neither well monitored by Low Tide Counts (Maclean & Austin 2006). Instead, the distributions of two species apparently increasing at low water, Teal and Redshank, are considered here. The winters of 2005/06 and 1998/99 are compared.

Mean site density of Teal has increased from 0.17 to 1.11 birds per ha between the two winters, meaning a mean winter count of around 500 is now possible. The reasons for this change are thought to be from increased movement on to the estuary from the surrounding grazing marshes and pools of Berney Marshes in response to increased numbers of wintering raptors on Halvergate Marshes, and also as a result of compensatory habitat installed after flood defence works in 1994/95 that now benefit wildfowl including Teal (J. Rowe *pers. comm.*). The species has visibly expanded into most of the count sectors north of the river channel (Figure 62.). In particular the area between Berney Marshes and Acle Marshes has seen mean winter Teal numbers increase from four to 209 birds at low water.

Most sectors at Breydon Water have undergone increases in Redshank density between the winters, with the exception of the intertidal area on the south bank west of Humberstone Marshes (Figure 62.). The overall change has been an increase in mean site density of 1.21 birds per ha. East of Humberstone Marshes, density has increased notably since 1998/99 when no Redshank occurred on the outermost feeding flats; to the north of the River Yare, birds are distributed more widely but still at higher density than in the earlier of the winters.

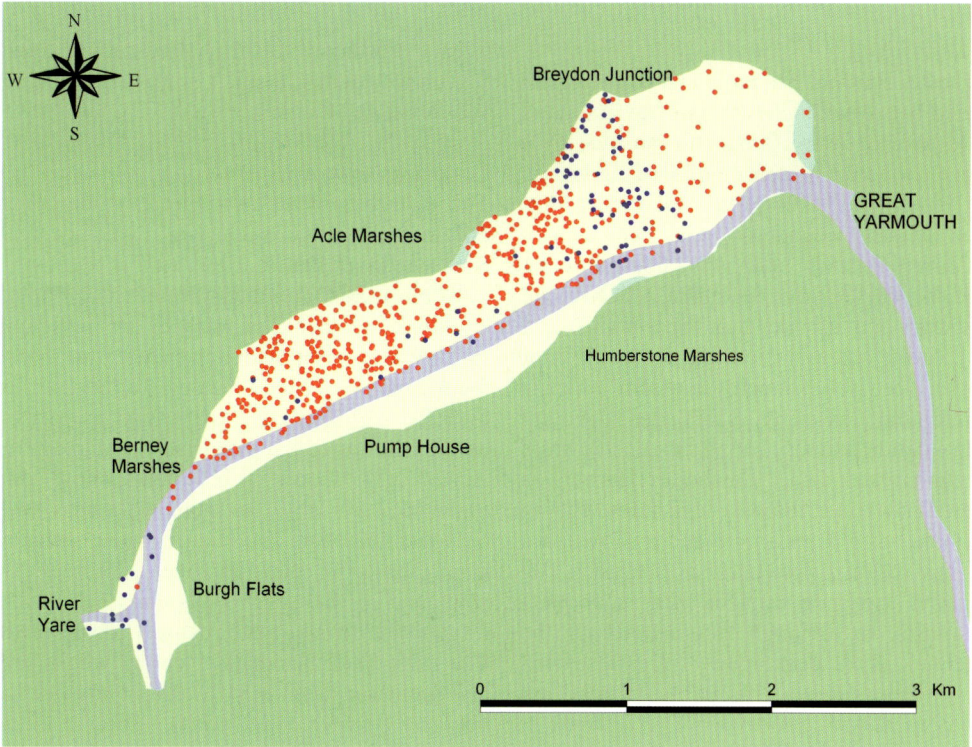

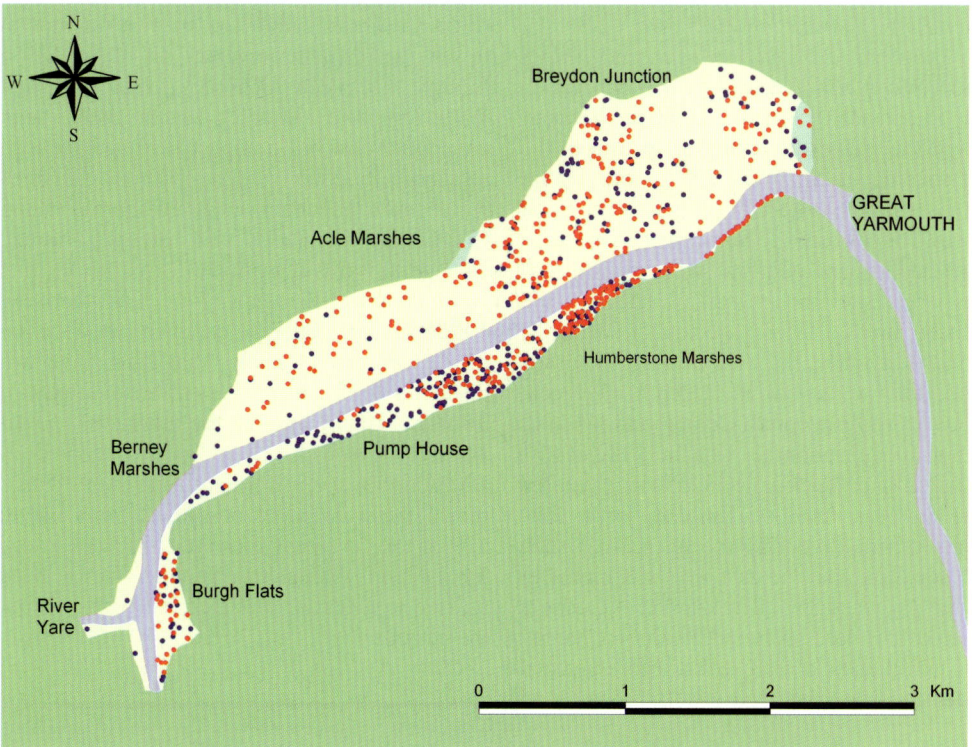

Figure 62. Low tide distribution of Teal (above) and Redshank (below: 1 dot = 3 birds) at Breydon Water for the winters of 1998/99 (blue) and 2005/06 (red). Yellow = intertidal; pale green = non-tidal; blue = subtidal.

CHICHESTER HARBOUR

Site description

Chichester Harbour is a large and complex site situated between Chichester and Havant and is linked to Langstone Harbour to the west by a channel along the north side of Hayling Island. The Harbour is a land-locked area of deep salt-water channels; bounded by mud banks which are covered twice daily by tides flowing through the narrow entrance. There are sandbanks and shingle near the entrance and much of the shore at the high-tide mark is of shingle. The river channels are muddy whereas the intertidal areas south of Thorney Island are much sandier, and also support extensive areas of eelgrass and algae. Chichester Harbour is covered by international legislation, being designated as a Ramsar site and a SPA and a also has national protection as a SSSI. The estuary is extremely popular with watersports enthusiasts, so although the majority of the shoreline is undeveloped with restricted access, those areas with public access are heavily used.

General bird distribution 2005/06

Area covered 2,515ha; Mean total birds 23,861; Mean bird density 9.5 birds per ha.

With 53 species recorded, Chichester Harbour showed the greatest diversity of species of the sites covered in 2005/06. Dunlin was by far the most numerous species, with a mean count of over 10,000 birds. Some species such as Oystercatcher and Curlew were widespread across the site though many species had preferred areas. Most species of wildfowl generally favoured the area north of Thorney Island known as Great Deep. Wigeon, however, also favoured Colner and Cutmill Creeks and Fishbourne Channel. Typically, Bar-tailed and Black-tailed Godwits favoured different areas from one another, with Bar-tailed Godwits favouring the open areas such as Pilsey Sand and the east side of Chichester Channel, whereas Black-tailed Godwits favoured the more sheltered Thorney Channel and Cutmill and Colner Creeks. More unusual species recorded included Black-throated and Great Northern Divers, Red-necked and Slavonian Grebes, Shag, Common Sandpiper, Whimbrel and Guillemot.

Comparative bird distribution

The distributions of two species undergoing different patterns of change, Shelduck and Knot, are investigated.

Core Count numbers from 2005/06 of Shelduck in Chichester Harbour are sufficient to exceed the national importance threshold, though there is a Medium Alert for this species over the period since classification as part of the Chichester & Langstone SPA (Maclean & Austin 2006). Low Tide Counts show that Shelduck are still widespread across the harbour area, but, as is immediately evident from Figure 63., there has been a decrease in abundance of the species from areas in which it was previously more densely distributed. The density of Shelduck over the whole site has decreased from 0.31 to 0.13 birds per ha between 1996/97 and 2005/06. There has not been an apparent retraction of Shelduck into restricted areas, rather the species remains at lower density throughout. In 1996/97, a triangle between Thorney Island, West Wittering and Middle Marsh held high concentrations, but numbers here had reduced markedly in 2005/06. In 2005/06, Shelduck were most abundant at Southbourne, along Chichester Channel, and along Thorney Channel towards Langstone.

In contrast, the density of Knot between 1996/97 and 2005/06 declined to a lesser degree. Figure 63. shows that Knot have two distinctly preferred areas; to the southwest of Thorney Island and at the northern end of Thorney Channel. These were occupied in 2005/06, but at a lower density than in 1996/7; mean winter averages decreased from 638 to 139 at the former. Other areas such as Pilsey Sands and Mengham Salterns have been utilised to a greater or lesser extent throughout the period.

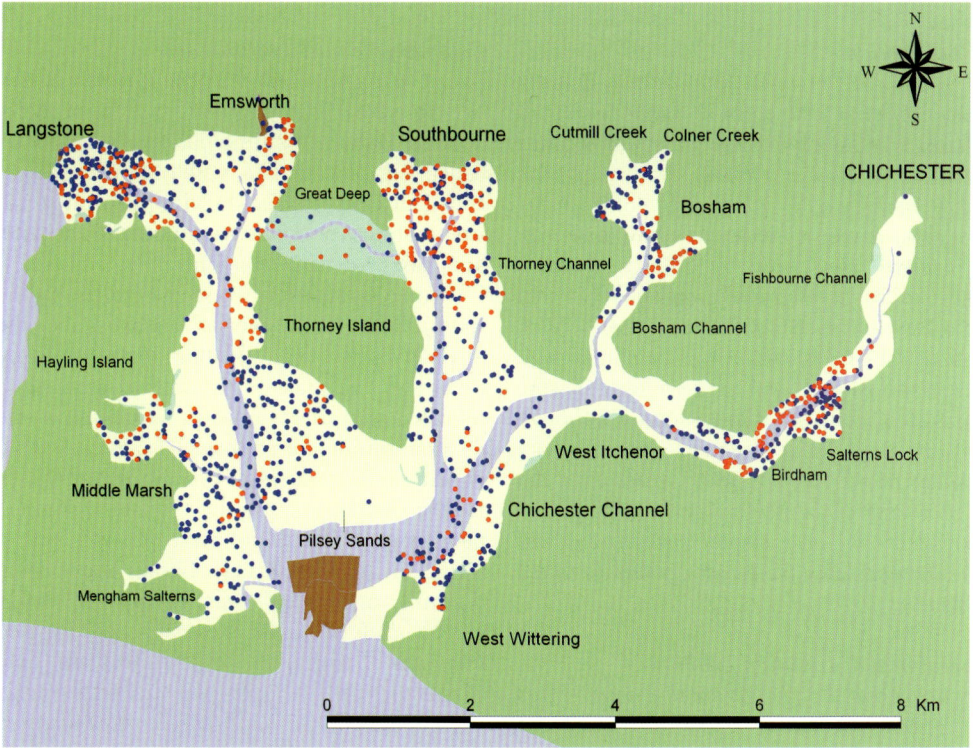

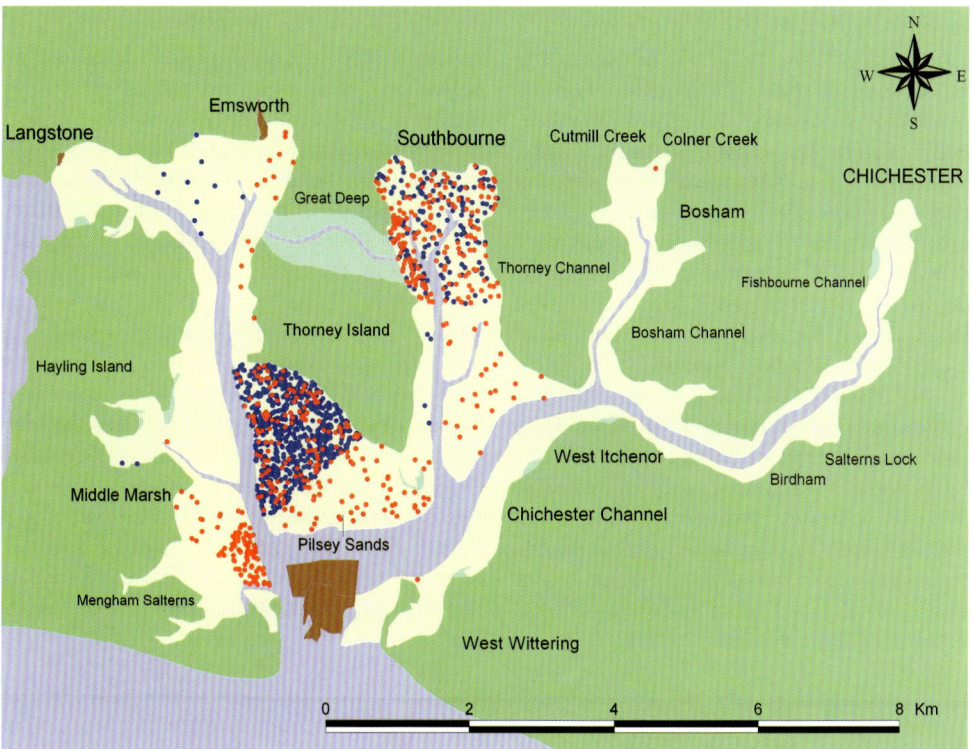

Figure 63. Low tide distribution of Shelduck (above) and Knot (below) at Chichester Harbour for the winters of 1996/97 (blue dots) and 2005/06 (red). Yellow = intertidal; pale blue = subtidal; pale green = intertidal. Brown areas not counted in later winter.

CLEDDAU ESTUARY

Site description

The WeBS site known as the Cleddau Estuary is the collective name for a series of small estuaries all opening into the waters of Milford Haven in southwest Wales. Although the estuary as a whole is not protected under legislation, some of the areas which make up the estuary are, with the Gann Estuary, Pembroke River and Pwllcrochan Flats and Angle Bay all being notified as SSSIs. Also, much of the site is within the Pembrokeshire Coast National Park. All but four sectors were covered for WeBS Low Tide Counts, although many of the intervening sections of rocky coast are not covered. Eleven years after the *Sea Empress* oil spillage, the area is still a major centre for oil transport and refining and this still represents the largest potential threat to this site.

General bird distribution 2005/06

Area covered 1,323 ha; Mean total birds 13,583; Mean bird density 10.3 birds per ha.

The Cleddau Estuary lacks the wide expanses of intertidal habitat associated with many other estuaries, and this is reflected in where the birds are concentrated. Distribution of birds on the Cleddau Estuary is fairly widespread, though Pembroke River, Angle Bay, Fowborough Point, Carew River and Gann Estuary are generally favoured areas by many species of birds, these being the more extensive areas of intertidal habitat. Thirty-eight species were recorded, with the most abundant being Lapwing, though counts of Golden Plover, Dunlin, Wigeon and Curlew were also above 1,000 birds. Unusually for a low tide count, good numbers of Snipe were recorded with a peak of 86 in December; this being a species that is often overlooked due to their secretive nature. Another notable count was of 42 Greenshank in January, with double figure counts also recorded in the other winter months. Finally, a Common Sandpiper was seen throughout the winter.

Comparative bird distribution

The Cleddau Estuary was last covered at low tide for WeBS in 1997/98, a gap of eight winters. During this time period, most species underwent little change, with only the sparsely abundant Grey Plover showing major decline. Distributions of Little Egret and Redshank, both having expanded, are investigated here.

The colonisation of southern Britain by the Little Egret is now well established, and the distinctive bird is a familiar sight on many estuaries. The scale of the rate of increase is reflected well in Figure 64., which shows the distribution of the species on the Cleddau Estuary in the winters of 1997/98 and 2005/06. In the earlier of the two winters, limited numbers of Little Egret were present, with a mean site count of nine birds. These were typically distributed evenly between creeks and channels of the outer estuary; only isolated individuals appeared further upstream than Cosheston Pill. Seven winters later, mean winter numbers of Little Egret had increased more than threefold to 40. Increased numbers of the species occupied not only the same creeks as previously at higher densities, but also had expanded throughout the river complex, appearing nearly as far upstream as Haverfordwest.

Redshank have also increased in mean winter density across the Cleddau Estuary, from 0.29 to 0.42 birds per ha over the study period. On the outer part of the complex, Angle Bay is an area of particularly high density increase (Figure 64.). It is possible that the area was avoided for reasons connected with the *Sea Empress* oil spill in the earlier winter, as Angle Bay was heavily oiled (Armitage *et al.* 1997). On the Western Cleddau, there has been limited local movement between sectors in the Fowborough Point / Landshipping Quay area. In general, more sectors have increased in Redshank density than have decreased, leading to the overall increase in mean winter density.

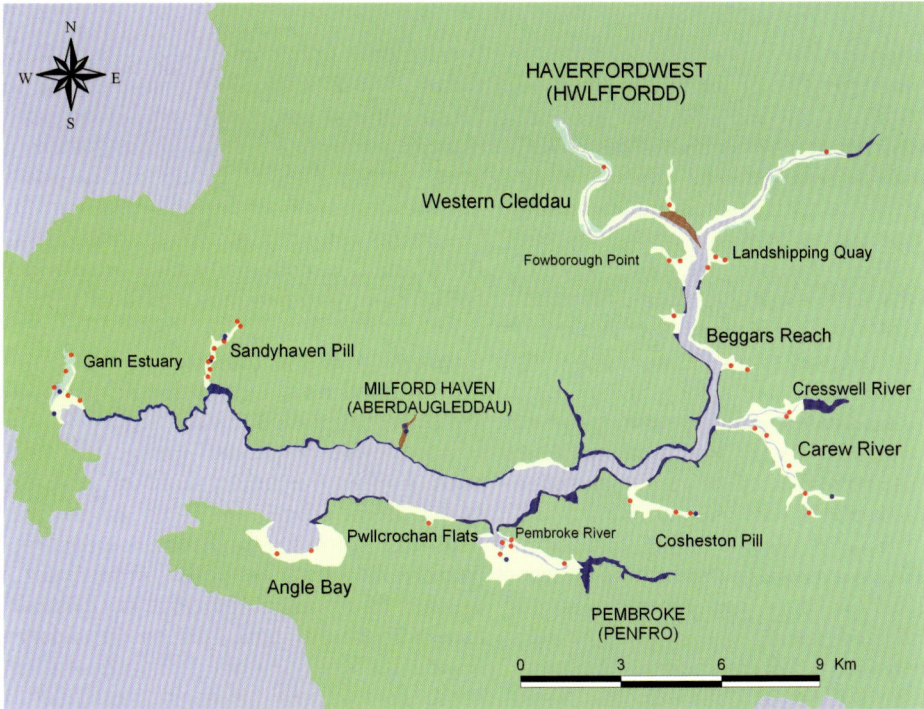

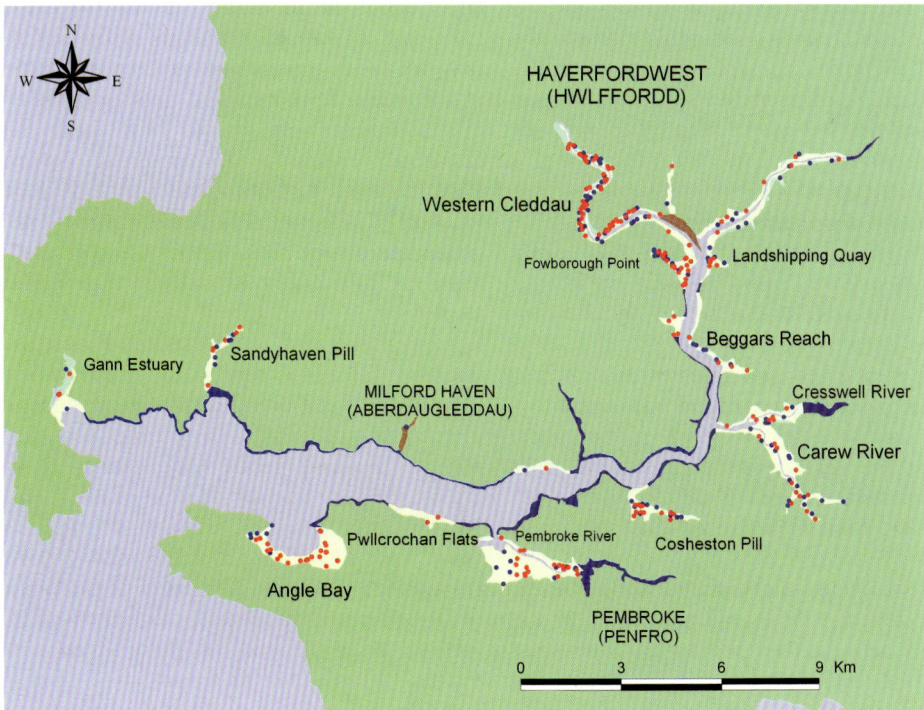

Figure 64. Low tide distribution of Little Egret (above) and Redshank (below: 1 dot = 3 birds) at the Cleddau Estuary for the winters of 1997/98 (blue dots) and 2005/06 (red). Yellow = intertidal; pale blue = subtidal; pale green = intertidal. Blue areas never surveyed. Grey areas not covered in earlier winter; brown in later winter.

LINDISFARNE

Site description

Lindisfarne forms one of the largest intertidal areas in northeast England. This site, as one of only two barrier beach systems within the UK, has an unusual structure. The majority of the site is sandy, although there are increasing amounts of silt in parts of Budle Bay and Fenham Flats. Several freshwater creeks traverse the flats at low tide. Saltmarsh exists between Goswick and Fenham, especially around the causeway to Holy Island, and along the southwestern shore of Budle Bay. Extensive sand dunes occur on several parts of the site, with dune slacks, dune heath and dune pasture also represented. The eastern shoreline of Holy Island is mainly rocky, with a few patches of shingle. There is a small harbour on Holy Island but no other industry is present. Recreational activities are generally water-based and occur mainly in Budle Bay, though beach recreation is widespread over the entire area, as are walking and birdwatching. Some grazing and hand gathering of mussels occurs, as does wildfowling, but this is strictly licensed. Wildlife conservation is in force, with the area protected by SPA and Ramsar status, and in 1997 a waterbird refuge was set up on the southern Fenham Flats.

General bird distribution 2005/06

Area covered 2,950 ha; Mean total birds 19,596; Mean bird density 6.6 birds per ha.
Counts were made in November and January only, but counters at this site still managed to record a great diversity of species, 52 different species representing the second widest array of species across sites covered this winter. Amongst the scarcer species recorded were Black-throated and Great Northern Diver, Red-necked and Slavonian Grebe, Goosander and Scaup.

Most of Lindisfarne supports some birds at low tide, but the most important areas of the site continue to be the intertidal areas between Beal Point and Ross Point (incorporating Fenham Flats and extending offshore to Holy Island), and Budle Bay. Lindisfarne is an important site for internationally important numbers of both Barnacle and Light-bellied

Brent Geese from the Svalbard populations, and also Icelandic breeding Pink-footed Geese. Fenham Flats and Budle Bay were areas especially notable for dense concentrations of Knot, Dunlin Golden Plover and Lapwing. Redshank and Oystercatchers also favoured these areas but were also concentrated around Holy Island, whilst other waders were widely present in varying densities around the site.

Comparative bird distribution

The distributions of Light-bellied Brent Goose and Grey Plover are discussed here.

As previously stated, Lindisfarne holds internationally important numbers of Light-bellied Brent Geese, and is by far the most important site in England for this population. Numbers present here are dependent on weather conditions in the other key wintering area in northern Denmark and so are subject to fluctuation between years. Numbers of Light-bellied Brent Geese present between 2001-02 and 2005-06 have, however, decreased markedly (Figure 65.), though of course this may be due to favourable weather conditions in Denmark. It is notable, however, that the distribution of birds in the two winters has apparently changed. Although the Fenham Flats area is still the favoured low water location, in 2005/06 there were also concentrations of birds further offshore to the south of Holy Island, as well as north of Beal Point.

Lindisfarne SPA includes Grey Plover as a interest feature due to international importance, yet Core Counts indicate a series of declines, sufficient to trigger Medium Alerts over a sustained period (Maclean & Austin 2006). The low tide distribution of the species has similarly changed, with an average 220 fewer individuals in 2005/06 than 2001/02 (Figure 65.). Despite differences in the timing of site coverage, it seems as though there has been a large decline in mean density on Fenham Flats. This has been only partially compensated by a more widespread distribution towards Holy Island, whilst within Budle Bay, mean Grey Plover density has altered little.

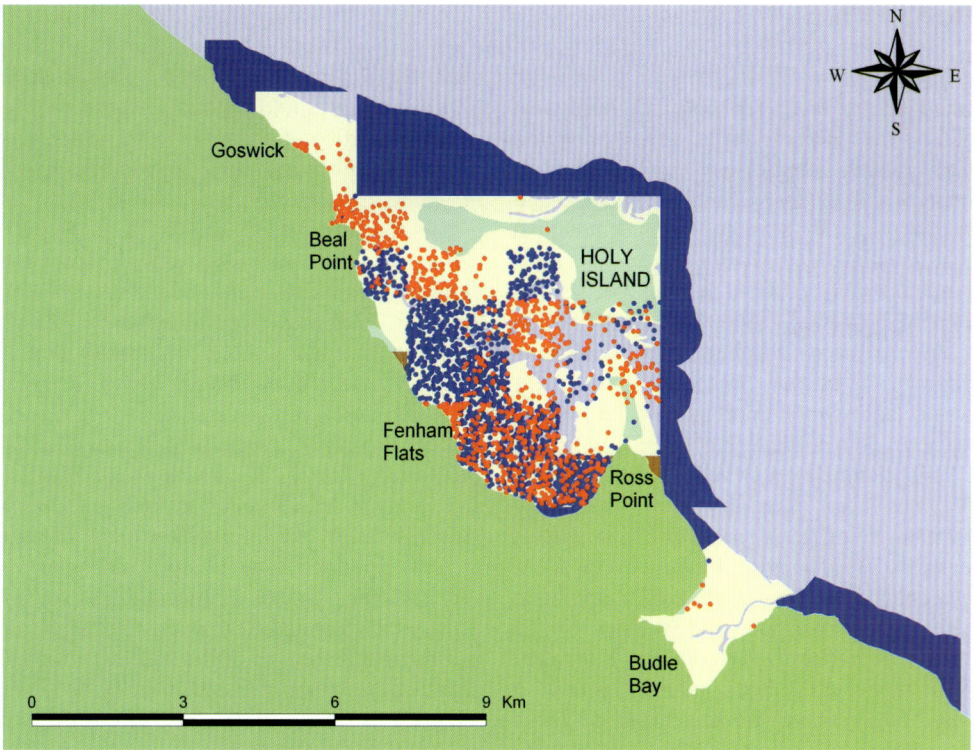

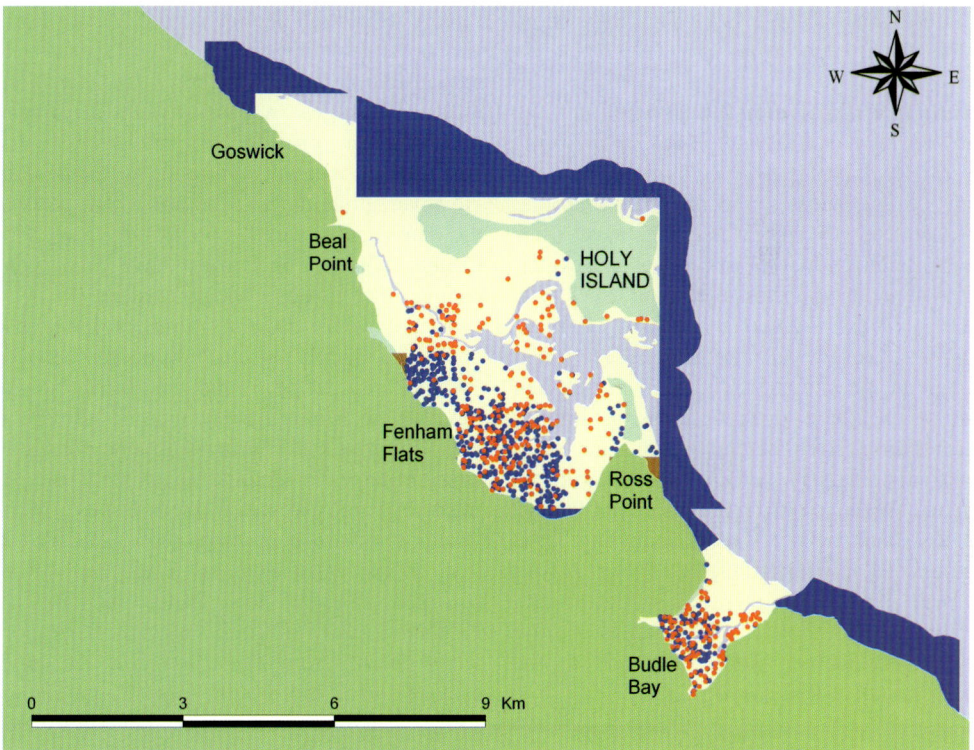

Figure 65. Low tide distribution of Light-bellied Brent Goose (above: 1 dot = 2 birds) and Grey Plover (below) at Lindisfarne for the winters of 2001/02 (blue dots) and 2005/06 (red). Yellow = intertidal; pale blue = subtidal; pale green = nontidal. Brown areas not covered in later winter; dark blue areas not covered in either winter.

MERSEY ESTUARY

Site description

Located on the Irish Sea coast of northwest England, the Mersey is a large, sheltered estuary which comprises large areas of saltmarsh and extensive intertidal sand- and mud-flats, along with limited areas of brackish marsh, reclaimed marshland, rocky shoreline and boulder clay cliffs. The Mersey has the second highest tidal range in the UK, which has created deep channels and sandbanks throughout the estuary. Since an improvement in water quality, the importance of the Mersey Estuary as both a wintering and staging post for large numbers of wildfowl and wading birds has been underlined by protective legislation, being designated as a SSSI and, as recently as 1995, as a Ramsar site and a SPA. Large conurbations on both banks dominate the site, with Liverpool and Widnes to the north and Birkenhead, Runcorn and Ellesmere Port to the south. Major industry is also a feature of the estuary with adjacent large docks and petrochemical plants; consequent pollution, plus habitat loss through expansion, are the primary concern.

General bird distribution 2005/06

Area covered 3,402 ha; Mean total birds 42,798; Mean bird density 12.6 birds / ha

Many of the marshes and sandbanks around the perimeter of the Mersey Estuary offer suitable habitat for some species at low water. Many abundant species including Shelduck, Pintail, Teal, Wigeon, Golden Plover, Lapwing, Curlew and Redshank are found throughout the site at varying density, though most species have discrete areas of highest concentration. Both Stanlow Banks and Ince Banks support many wildfowl species, such as Teal and Wigeon, plus high densities of Dunlin and Redshank. Many waders including large numbers of Lapwing and Golden Plover favour the large expanse of intertidal habitat around Weston Point in the east. Other areas of high bird density include Rock Park and at the estuary mouth at Wallasey.

Comparative bird distribution

When the Mersey was last covered at low tide for WeBS, in 1996/97, numbers of Wigeon were sufficient to exceed the national importance threshold, but this was no longer the case in 2005/06. The High Alert identified for the species at the site (Maclean *et al.* 2005) indicates a decline that may involve site factors, because the species is increasing at a national level. It is immediately evident that there has been a retraction of the species from areas in which it was previously densely distributed (Figure 66.). In both winters, Wigeon were most abundant along the south bank of the river, though densities in the earlier winter were far greater than in 2005/06. Most of the sectors counted between Eastham and Runcorn held Wigeon in high densities in 1996/97, but by the later winter many of these sectors supported few, if any, birds. On one sector of the marshes at Ince Banks, Wigeon density has declined from over 22 birds per ha to a complete absence of the species. The peak count of the sector at Eastham in 1996/97 was recorded as 6,850, the highest across the site; by 2005/06 the figure was reduced to just 60. Pintail are also in decline on the Mersey, so it is possible that common factors are responsible. Some potential explanations include changes in wastewater treatment, accretion or erosion of salt marsh and sustained movements to other wetland sites. However, it should also be noted that parts of the site not covered by WeBS Low Tide Counts (Figure 66.) may have undergone changes that have not been recorded.

Mean winter Dunlin numbers have also decreased sharply on low tide counts between 1996/97 and 2005/06, and the distribution of birds on the estuary has changed in this time as well (Figure 66.). The most notable change in distribution is a decline in density on the intertidal area west of Ince Banks. Mean winter numbers here were some 20,000 less than in the earlier winter. Some birds were recorded in new low water locations at Eastham Locks and Rock Park, but many more were absent from flats both on the inner estuary and at the mouth. Core Count figures show Dunlin numbers fluctuating but increasing (Maclean & Austin 2006), so it is to be hoped that this low tide pattern is not a harbinger of new detrimental changes.

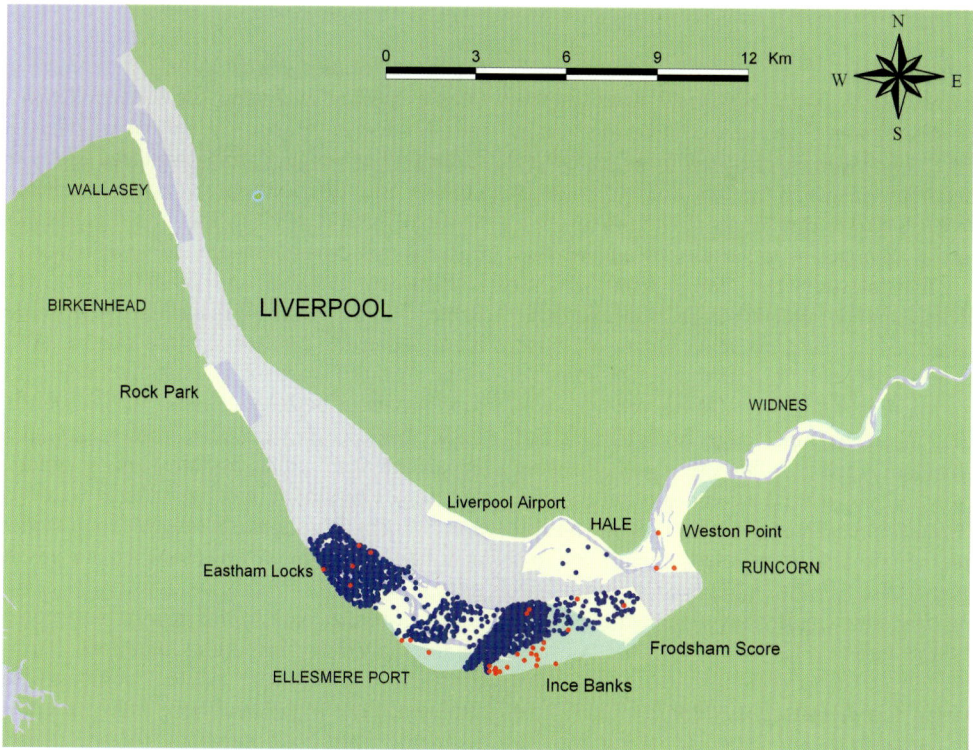

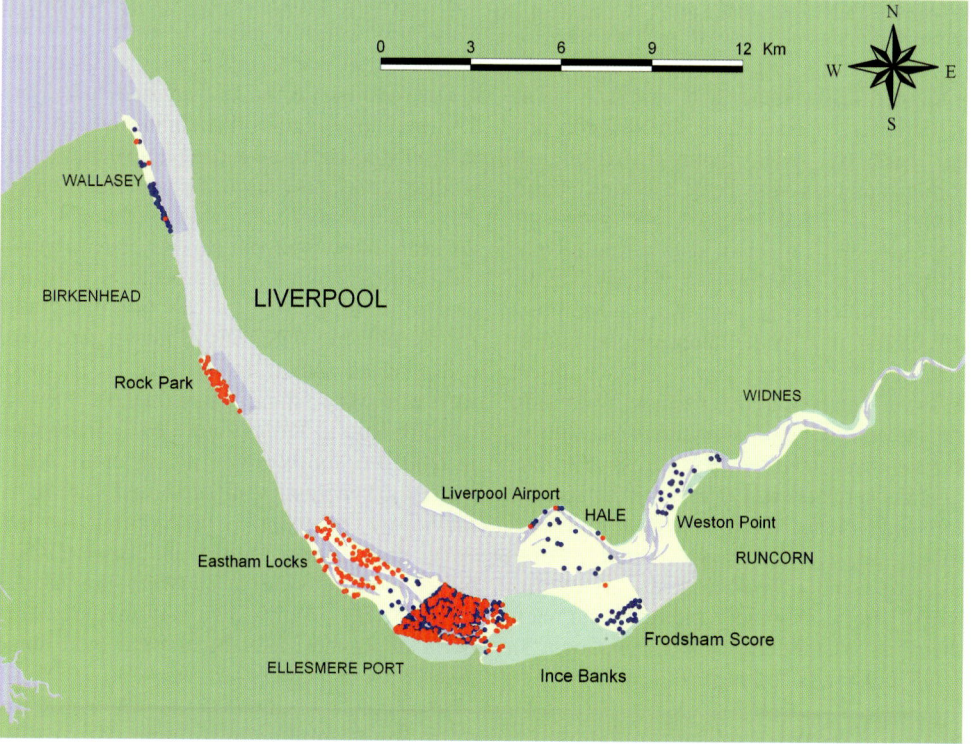

Figure 66. Low tide distribution of Wigeon (above: 1 dot = 5 birds) and Dunlin (below: 1 dot = 50 birds) at the Mersey Estuary for the winters of 1996/97 (blue dots) and 2005/06 (red). Yellow = intertidal; pale blue = subtidal; pale green = intertidal. Grey areas not counted in either winter.

MONTROSE BASIN

Site description

Montrose Basin is an enclosed estuary of the South Esk, about 3 km across covering nearly 985 ha. The basin is separated from the sea by a broad spit on which the town of Montrose is situated; the river discharges to the sea through a narrow channel at the southern end of this spit. The intertidal flats range from sand to mud and shingle and there are also extensive mussel beds. Eelgrass and algae are also present and provide a food source for some of the wintering wildfowl. There are areas of saltmarsh on the inner edge of the basin and freshwater grazing fields nearby. Pressure from wildfowling used to be heavy on this site but has been restricted since 1981 when a local nature reserve was established; this has led to a dramatic rise in the numbers of waterfowl using the site, particularly Pink-footed and Greylag Geese.

General bird distribution 2005/06

Area covered 801 ha; Mean total birds 10,001; Mean bird density 12.5 birds per ha.

Montrose Basin is a very important site for many wildfowl species, many of which occur here in nationally important numbers. The most numerous species here are Wigeon and Eider, both of which occur in nationally important numbers with peak low tide counts of 4,849 and 2,833 respectively. Although perhaps not well reflected in low tide counts, Redshank are present in internationally important numbers as are Knot and Pink-footed Geese, though the latter species mainly uses the site for roosting and so may be absent or in lower numbers at other times. With most of the site consisting of intertidal areas, wading birds are distributed widely across the site, though Oystercatchers typically preferred the eastern side of the basin, whereas Redshank favoured the south western corner. Both Goosander and Red-breasted Merganser occurred here in nationally important numbers at the time of the last Low Tide Count in 1997/98, but this is no longer the case.

Comparative bird distribution

Of the species evaluated for WeBS Alerts at the Montrose Basin, none have declined sufficiently to trigger alerts (Maclean & Austin 2006). The distributions of two species of national importance at the site, undergoing different patterns of change, are investigated.

Between 1997/98 and 2005/06, the mean winter density of Wigeon at low water increased from 2.4 to 4.2 birds per ha. As is clear from Figure 67., this increase was brought about both by changes in bird density on sectors used in the earlier winter, as well as expansion into areas not previously holding Wigeon. Prominent amongst these were the area west of Rossie Mills toward The Lurgies, and bordering the town of Montrose in the east.

By contrast, in 1997/98 Knot were present in internationally important numbers, but by 2005/06, their mean winter numbers had dropped. Although they were still present in nationally important numbers, the site density had decreased from 1.97 to 0.67 birds per ha. This low water decline, however, was not matched by Core Count figures and were not severe enough to have triggered any alerts. This perhaps suggests that either birds roosting at Montrose Basin now feed elsewhere, or that feeding areas now used are more difficult to survey accurately; the latter is thought to be the more likely by local experts, particularly as the geomorphology of the basin is known to change consistently. The low tide distribution of Knot in 1997/98 shows the birds concentrated into three discrete areas; to the west of Rossie Island, by Steinshell Burn and north of Rossie Mills (Figure 67.). In contrast, in 2005/06, birds were much more widely distributed across the site, with areas not previously frequented now being utilised. Areas such as near Tayock and Tayock Bridge in the north of the basin were not used in 1997/98; also in the previously favoured areas north of Rossie Mills, birds were more thinly distributed in the later winter.

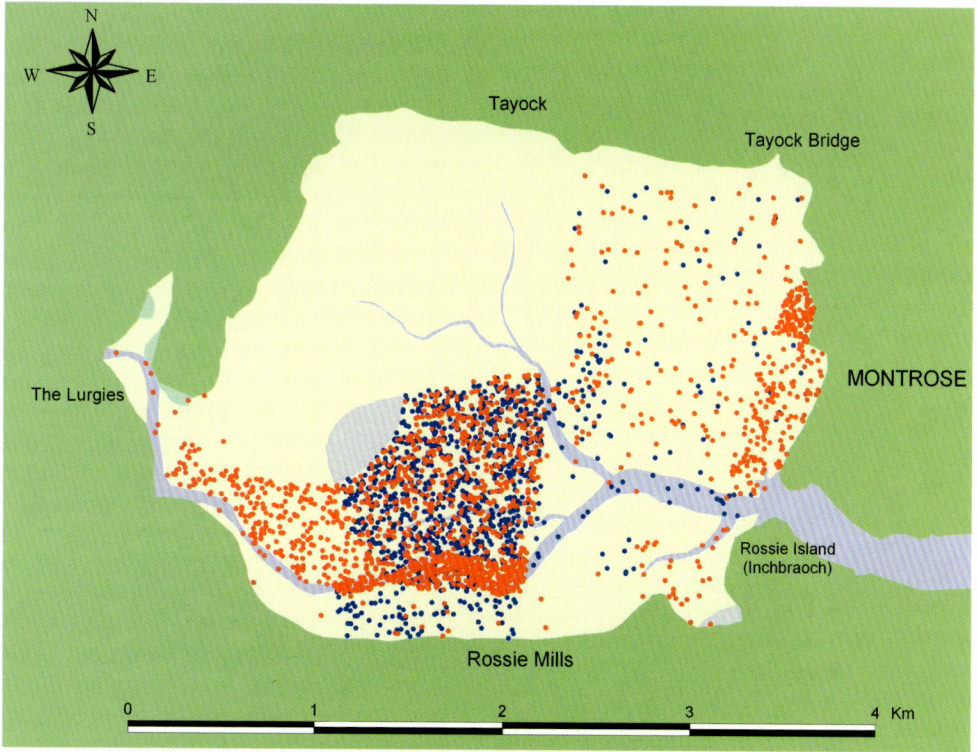

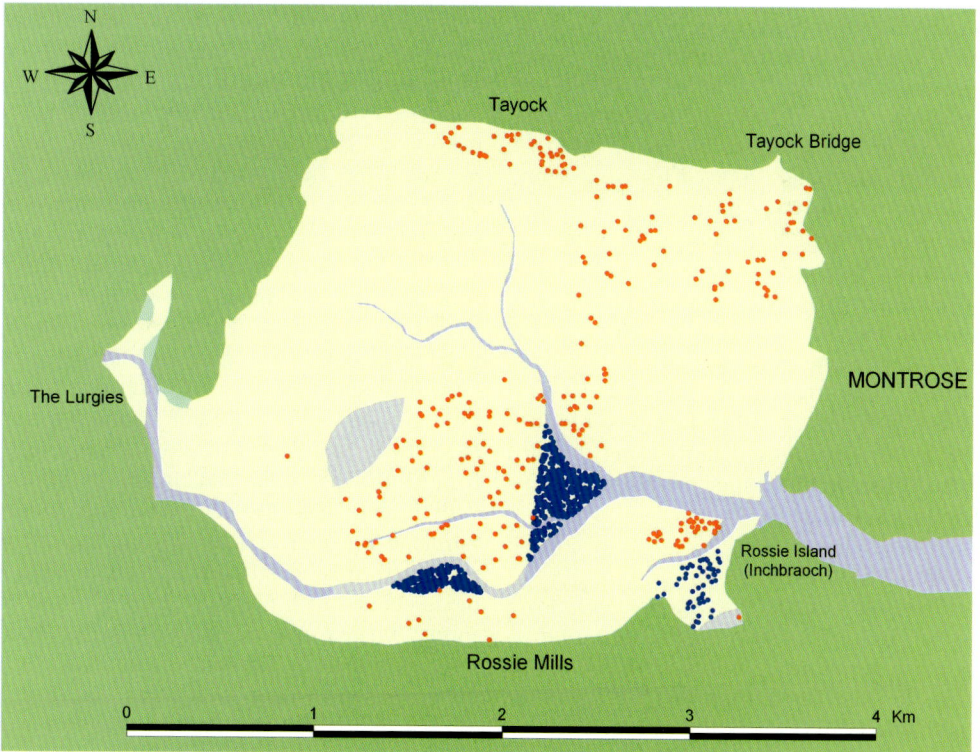

Figure 67. Low tide distribution of Wigeon (above: 1 dot = 2 birds) and Knot (below: 1 dot = 2 birds) at Montrose Basin for the winters of 1997/98 (blue dots) and 2005/06 (red). Yellow = intertidal; pale blue = subtidal; pale green = intertidal. Grey areas not counted in later winter.

MORECAMBE BAY

Site description

Situated on the north west coast of England, Morecambe Bay is a huge tidal embayment, draining the Lake District through the rivers Leven, Lune and Kent, and parts of the Bowland Fells in Lancashire through the River Wyre. In the west of the bay, Walney Island shelters an area of rich intertidal mud known as South Walney & Piel Channel Flats. The site is characterised by extensive invertebrate-rich sand and mud flats, stretching as far as 7 km offshore, with localised rocky outcrops ('skears') often providing suitable mussel *Mytilus edulis* beds. Indeed the flats represent the largest continuous area of such habitat in the UK; this factor, in addition to the huge tidal range in the bay, has made comprehensive WeBS Low Tide Counts impossible in the past, though mid-tide counts have occurred in parts of the bay. However, the site is of rare importance, supporting ten different species of international importance during the winter, as well as providing crucial breeding and passage habitat for other species of waterbirds. For this reason, Natural England funded a unique low tide aerial survey of the entire bay for the winter of 2005/06, in addition to volunteer support covering the four major estuaries feeding the bay using standard count methods (Banks 2006). This is therefore the first attempt to describe low water distribution of waterbirds across the whole of Morecambe Bay, and comparative distribution analysis is not possible. A huge combined area of 42,795 ha was surveyed. A general point about Morecambe Bay is that due to tidal patterns, birds may have finished feeding before low water. It is therefore possible that these surveys overlooked birds using the bay that had returned to roost or pre-roost areas by the lowest low tide.

i. AERIAL SURVEY

Area covered 36,400 ha; Mean total birds 20,816; Mean bird density 0.88 birds per ha.

It is obvious that using a low-level aircraft to survey this site has advantages; the figure of 36,400 ha covered dwarfs low water coverage of anywhere else in the UK. However, there are associated disadvantages, the foremost being that identification can be difficult and that numbers can be greatly underestimated.

This is particularly true of small waders such as Knot and Dunlin that congregate in tight flocks. Subsequently, waders and wildfowl were often categorised into size classes. This problem is not overly detrimental to distribution surveys, as species density is relative and 'hotspots' of distribution still emerge. However, some species, such as the distinctive Oystercatcher, were readily identifiable and recorded easily. The average winter distribution of the species is presented (Figure 68.), from which it is evident that although the species was widespread, not all of the sand and mud flats were selected in equal proportion. Areas close to the shore at Newbiggin, Warton Sands and Cockerham Sands held the highest densities of Oystercatcher; the mouth of the Kent and the outermost flats the lowest. The distribution of the species at low water superficially corresponds to knowledge of roost sites, in that highest concentrations of Oystercatchers seem to occur near to the largest roosts (R. Horner, *pers. comm.*). Although the distribution is representative, the mean winter average for Oystercatcher at low tide in the parts of Morecambe Bay surveyed from the air comprised 10,253 individuals. This figure is merely a fifth of the five-year site mean recorded on Core Counts illustrating how aerial surveys can underestimate numbers. Unidentified small waders (principally Knot and Dunlin) were on average the next most abundant species counted (a mean of 5,961 over the winter). These were patchily distributed across the intertidal habitat, with particular aggregations at Cockerham Sands, Preesall Sands and the Kent Channel toward Hest Bank. Curlew and Lapwing typically fringed the intertidal and saltmarsh sectors, with these species also detected in surrounding farmland on approach to the bay. Shelduck, another readily identifiable species, tended to be associated with intertidal areas close to the shore, whereas the equally distinctive Eider were most densely distributed around the south tip of Walney Island, with more surprisingly advanced into the bay in the channels near Warton Sands.

ii. KENT ESTUARY

Area covered 1,106 ha; Mean total birds 3,977; Mean bird density 3.6 birds per ha.

The Kent Estuary, a shallow-banked river channel in the north east of the bay, supported 21 different species at low water, a relatively low diversity of waterbirds. Three species of international importance for Morecambe Bay SPA were however present in noteworthy densities; Pintail, Oystercatcher and Dunlin. The most numerous was Dunlin, with a winter average of 1,522 birds. Figure 68. shows that these were almost exclusively concentrated at the mouth of the river, and never above the viaduct at Arnside [note that the southern extent of the distribution marks the boundary of the count sector and not necessarily the limit of Dunlin]. North of the bridge contained few waders except for some scattered Redshank and more densely distributed Curlew. These species were found in greater density near Grange-Over-Sands at the mouth of the river; these sectors were most heavily used overall, with Oystercatcher and Shelduck also exploiting the feeding flats. The latter was more thinly spread north of the Arnside viaduct. Sectors near the mouth of the river were additionally used by wildfowl such as Wigeon and Pintail, which were exclusively found in this area, owing to accreting saltmarsh at Grange-Over-Sands. Peak low tide counts of Pintail exceeded 1,000 birds, which exceeds the international importance threshold for the species. Individuals of other species, including Red-breasted Merganser and Spotted Redshank, were recorded on the estuary.

iii. LEVEN ESTUARY

Area covered 1,342 ha; Mean total birds 1,668; Mean bird density 1.2 birds per ha.

Of the riverine branches surveyed as part of Morecambe Bay in 2005/06, the Leven Estuary supported waterbirds in the lowest overall density (1.2 birds per ha), despite its large size and extensive saltmarsh and intertidal habitat. The mean winter count for the site was greatest for Redshank, whilst densities of Dunlin and Oystercatcher were identical to that of Redshank at 0.21 birds per ha. Distributions of these species were broadly similar, with least birds towards the river head and most in the middle and outermost reaches. It is the distribution of Redshank that is shown here (Figure 69.). The suite of common wildfowl found on most estuaries in the UK were present: Shelduck, Wigeon, Teal and Mallard, at varying densities. Pintail were much scarcer than on the Kent, the winter average for the Leven totalling just eight. Relatively low numbers of godwits, plovers, Knot and Curlew were registered. Lapwing and Golden Plovers may have been underrepresented, as many were seen in surrounding farmland; it is possible the estuary is also used but not during periods of observation. In general, highest bird densities were on the flats at the river mouth near Ulverston and on the east of the middle reaches.

iv. LUNE ESTUARY

Area covered 1,037 ha; Mean total birds 13,140; Mean bird density 12.7 birds per ha.

In contrast to the Leven, the Lune Estuary held the highest mean bird density of the estuarine areas surveyed in Morecambe Bay in 2005/06. For the purposes of this survey, the Lune Estuary also included some areas of foreshore at Middleton, Morecambe and Half Moon Bay not covered by aerial survey. The latter contained an estimated 14,000 Knot in December 2005, resulting in a high mean species density (6.41 birds per ha) and contributing to the high bird density of the site as a whole. Also at high density were the common winter plovers Lapwing and Golden Plover. The distribution of the former was extensive on the estuarine sectors, with the area at Conder Green especially heavily populated (Figure 69.). The latter was more restricted to the lower reaches of the river, particularly north of the channel to the east of Bazil. Oystercatcher were recorded on the coastal sectors, principally at Half Moon Bay and Morecambe where rocky skears are exposed during low tide periods. Of the wildfowl, a mean count of 130 Mute Swan was notable, the equivalent figure of Wigeon totalling over 1,000 birds. Comparatively few Shelduck appeared on the Lune relative to the other estuaries around Morecambe Bay.

v. SOUTH WALNEY & PIEL CHANNEL FLATS

Area covered 2,231 ha; Mean total birds 13,347; Mean bird density 6.0 birds per ha.

The area known as South Walney & Piel Channel Flats contains a variety of habitats, including the sheltered intertidal mudflats along Piel Channel, the more exposed flats on the west of Walney Island, the limited saltmarsh in the north towards Vickerstown, the large non-tidal waterbody of Cavendish Dock in Barrow-in-Furness and the rocky skears around Foulney Island. Subsequently a diverse array of waterbird species is typically recorded at low tide; in 2005/06, 35 species, including 14 waders, were discovered. At high densites were Tufted Duck and Coot, both restricted to Cavendish Dock but present in high mean winter numbers. Elsewhere, Oystercatcher were found at high densities on most intertidal sectors apart from South Walney and Roosecoote Sands. The latter is important for Oystercatchers and other waders, especially in cold weather (J. Sheldon, *pers. comm.*), and the lack of Oystercatchers recorded probably reflects the single survey of the sector (in November); redistribution in response to raptor disturbance is common and may have contributed to this pattern. In particular, the flats and skears around Foulney Island were strongly favoured (Figure 70.), as was the area in the north west of the channel. Roosecoote Sands did, however, support Shelduck and Redshank at relatively high densites, species which were more thinly scattered on other intertidal count sectors. Knot were limited to the area west of Piel Island and south of Roosecoote Sands, while in contrast Dunlin were thinly and widely spread across the sheltered feeding flats. Curlew shared a similar though less dense distribution to Oystercatcher, with the Foulney Island and Blackamoor Ridge areas especially important.

These locations also held large numbers of Eider in a small area, meaning this species was at the highest density recorded (7.4 birds per ha). It is likely that individuals of these species feed on the shellfish beds found here.

vi. WYRE ESTUARY

Area covered 679 ha; Mean total birds 6,359; Mean bird density 9.4 birds per ha.

The River Wyre reaches Morecambe Bay at the bisection of Fleetwood and Knott End-On-Sea, having run through extensive areas of intertidal and saltmarsh. The west bank is industrialised to an extent, though the surrounding conurbations rarely encroach on to the estuary. By contrast, the east bank is characterised by lowland agricultural habitat. This may help to explain why at a mean 3.46 birds per ha and winter average of 2,348, Lapwing represents the most abundant species recorded at low water in 2005/06. It was most densely distributed between Skippool Marsh and Little Singleton. The Skippool Marsh area was important for other species of wader too, including Dunlin and Golden Plover. Oystercatcher, on the other hand, were aggregated in sectors at the mouth of the river, whilst Curlew (Figure 70.) were present on all sectors. Their preferred areas were in the vicinity of Little Singleton, quite far upriver but amongst areas of farmland upon which they may also feed. Wildfowl recorded included Pink-footed Goose, Shelduck, Wigeon, Teal and Mallard, all at similar densities (0.26 – 0.38 birds per ha).

Aircraft of type used for aerial survey of Morecambe Bay (Alex Banks)

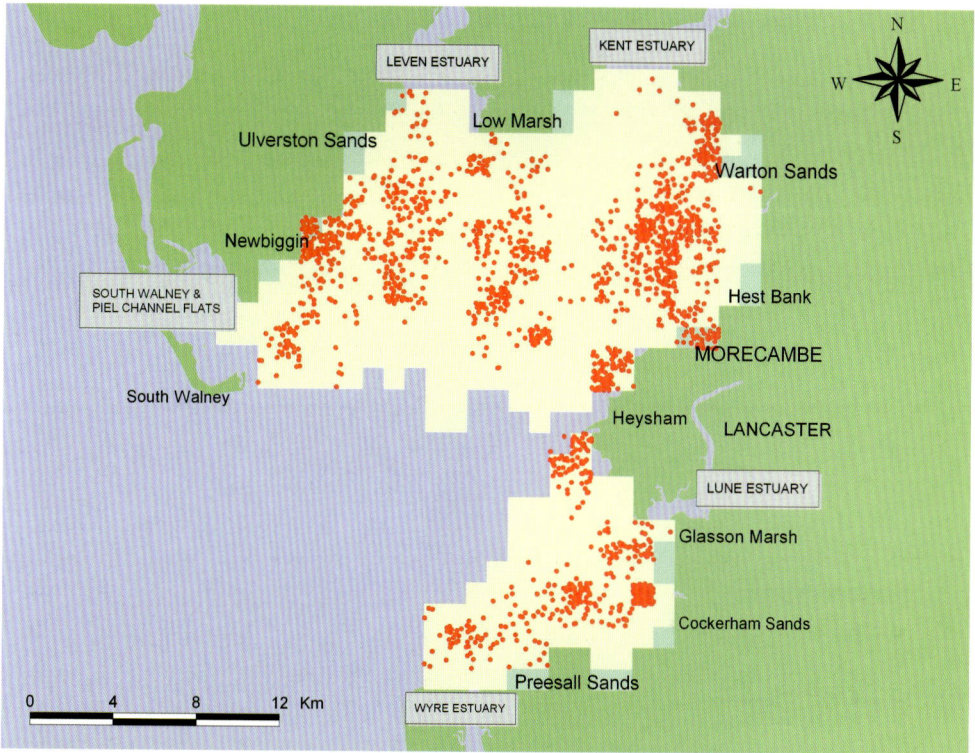

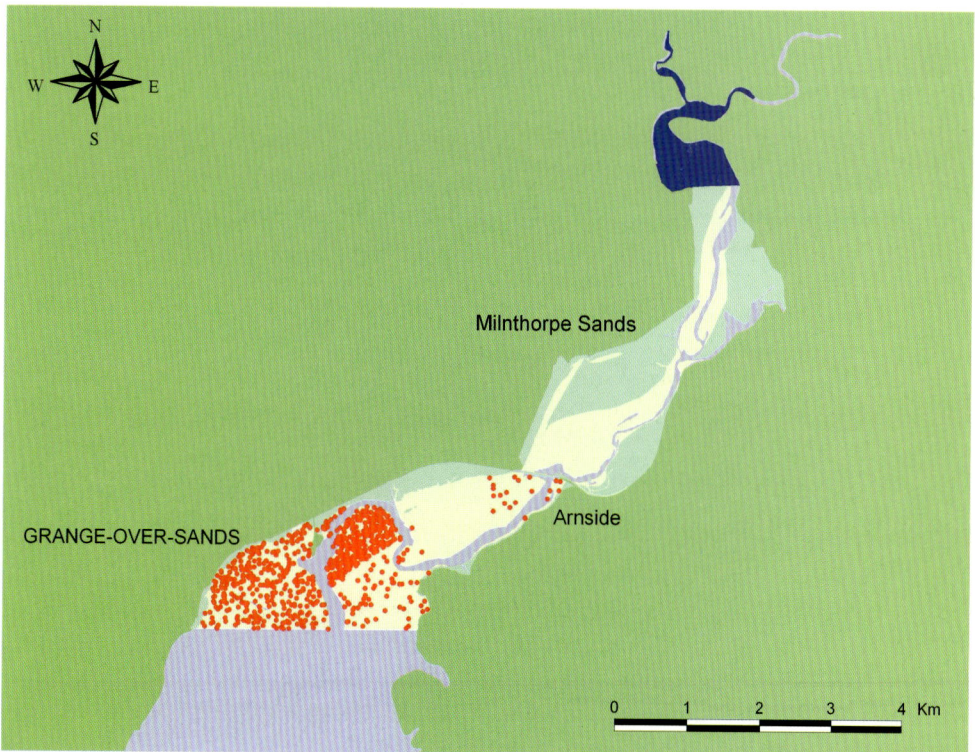

Figure 68. Low tide distribution of Oystercatcher (above: 1 dot = 5 birds) on aerial survey of Morecambe Bay in 2005/06, and Dunlin on standard survey of the Kent Estuary (below: 1 dot = 2 birds). Yellow = intertidal; pale blue = subtidal; pale green = intertidal. Dark blue area not covered.

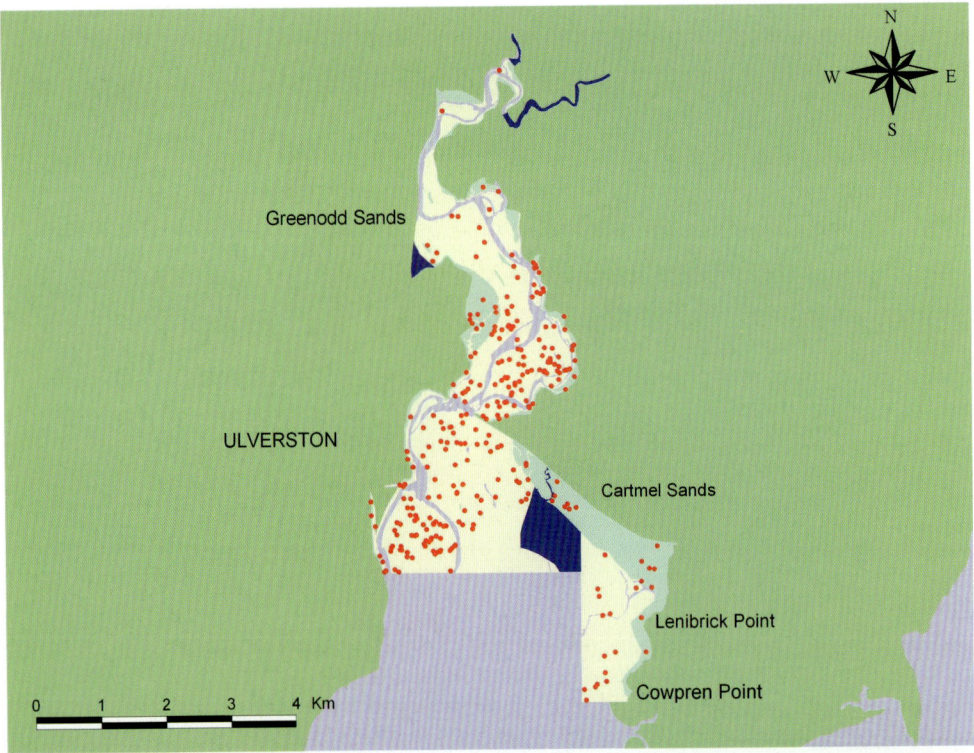

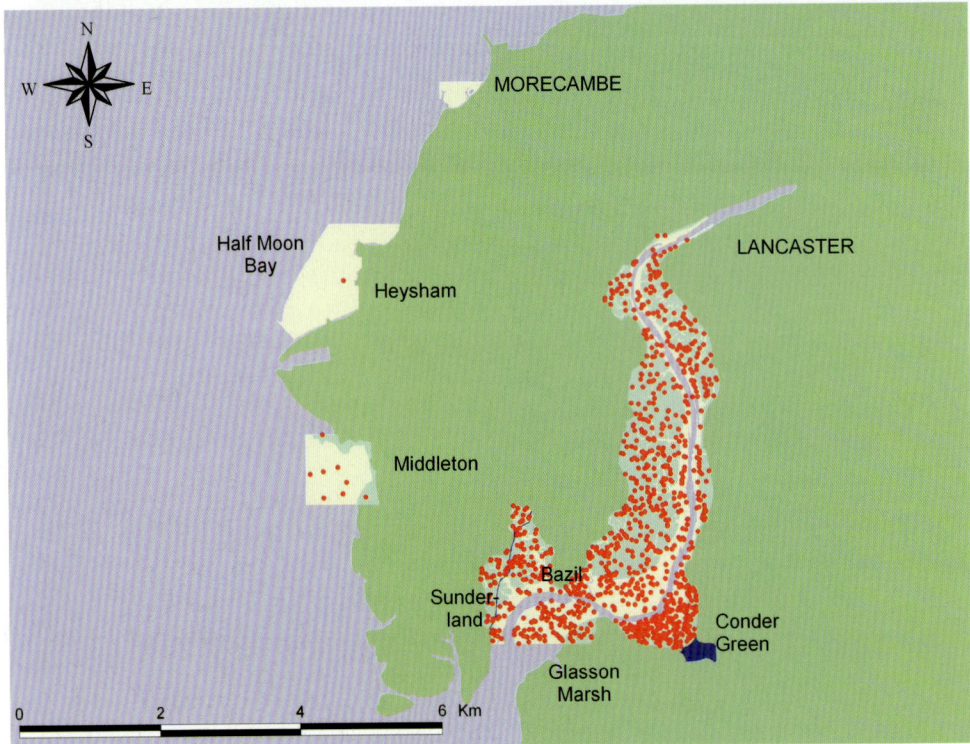

Figure 69. Low tide distribution of Redshank (above) on standard survey of the Leven Estuary in 2005/06, and Lapwing (below: 1 dot = 5 birds) on similar survey of the Lune Estuary. Yellow = intertidal; pale blue = subtidal; pale green = intertidal. Dark blue areas not covered.

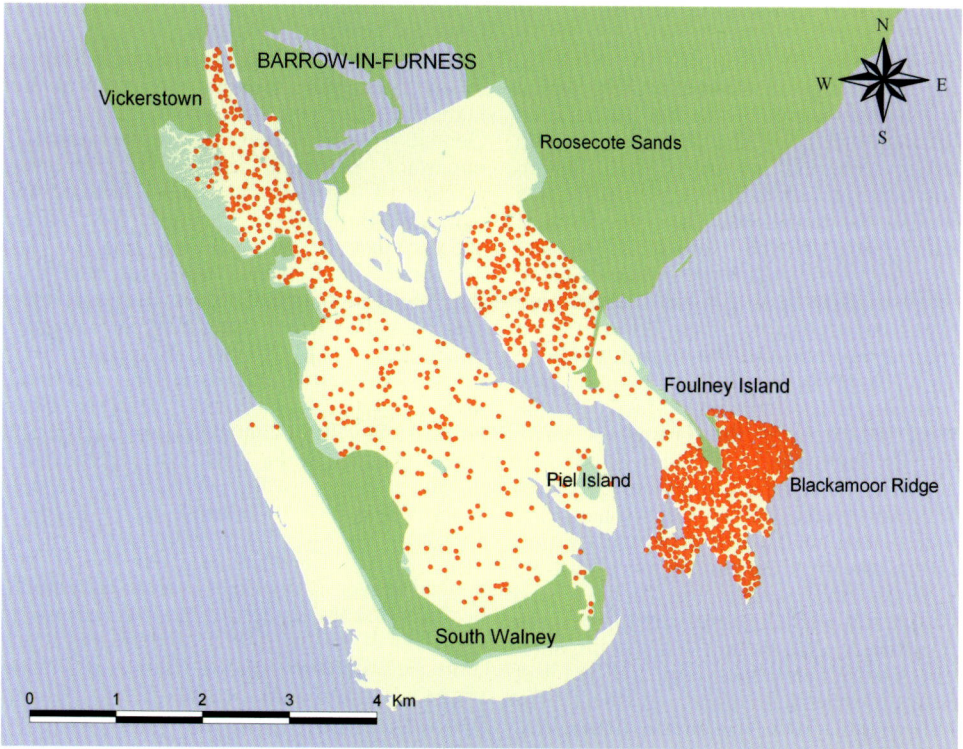

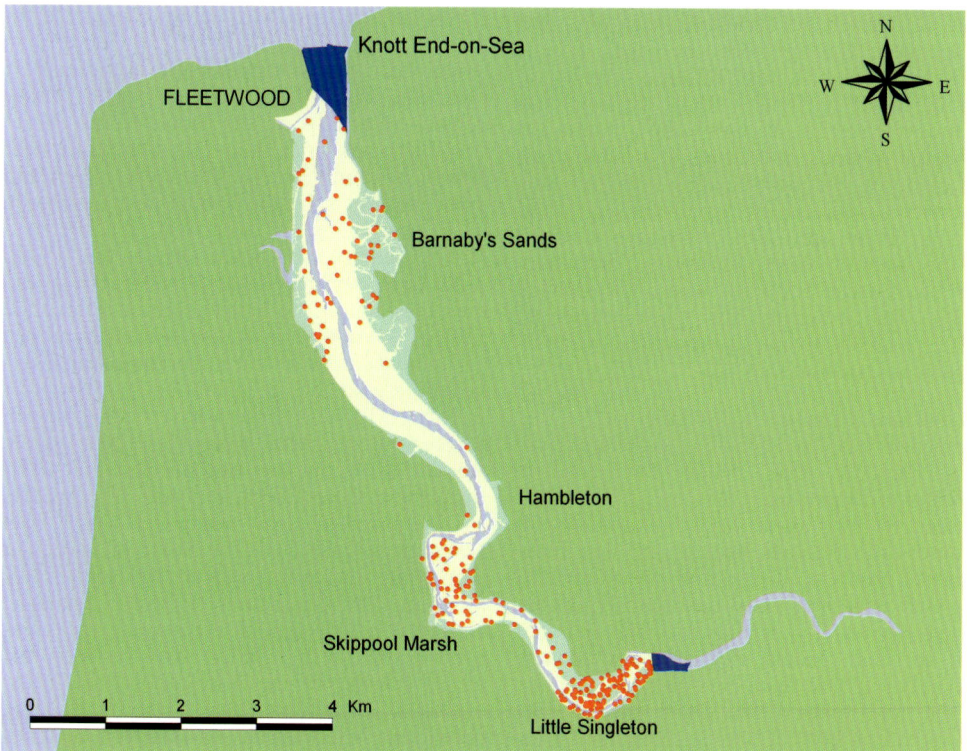

Figure 70. Low tide distribution of Oystercatcher (above: 1 dot = 3 birds) on standard survey of South Walney & Piel Channel Flats in 2005/06, and Curlew (below) on similar survey of the Wyre Estuary. Yellow = intertidal; pale blue = subtidal; pale green = intertidal. Dark blue areas not covered.

STOUR & ORWELL ESTUARIES

Site description

The Stour is a long and straight estuary, which forms the eastern end of the border between Suffolk and Essex. The estuary's mouth converges with that of the Orwell, which extends from Ipswich to Felixstowe, as the two rivers enter the North Sea. The outer Stour is becoming sandier and substrates become progressively muddier further upstream. There are seven shallow bays along the estuary and sharply rising land or cliffs, covered with ancient coastal woodland and agricultural land, border much of its length leaving little room for saltmarsh development. Much of the intertidal substrate of the Orwell is fairly muddy. In mitigation for the latest port development, both the north and south shores of the lower reaches of the estuary have had soft silts placed behind stiff clay bunds within the intertidal areas, changing the substrate again. Additionally, suspended silts have been placed directly into the water column of both lower estuaries, also as part of the mitigation process. Long stretches of farmland and wet meadow are situated along the mid-estuary, the latter providing roost sites for waterbirds. Nature conservation in the area includes the Stour & Orwell Estuaries Ramsar site and SPA, with management by the RSPB, Woodland Trust, Essex Wildlife Trust and Suffolk Wildlife Trust. Some sailing and shooting occurs, though the continued expansion of dock operations and subsequent land claim of important feeding areas remains a concern. Recently, bait-digging and recreational pressure (especially dog walking) on the Orwell have increased, and may influence bird distribution where disturbance results. The estuaries are here considered together as a single unit to reflect the extent of the SPA designation.

General bird distribution 2005/06

Areas covered 1,627/1,227 ha; Mean total birds 38,605/17,953; Mean bird density 23.7/14.6 birds per ha.

Between the two sites, the Stour and Orwell Estuaries support ten species in nationally important numbers. In addition, the Stour also surpassed national passage threshold for Ringed Plover. In keeping with the national trend for the species, Avocet now achieve nationally important numbers on both estuaries for the first time. Little Egret too is now widespread on both estuaries, testament to the meteoric rate of increase nationwide. As in the winter of 2004/05, many areas of the Orwell were favoured by many species, areas such as near Nacton, Jill's Hole, Mulberry Middle, Trimley Marshes and Loompit Lake. The more

sheltered intertidal habitats and series of bays on the Stour attract many species in greater numbers than its neighbour. Black-tailed Godwits occur here in internationally important numbers, favouring the inner reaches of the estuary around Jacques and Holbrook Bays west to Seafield Bay. Other species such as Grey Plover, Dunlin and Turnstone also occur here in nationally important numbers that do not reach equivalent levels on the Orwell. Single figure counts of Greenshank and Spotted Redshank were present on both estuaries, while other more unusual species recorded included Great Northern Diver, Slavonian Grebe, Black Brant, Scaup and Jack Snipe.

Comparative bird distribution

Comparisons with the winter of 1996/97 are displayed for Dark-bellied Brent Goose and Ringed Plover, two species that are undergoing declines of concern according to WeBS Alerts (Maclean & Austin 2006). Medium and short-term declines have led to Medium Alerts being triggered for Dark-bellied Brent Goose matching those occurring regionally and nationally and suggesting that adverse local conditions are less likely to be responsible for the triggering of these alerts. Low Tide Counts reveal different patterns on the two estuaries. Over the period in question, mean winter density decreased on the Stour, from 0.42 to 0.34 birds per ha (Figure 71.). On the Orwell, however, Loompit Lake, Trimley Marshes, and other areas of sub-tidal habitat near the mouth have seen increases over time, leading to a higher mean density of the species in 2005/06 (0.53 birds per ha) than in 1996/97 (0.33), and thus contributing to a net increase across the SPA as a whole; this is due to changes in use and management of the hinterland, which has benefited wildfowl in general.

Ringed Plover numbers have declined more severely to trigger a High Alert (Maclean & Austin 2006), but low tide distributions differ from these Core Count analyses. The species has increased in mean winter density on both estuaries (Figure 71.), the Stour in particular witnessing on average over 100 individuals more in 2005/06 than in 1996/97. Bathside, Holbrook and Seafield Bays were the chief beneficiaries of this change. This suggests that at least some of the birds occurring in the estuaries at low tide may well be roosting outwith the site, perhaps on Hamford Water to the south.

The Stour & Orwell Estuaries are covered by Suffolk Wildlife Trust under contract to Harwich Haven Authority. These data are generously made available to The Wetland Bird Survey.

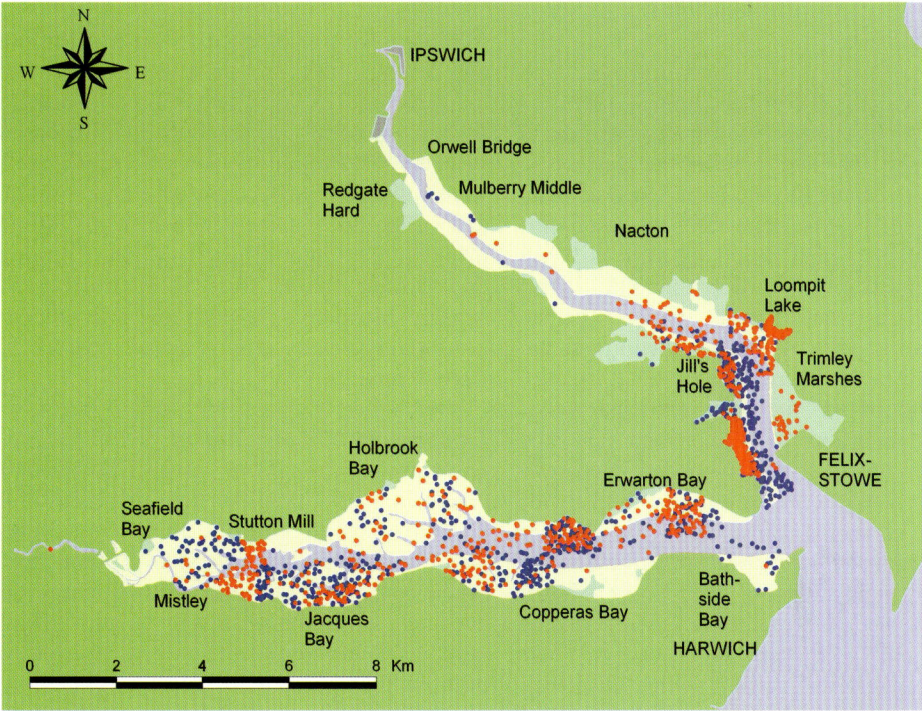

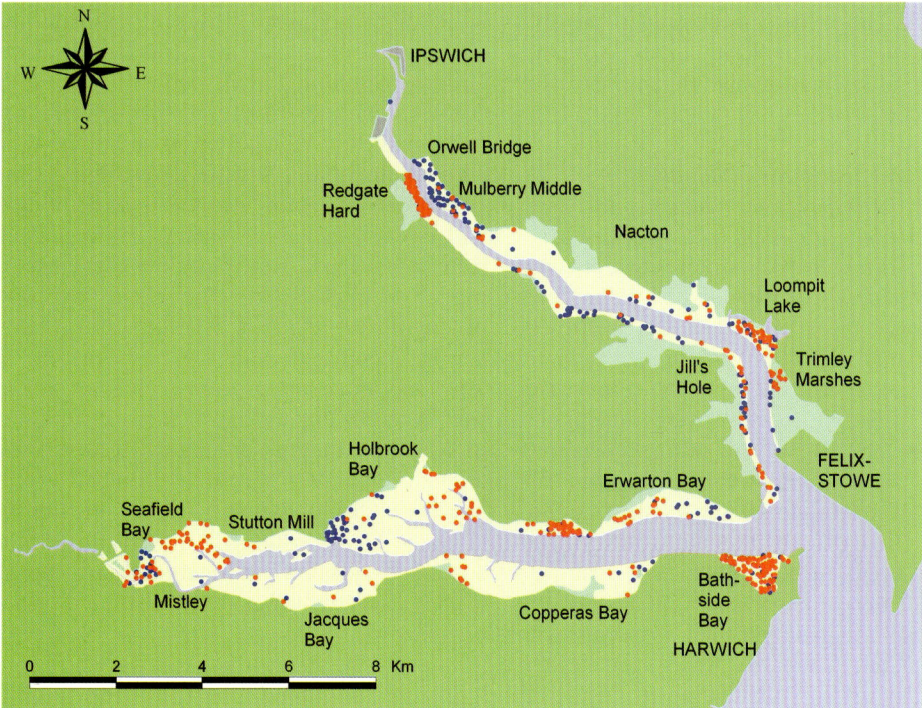

Figure 71. Low tide distribution of Dark-bellied Brent Goose (above: 1 dot = 2 birds) and Ringed Plover (below) on the Stour & Orwell Estuaries for the winters of 1996/97 (blue dots) and 2005/06 (red). Yellow = intertidal; pale blue = subtidal; pale green = intertidal. Grey areas not counted in earlier winter.

STRANGFORD LOUGH

Site description

Strangford Lough is a large shallow sea lough on the east coast of Northern Ireland, protected as a SPA, a Marine Nature Reserve, and a Ramsar Site. The site includes the Narrows, a deep rocky channel to the Irish Sea. The main body of the lough is sheltered to the east by the Ards Peninsula, and is fed by various rivers and tributaries. Downpatrick and Newtownards are the largest human habitations nearby. Within the lough there are numerous rocky outcrops and small islands. The north of the lough in particular holds extensive intertidal mud and sand flats and there are countless other bays and inlets, and large expanses of open water, providing a wide diversity of habitat. Since 2001, mobile gear fishing has been banned in Strangford Lough to allow populations of the Horse Mussel *Modiolus modiolus* to recover. Static fishing and catching of crustaceans still occurs. There is some recreational activity within the lough, including sailing. Despite the enormity of Strangford Lough, dedicated counters are able to count along the majority of its shoreline, and do so at low tide annually – an impressive achievement.

General bird distribution 2005/06

Area covered 4,325 ha; Mean total birds 44,396; Mean bird density 8.86 birds per ha.

As with most winters of survey, counter effort was rewarded with a wide diversity of species (51 different types of waders and wildfowl), many at high densities and in mean winter numbers exceeded only by the Blackwater Estuary. Certain species were ubiquitous, found on most if not all sectors containing intertidal habitat; unsurprisingly the dominant species were waders, especially Curlew, Dunlin, Oystercatcher and Redshank, Light-bellied Brent Geese, which occur in internationally important numbers at Strangford Lough. Shelduck, a species with similar status, was also widespread though most densely concentrated in the north where the feeding flats are most favourable. The lough is one of only two sites in Northern Ireland holding internationally important numbers of Bar-tailed Godwits; these birds were found at Castle Espie and on the north east shoreline. Other abundant wildfowl included Teal and Wigeon, whilst Lapwing and Golden Plover were fairly widely and densely distributed. At much lower density but also widely scattered were an average of 36 Greenshank.

Comparative bird distribution

Distribution data from Low Tide Counts undertaken in 1995/96 are displayed for comparison with bird distribution ten years later in 2005/06, for Pintail and Knot, both species occurring in important numbers. In keeping with the national trend, Pintail numbers have been steadily increasing over the past ten years. This shows with a comparison of the mean winter count for the two winters; in 1995/96 the figure was 118, compared to 631 in 2005/06. Around Strangford Lough, birds were confined to the northern areas, with distinct concentrations in both years at Ardmillan Bay, Mount Stewart and in the very northern tip around Newtownards (Figure 72.).

In contrast, Knot numbers peaked in the mid 1990s (Maclean & Austin 2006) but since then have seen a steady decline sufficient to trigger a High Alert. The mean low tide density for the winter of 2005/06 was 1.3 birds per ha, compared with 1.9 birds per ha in 1995/96, reflecting a decline in Knot at the site. As with Pintail, the main concentrations of Knot are in the northern bays, especially around Castle Espie, Comber and Newtownards (Figure 72.). Previously favoured areas on the east side of the lough, such as Greyabbey and Mount Stewart, appear to be used by fewer birds than ten years previously.

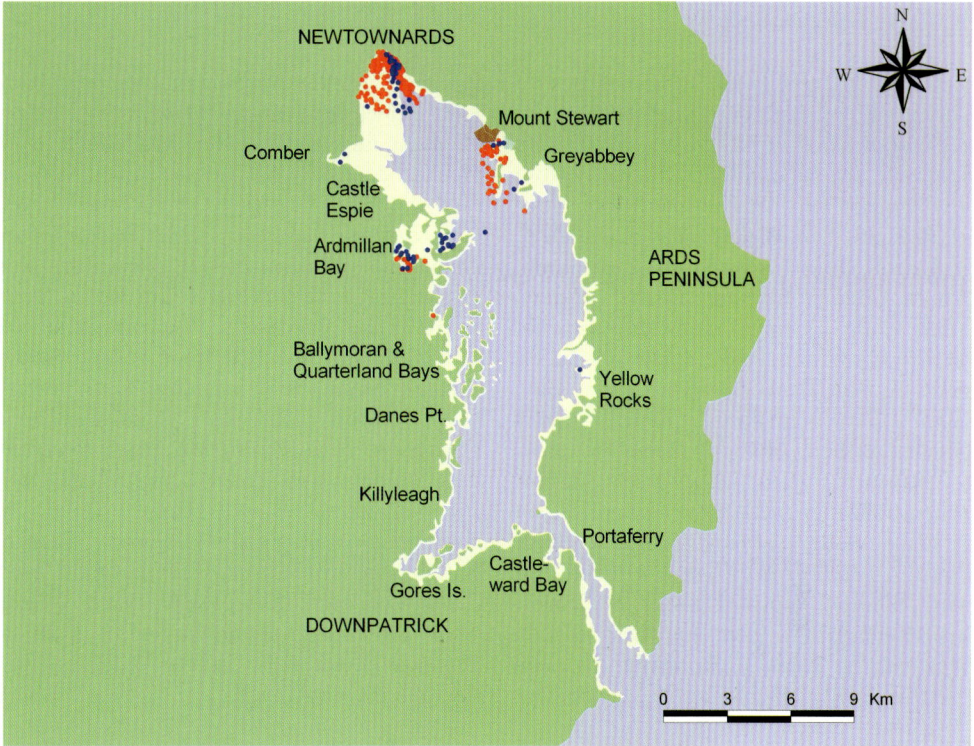

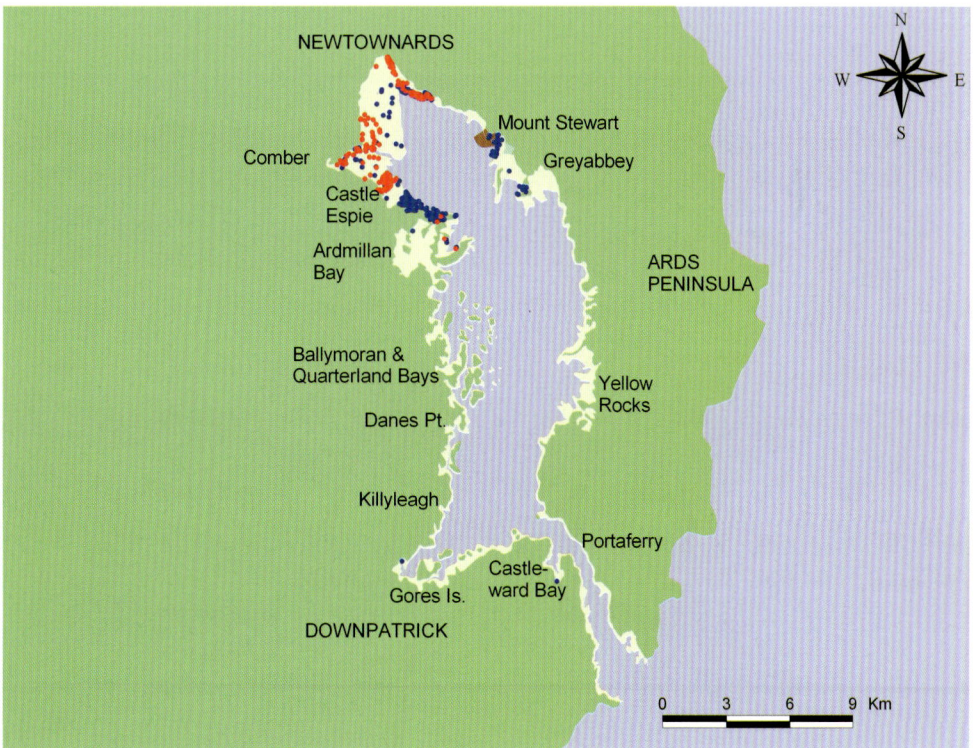

Figure 72. Low tide distribution of Pintail (above: 1 dot = 2 birds) and Knot (below: 1 dot = 30 birds) at Strangford Lough for the winters of 1995/96 (blue dots) and 2005/06 (red). Yellow = intertidal; pale blue = subtidal; pale green = nontidal. Brown area not counted in later winter.

ACKNOWLEDGEMENTS

We are very grateful to the following people and organisations that contributed to the Low Tide Count scheme in the winter of 2005/06. Apologies to anyone omitted accidentally from the list. Rachel Bain, Jean Barnard, John Best, Peter Beukers, Dot Blakey, David Callahan, Alex Carroll, Eve Catlett, Carl Clee, Chris Cockburn, Barry Collins, Declan Coney, Jonathan Copp, David Cousins, Jim Cowlin, Jason Crook, Adrian Dally, Anne de Potier, Frances Donnan, Jack Donovan, Colin Gay, Rosalyn Gay, Stephen Gilbert, Andy Goodman, Anthony Harbott, Annie Haycock, Bob Haycock, Will Hayward, Alastair Hearmon, John Hendrie, Martin Henry, Jeremy High, Jane Hodges, Mike Hodgson, Steve Holloway, Chas Holt, Robin Horner, Ed Hunter, Philip Johnston, Mervyn Jones, Juliette Kerr, H.J.A. Lee, Jon Lees, Paddy Livingstone, Ralph Loughlin, Euan MacAlpine, Paddy Mackie, Kerry Mackie, Ilya Maclean, Seamus Magouran, John Marchant, T.H. Mawdsley, Kevin Mawhinney, Hilary Mayne, Craig McCoy, Neil McCulloch, Niall McCutcheon, Roy McGregor, James McNair, Russell Neave, Geoffrey Orton, Tony Parker, Terry Paton, Brian Pavey, Colin Peake, Trevor Price, Eric Rainey, Neil Ravenscroft, Neil Rawlings, Geoff Robinson, Jim Rowe, Peter Royle, L.H. Sanderson, Richard Schofield, Jack Sheldon, John Shillitoe, Mark Smart, Graham Smith, Jeff Stenning, Len Stewart, Matt Sutton, Adrian Thomas, David Thompson, John Thorogood, Hugh Thurgate, Jack Torney, Bob Treen, Derek Tutt, Chris Tyas, Andrew Upton, Colin Wells, Jo Whatmough, Stephen Wilkinson, Richard Williamson, Jim Wilson, H. Wingfield-Hayes, C.W. Woodburn, Ken Wright, Mick Wright, Allan Drewitt (Natural England), Staff at Ravenair (Liverpool), Suffolk Wildlife Trust, Harwich Haven Authority.

Knot (John Bowers)

References

Armitage, M.J.S., Burton, N.H.K., Rehfisch, M.M. & Clark, N.A. 1997. *The Abundance and Distribution of Waterfowl within Milford Haven after the Sea Empress Oil Spill. Interim Report January 1997.* BTO Report No: 173 (Report of work carried out by the BTO under contract to the Countryside Council for Wales and the Sea Empress Environmental Evaluation Committee).

Banks, A.N. 2006. *Surveying waterbirds in Morecambe Bay for the Wetland Bird Survey (WeBS), Low Tide Count Scheme.* BTO Report No: 443. BTO, Thetford.

Banks, A.N., Bolt, D., Bullock, I., Haycock, B., Musgrove, A., Newson, S., Fairney, N., Sanderson, W., Schofield, R., Smith, L., Taylor, R. & Whitehead, S. 2005. *Ground and aerial monitoring protocols for in shore Special Protection Areas.* Countryside Council for Wales Marine Monitoring Report No: 11, 89pp.

Banks, A.N., Collier, M.P., Austin, G.E., Hearn, R.D. & Musgrove, A.J. 2006. *Waterbirds in the UK 2004/05: The Wetland Bird Survey.* BTO/WWT/RSPB/JNCC, Thetford.

Bibby, C.J., Burgess, N.D., Hill, D.A. & Mustoe, S. 2000. *Bird Census Techniques. Second Edition.* Academic Press, London.

Bill, D.I. & Hollins, J.R. 1989. Report on Portsmouth Harbour low water waterbird counts - Winter 1988/89. Unpublished.

BOURC. 1999. British Ornithologists' Union Records Committee: 25th Report (October 1998). *Ibis* 141: 175-180.

Buckland, S.T., Anderson, D., Burnham, K., Laake, J., Borchers, D. & Thomas, L. 2001. *Introduction to distance sampling: estimating abundance of biological populations.* Oxford University Press, Oxford.

Buckland, S.T., Anderson, D.R., Burnham, K.P., Laake, J.L., Borchers, D.L. & Thomas, L. (editors). 2004. *Advanced Distance Sampling.* Oxford University Press.

Burton, N.H.K., Musgrove, A.J., Rehfisch, M.M., Sutcliffe, A. & Waters, R.J. 2003. Numbers of wintering gulls in the United Kingdom, Channel Islands and Isle of Man: a review of the 1993 and previous Winter Gull Roost Surveys. *British Birds* 96: 376-401.

Cranswick, P.A., Kirby, J.S., Salmon, D.G., Atkinson-Willes, G.L., Pollitt, M.S. & Owen, M. 1997. A history of wildfowl counts by The Wildfowl & Wetlands Trust. *Wildfowl* 47: 217-230.

Crowe, O. 2005. *Ireland's Wetlands and their Waterbirds: Status and Distribution.* BirdWatch Ireland, Newcastle, Co. Wicklow.

Danielsen, F., Skov, H. & Durnick, J. 1993. Estimates of the wintering population of Red-throated Diver *Gavia stellata* and Black-throated Diver *Gavia arctica* in northwest Europe. *Proceedings of the 7th Nordic Congress of Ornithology, 1990.* pp18-24.

Dean, B.J., Webb, A., McSorley, C.A., Schofield, R.A. & Reid, J.A. 2004. *Surveillance of wintering seaducks, divers and grebes in UK inshore areas: Aerial surveys and shore-based counts 2003/04.* JNCC report 357.

Fraser, P.A., Rogers, M.J. & the Rarities Committee. 2007. Report on rare birds in Great Britain in 2005. *British Birds* 100, 16-61.

Fox, A.D. & Francis, I. 2004. *Report of the 2003/2004 national census of Greenland White-fronted Geese in Britain.* Greenland White-fronted Goose Study, Kalø.

Gilbert, G., Gibbons, D.W. & Evans, J. 1998. *Bird Monitoring Methods.* RSPB, Sandy.

Griffin, L.R. & Mackley, E.R. 2004. *WWT Svalbard Barnacle Goose Project Report 2003-04.* WWT Internal report, Slimbridge.

Hastie, T. & Tibshirani, R. 1990. *Generalized Additive Models.* Chapman & Hall, London.

Heubeck, M. & Mellor, M. 2005. *SOTEAG ornithological monitoring programme: 2004 summary report.* SOTEAG, Aberdeen.

Holmes, J.S., Marchant, J., Bucknell, N., Stroud, D.A. & Parkin, D.T. 1998. The British List: new categories and their relevance to conservation. *British Birds*, 92: 2-11.

Holmes, J.S. & Stroud, D.A. 1995. Naturalised birds: feral, exotic, introduced or alien? *British Birds*, 92: 2-11.

Kershaw, M. & Cranswick, P.A. 2003. Numbers of wintering waterbirds in Great Britain, 1994/1995-1998/1999: I. Wildfowl and selected waterbirds. *Biological Conservation* 111: 91-104.

Kirby, J.S., Salmon, D.G., Atkinson-Willes, G.L. & Cranswick, P.A. 1995. Index numbers for waterbird populations, III. Long-term trends in the abundance of wintering wildfowl in Great Britain, 1966/67 to 1991/92. *Journal of Applied Ecology* 32: 536-551.

Maclean, I.M.D., Austin, G.E., Mellan, H.J. and Girling, T. 2005. *WeBS Alerts 2003/2004: Changes in numbers of wintering waterbirds in the United Kingdom, its Constituent Countries, Special Protection Areas (SPAs) and Sites of Special Scientific Interest (SSSIs).* BTO Research Report No. 416 to the WeBS partnership. BTO, Thetford.

Maclean, I.M.D., & Austin, G.E. 2006. *WeBS Alerts 2004/2005: Changes in numbers of wintering waterbirds in the United Kingdom, its Constituent Countries, Special Protection Areas (SPAs) and Sites of Special Scientific Interest (SSSIs).* BTO Research Report No. 458 to the WeBS partnership. BTO, Thetford.

Musgrove, A.J., Langstone, R.H.W., Baker, H. & Ward, R.M. 2003. *Estuarine Waterbirds at Low Tide: the WeBS Low Tide Counts 1992/93 to 1998/99.* WSG/BTO/WWT/RSPB/JNCC, Thetford.

Prŷs-Jones, R.P., Underhill, L.G. & Waters, R.J. 1994. Index numbers for waterbird populations. II Coastal wintering waders in the United Kingdom, 1970/71-1990/91. *Journal of Applied Ecology* 31: 481-492.

Ramsar Convention Bureau. 1988. *Convention on Wetlands of International Importance especially as Waterfowl Habitat.* Proceedings of the third meeting of the Conference of the Contracting Parties, Regina, Canada, 1987. Ramsar, Switzerland.

Rehfisch, M.M, Austin, G.E., Armitage, M.J.S., Atkinson, P.W., Holloway, S.J., Musgrove, A.J. & Pollitt, M.S. 2003. Numbers of wintering waterbirds in Great Britain and the Isle of Man (1994/1995-1998/1999): II. Coastal waders (Charadrii). *Biological Conservation* 112: 329-341.

Rose, P.M. & Scott, D.A. 1997. *Waterfowl Population Estimates - Second Edition.* Wetlands International Publ. 44, Wageningen, The Netherlands.

Rose, P.M. & Stroud, D.A. 1994. Estimating international waterfowl populations: current activity and future directions. *Wader Study Group Bulletin* 73: 19-26.

Rowell, H.E. & Spray, C.J. 2004. The Mute Swan *Cygnus olor* (Britain and Ireland populations) in Britain and Northern Ireland

1960/61 – 2000/01. Waterbird Review Series, The Wildfowl & Wetlands Trust/Joint Nature Conservation Committee, Slimbridge.

Rowell, HE. 2005. *The 2004 Icelandic-breeding Goose Census.* The Wildfowl & Wetlands Trust/Joint Nature Conservation Committee, Slimbridge.

Sangster, G., Collinson, M.J., Helbig, A.J., Knox, A.G. & Parkin, D.T. 2005. Taxonomic recommendations for British birds: third report. Ibis 147; 821-826.

Simpson, J. & Maciver, A. 2005. *Population and distribution of Bean Geese in the Slamannan area 2004/2005.* Report to Bean Goose Action Group.

Smith, L.E., Hall, C., Cranswick, P.A., Banks, A.N., Sanderson, W.G. & Whitehead, S. 2007. *The status of Common Scoter Melanitta nigra in Welsh waters and Liverpool Bay, 2001-06.* Welsh Birds 5: 4-28.

Söhle, I., Wilson, L.J., Dean, D.J., O'Brien, S.H., Webb A. & Reid, J.B. 2006. Surveillance of wintering seaducks, divers and grebes in UK onshore ares: Aerial surveys and shore-based counts 2005/06. JNCC, Peterborough.

Underhill, L.G. 1989. *Indices for waterbird populations.* BTO Research Report 52.

Underhill, L.G. & Prŷs-Jones, R. 1994. Index numbers for waterbird populations. I. Review and methodology. *Journal of Applied Ecology*, 31: 463-480.

Vinicombe, K., Marchant, J. & Know, A. 1993. Review of status and categorization of feral birds on the British List. *British Birds*, 75: 1-11.

Waltho, C.M. 2004. *Firth of Clyde Eider News: No 5. August 2004.* Private report

Ward, R.M., Cranswick, P.A., Kershaw, M., Austin, G., Brown, A.W., Brown, L.M., Coleman, J.T., Chisholm, H. & Spray, C. 2007. *National Mute Swan Census 2002.* WWT, Slimbridge.

Way, L.S., Grice, P., MacKay, A., Galbraith, C.A., Stroud, D.A. & Pienkowski, M.W. 1993. *Ireland's internationally important bird sites: a review of sites for the EC Special Protection Area network.* JNCC, Peterborough, 231 pp.

Wetlands International. 2006. *Waterbird Population Estimates - Forth Edition.* Wetlands International, Wageningen, The Netherlands.

Glossary

The terms listed below are generally restricted to those that have been adopted specifically for use within WeBS or more widely for monitoring.

1% criterion The criterion identifies sites as being of *international importance* if at least 1% of the *waterbirds* of a particular migratory flyway or population regularly make use of a site during their annual cycle. The term thus relates to the proportion (1%) that is used as a criterion of site selection. First used in the Ramsar Convention, the 1% criterion is used widely in assessment of site importance.

1% threshold This logically derives from the *1% criterion* and relates to the number of birds that are used as the nominal 1% of the population for the purposes of site selection. Thus, an international population of 75,215 Shelduck has a derived 1% threshold (adopting rounding conventions) of 750.

African-Eurasian Migratory Waterbird Agreement (AEWA) An independent international treaty developed under the Convention on the Conservation of Migratory Species of Wild Animals (*'Bonn Convention'*). Parties to the Agreement are called upon to engage in a wide range of conservation actions addressing key issues such as species and habitat conservation, management of human activities, research and monitoring, education and information, and implementation.

All-Ireland Comprises the whole island of Ireland (Northern Ireland and the Republic of Ireland).

Autumn For waders, autumn comprises July to October inclusive. Due to differences in seasonality between species, a strict definition of autumn is not used for wildfowl.

British Trust for Ornithology (BTO) The BTO is a well-respected organisation, combining the skills of professional scientists and volunteer birdwatchers to carry out research on birds in all habitats and throughout the year. Data collected by the various surveys form the basis of extensive and unique databases, which enable the BTO to objectively advise conservation bodies, government agencies, planners and scientists on a diverse range of issues involving birds.

Complex site A *WeBS site* that consists of two or more *WeBS sectors*.

Core Counts The fundamental WeBS counts that monitor all types of wetlands throughout the UK once per month on, or as near as possible to, pre-selected *priority dates*. Used to determine population estimates and trends and identify important sites.

Great Britain The countries of England Scotland and Wales (excludes the Channel Isles and the Isle of Man).

Incomplete counts When presenting counts of an individual species, a large proportion of the number of birds was suspected to have been missed, *e.g.* due to part coverage of the site or poor counting conditions, or when presenting the total number of birds of all species on the site, a significant proportion of the total number was missed.

I-WeBS An independent but complementary scheme operating in the Republic of Ireland to monitor non-breeding *waterbirds*, organised by BirdWatch Ireland, the National Parks and Wildlife Service (Ireland) and The *Wildfowl & Wetlands Trust*.

Joint Nature Conservation Committee (JNCC) JNCC is the statutory body constituted by the Environmental Protection Act 1990 to be responsible for research and advice on nature conservation at both UK and international levels. The committee is established by Natural England, Scottish Natural Heritage and the Countryside Council for Wales, together with independent members and representatives from the Countryside Commission and Northern Ireland, and is supported by specialist staff.

Local Organiser Person responsible for coordinating counters and counts at a local level, normally a county or large estuary, and the usual point of contact with the *WeBS office*.

Low Tide Counts (LTC) WeBS counts made at low tide to assess the relative importance of different parts of individual estuaries as feeding areas for intertidal *waterbirds*.

Priority date Pre-determined dates published by the *WeBS Office* to aid coordination of surveys. Counters are asked to count on, or as near as possible to, priority dates to minimise the risk of missing birds or double counting.

Royal Society for the Protection of Birds (RSPB) The RSPB is the charity that takes action for wild birds and the environment in the UK. The RSPB is the national BirdLife partner in the UK.

Spring For waders, spring comprises April to June inclusive. Due to differences in seasonality between species, a strict definition of spring is not used for wildfowl.

United Kingdom *Great Britain* and Northern Ireland (excludes the Channel Isles and the Isle of Man).

Waterbirds WeBS follows the definition adopted by *Wetlands International*. This includes a large number of families, those occurring regularly in the UK being divers, grebes, cormorants, herons, storks, ibises and spoonbills, wildfowl, cranes, rails, waders, gulls and terns.

WeBS count unit The area/boundary within which a count is made. The generic term for *WeBS sites, WeBS sub-sites* and *WeBS sectors*.

WeBS Office Main administrative centre for the day-to-day running of WeBS and main point of contact for information or data pertaining to WeBS (see *Contacts* section).

WeBS sector The unit of division of large *sites* into areas that can be counted by one person in a reasonable time period. They are often demarcated by geographic features to facilitate recognition of the boundary by counters. The finest level at which data are recorded.

WeBS site A biologically meaningful area that represents a discrete area used by *waterbirds* such that birds regularly move within but only occasionally between sites. The highest level at which count data are stored.

WeBS sub-site A grouping of *sectors* within a *site* to facilitate coordination. In most cases, sub-sites also relate to biologically meaningful units for describing *waterbird* distribution.

WeBS Year Defined as July to June inclusive the WeBS Year is centred on the time when most *waterbird* species are present in their largest number, during *winter*. Counts during *autumn* passage and *spring* passage the following calendar year are logically associated with the intervening *winter*.

Wetlands International A leading global non-profit organisation whose mission is to sustain and restore wetlands, their resources and biodiversity for future generations through research, information exchange and conservation activities, worldwide.

Wildfowl & Wetlands Trust (WWT) Founded by Sir Peter Scott in 1946, WWT is the largest international wetland conservation charity in the UK. WWT works to conserve wetlands and their biodiversity, focusing particularly on waterbirds and their habitats, and seeks to raise awareness of the value of wetlands, the threats they face and the actions needed to save them. WWT has nine visitor centres throughout the UK.

Winter For waders, winter comprises November to March inclusive. Due to differences in seasonality between species, a strict definition of winter is not used for wildfowl.

Appendices

APPENDIX 1.   INTERNATIONAL AND NATIONAL IMPORTANCE

Any site recognised as being of international ornithological importance is considered for classification as a Special Protection Area (SPA) under the EC Directive on the Conservation of Wild Birds (EC/79/409), whilst a site recognised as an internationally important wetland qualifies for designation as a Ramsar site under the Convention on Wetlands of International Importance especially as Waterfowl Habitat. Criteria for assessing the international importance of wetlands have been agreed by the Contracting Parties to the Ramsar Convention on Wetlands of International Importance (Ramsar Convention Bureau 1988). Under criterion 6, a wetland is considered internationally important if it regularly holds at least 1% of the individuals in a population of one species or subspecies of waterbird, while criterion 5 states that any site regularly supporting 20,000 or more waterbirds also qualifies. Britain and Ireland's wildfowl belong, in most cases, to the northwest European population and the waders to the east Atlantic flyway population (Wetlands International 2006).

A wetland in Britain is considered nationally important if it regularly holds 1% or more of the estimated British population of one species or subspecies of waterbird, and in Northern Ireland important in an all-Ireland context if it holds 1% or more of the estimated all-Ireland population.

The 1% thresholds for British, all-Ireland and international waterbird populations, where known, are listed in Table A1. Thus, any site regularly supporting at least this number of

birds potentially qualifies for designation under national legislation, or the EC Birds Directive or Ramsar Convention. The international population for each species and subspecies is also specified in the table. However, it should be noted that, where 1% of the national population is less than 50 birds, 50 is normally used as a minimum qualifying threshold for the designation of sites of national or international importance.

It was agreed at the meeting of the Ramsar Convention in Brisbane that population estimates will be reviewed by Wetlands International every three years and 1% thresholds revised every nine years (Rose & Stroud 1994; Ramsar Resolution VI.4). 1% thresholds have not been derived for introduced species since protected sites would not be identified for these birds.

Sources of qualifying levels represent the most up-to-date figures following recent reviews: for wildfowl in Britain see Kershaw & Cranswick (2003); for waders in Britain see Rehfisch et al. (2003); for gulls in Britain see Burton et al. (2003); for all-Ireland importance for divers see Danielsen et al. (1993) and for other waterbirds see Whilde (in prep.) cited in Way et al. (1993). International criteria follow Wetlands International (2006).

It should be noted that for some populations, where the British total is the international total, the precise figure given for the estimates may differ because of different rounding conventions applied in the relevant publications.

Table A1. 1% thresholds for national and international importance

	Great Britain	all-Ireland	International	Subspecies/Population
Mute Swan: *British*	375	n/a	320	Britain
Irish	n/a	100	100	Ireland
Bewick's Swan	81	*25	200	*bewickii*, NW Europe (non-br)
Whooper Swan	57	100	210	Iceland (br)
Bean Goose: *Taiga*	*4	+	800	*fabalis*
Pink-footed Goose	2,400	+	2,700	Greenland, Iceland (br)
European White-fronted Goose	58	+	10,000	*albifrons*, Baltic-North Sea
Greenland White-fronted Goose	209	140	270	*flavirostris*
Greylag Goose: *Iceland*	819	*40	870	*anser*, Iceland (br)
Hebrides/N Scotland	90	n/a	100	*anser*, NW Scotland
Barnacle Goose: *Greenland*	450	75	560	E Greenland (br)
Svalbard	220	+	270	Svalbard (br)

	Great Britain	all-Ireland	International	Subspecies/Population
Dark-bellied Brent Goose	981	+	2,000	*Bernicla*, W Siberia (br)
Light-bellied Brent Goose: *Canada*	+	200	260	*hrota*, Ireland (non-br)
Svalbard	*30	+	70	*hrota*, Svalbard, N Greenland (br)
Shelduck	782	70	3,000	NW Europe (br)
Wigeon	4,060	1,250	15,000	NW Europe (non-br)
Gadwall	171	+	600	*strepera*, NW Europe (br)
Teal	1,920	650	5,000	NW Europe (non-br)
Mallard	3,520	500	**20,000	*platyrhynchos*, NW Europe (non-br)
Pintail	279	60	600	NW Europe (non-br)
Garganey	+	+	**20,000	W Africa (non-br)
Shoveler	148	65	400	NW & C Europe (non-br)
Red-crested Pochard	+	+	500	C Europe & W Mediterranean
Pochard	595	400	3,500	NE & NW Europe (non-br)
Tufted Duck	901	400	12,000	NW Europe (non-br)
Scaup	76	*30	3,100	*marila*, W Europe (non-br)
Eider	730	*20	12,850	*mollissima*, NW Europe[1]
Long-tailed Duck	160	+	**20,000	W Siberia, N Europe (br)
Common Scoter	500	*40	16,000	*nigra*
Velvet Scoter	*30	+	10,000	*fusca*, Baltic, W Europe (non-br)
Goldeneye	249	110	11,500	*clangula*, NW & Central Europe (non-br)
Smew	*4	+	400	NW & C Europe (non-br)
Red-breasted Merganser	98	*20	1,700	NW & C Europe (non-br)
Goosander	161	+	2,700	*merganser*, NW Europe[2]
Red-throated Diver	49	*10	3,000	NW Europe (non-br)
Black-throated Diver	*7	*1	3,750	*arctica*
Great Northern Diver	*30	?	50	NW Europe (non-br)
Little Grebe	78	?	4,000	*ruficollis*
Great Crested Grebe	159	*30	3,600	*cristatus*
Red-necked Grebe	*2	?	5,10	*grisegena*, NW Europe (non-br)
Slavonian Grebe	*7	?	55	*auritus*, NW Europe (large billed)
Black-necked Grebe	*1	?	2,200	*nigricollis*, Europe, N Africa
Cormorant	230	?	1,200	*carbo*, NW Europe
Shag	?	?	2,000	*aristotelis*
Little Egret	?	?	1,300	*garzetta*, W Europe, NW Africa
Grey Heron	?	?	2,700	*cinerea*, W Europe, NW Africa (br)
Moorhen	7500	?	**20,000	*chloropus*, Europe, N Africa (br)
Coot	1,730	250	17,500	*atra*, NW Europe (non-br)
Oystercatcher	3,200	500	10,200	*ostralegus*, Europe, NW Africa
Avocet	*35	+	730	W Europe (br)
Ringed Plover: *winter*	330	125	730	*hiaticula*, Europe & N Africa (non-br)
passage	300			
Golden Plover	2,500	2,000	9,300	*altifrons*, Iceland & Faeroes, E Atlantic[3]
Grey Plover	530	*40	2,500	E Atlantic (non-br)
Lapwing	**20,000	2,500	**20,000	Europe (br)
Knot	2,800	375	4,500	*islandica*
Sanderling: *winter*	210	*35	1,200	E Atlantic, W & S Africa (non-br)
passage	300			
Purple Sandpiper	180	*10	750	*maritima*, E Atlantic
Dunlin: *winter*	5,600	1,250	13,300	*alpina*, W Europe (non-br)[4]
passage	2,000			
Ruff	*7	+	?	W Africa (non-br)
Jack Snipe	?	250	?	NE Europe (br)
Snipe	?	?	**20,000	*gallinago*, Europe (br)
Woodcock	?	?	**20,000	Europe (br)
Black-tailed Godwit	150	90	470	*islandica*
Bar-tailed Godwit	620	175	1,200	*lapponica*
Whimbrel	+	+	6,800	*islandicus*
Curlew	1,500	875	8,500	*arquata*
Spotted Redshank	+	+	900	Europe (br)
Redshank	1,200	245	2,800	*brittanica*[5]
Greenshank	*6	*9	2,300	Europe (br)
Green Sandpiper	?	?	17,000	Europe (br)
Common Sandpiper	?	?	17,500	N, W & C Europe (br)
Turnstone	500	225	1,500	*interpres*, NE Canada, Greenland (br)

Table A1. continued

	Great Britain	all-Ireland	International	Subspecies/Population
Little Gull	?	?	1,230	N, C & E Europe (br)
Black-headed Gull	19,000	?	**20,000	N & C Europe (br)
Common Gull	9,000	?	**20,000	*canus*
Lesser Black-backed Gull	500	?	5,500	*graellsii*
Herring Gull	4,500	?	5,900	*argentatus*[6]
Great Black-backed Gull	400	?	4,400	NE Atlantic
Kittiwake	?	?	**20,000	*tridactyla*, E Atlantic (br)
Sandwich Tern	?	?	1,700	*sandvicensis*, W Europe (br)
Common Tern	?	?	1,900	*hirundo*, S, W Europe (br)
Little Tern	?	?	490	*albifrons*, W Europe (br)
Black Tern	?	?	7,500	*niger*

?   *Population size not accurately known.*
+   *Population too small for meaningful figure to be obtained.*
*   *Where 1% of the British or all-Ireland wintering population is less than 50 birds, 50 is normally used as a minimum qualifying level for national or all-Ireland importance respectively.*
**   *A site regularly holding more than 20,000 waterbirds qualifies as internationally important by virtue of absolute numbers.*

1   The degree of interchange of UK Eiders with birds on the continent is unclear, and although Wetlands International (2006) has recommended that birds in Britain and Ireland should be treated as a separate biogeographical population, a recent review of available data by DEFRA's SPA and Ramsar Scientific Working Group has found limited evidence to support this conclusion, and recommended that for site-selection purposes, British Eider continue to be considered as a component of the four groups of the Northwest European groups of the race *mollissima* with an international 1% threshold of 15,500. It is hoped that future genetic studies will help clarify the situation.

2   Although Wetlands International (2006) considers Goosanders breeding in Scotland, northern England and Wales to be a discrete population, a recent review of available data by DEFRA's SPA and Ramsar Scientific Working Group has found limited evidence to support this conclusion for the time being, and recommended that for site-selection purposes, British Goosanders continue to be considered as a component of the NW and C European population of Goosander, with an international 1% threshold of 2,700.

3   Three populations of Golden Plover listed by Wetlands International (2006) overlap in the UK in winter. Draft guidelines from Ramsar suggest that the largest of the three thresholds (*i.e.* that for *altifrons*, Iceland & Faeroes, E Atlantic) should be used for site-selection purposes.

4   Whilst several populations of Dunlin occur in the UK at different times of the year, most wintering birds are referable to the listed population.

5   Three populations of Redshank listed by Wetlands International (2006) overlap in the UK in winter: *totanus* E Atlantic (non-br), *robusta* and *brittanica*. Most *totanus* winter outside the UK but the other populations are known to occur widely. Draft guidelines from Ramsar suggest that the larger of the two thresholds (*i.e.* that for *brittanica*) should be used for site-selection purposes.

6   Two populations of Herring Gull overlap in the winter in the UK; *argentatus* and *argenteus*. Draft guidelines from Ramsar suggest that the larger of the two thresholds, *i.e.* that for *argentatus*, should be used for site-selection purposes.

APPENDIX 2.   LOCATIONS OF PRINCIPAL WeBS COUNT SITES

Table A2 provides details of principal WeBS sites that are mentioned in the Principal Sites table (Table 6.). Sites are listed alphabetically, with details of the Ordnance Survey 1-km square that the centre of the sites falls into.

Numbers following Principal Core Count sites refer to the sites' location in Figure A1. Details of all sites covered by WeBS are available from the www.bto.org/webs or the WeBS Office, (see *CONTACTS*).

Table A2. Details for Principal Sites mentioned in Table 6. Numbers refer to the sites' location in figure A1.

Site	1-km sq	No.
Abberton Reservoir	TL9717	103
Alde Complex	TM4257	98
Alt Estuary	SD2903	77
Arun Valley	TQ0314	121
Balnakeil Bay	NC3869	9
Balnakiel Farm	NC3968	10
Balranald Nat Res	NF7169	22
Beaulieu Estuary	SZ4297	129
Belfast Lough	IJ3983	141
Blackwater Estuary	TL9307	105
Blyth Estuary	TM4675	97
Branahuie Saltings	NB4631	16
Breydon Water & Berney Marshes	TG4706	96
Burry Inlet	SS5096	85
Cameron Reservoir	NO4711	51
Carlingford Lough	IJ1814	135
Carmarthen Bay	SN2501	84
Carsebreck & Rhynd Lochs	NN8609	38
Chew Valley Lake	ST5659	118
Chichester Harbour	SU7700	123
Cleddau Estuary	SN0005	83
Colne Estuary	TM0614	104
Cotswold Water Park (West)	SU0595	117
Cromarty Firth	NH7771	20
Crouch-Roach Est	TQ9895	107
Deben Estuary	TM2942	99
Dee Estuary (England & Wales)	SJ2675	79
Dee Flood Meadows	SJ4059	80
Dengie Flats	TM0302	106
Dornoch Firth	NH7384	18
Duddon Estuary	SD2081	70
Dundrum Bay	IJ4135	136
Dungeness GPs	TR0619	113
Dyfi Estuary	SN6394	82
Eden Estuary	NO4719	50
Exe Estuary	SX9883	133
Fleet and Wey	SY6976	132
Forth Estuary	NT2080	54
Hamford Water	TM2225	102
Hickling Broad	TG4221	94
Holburn Moss	NU0536	60
Hornsea Mere	TA1947	73
Hule Moss	NT7149	58
Humber Estuary	TA2020	74
Inner Firth of Clyde	NS3576	35
Inner Moray and Inverness Firth	NH6752	21
Island of Egilsay	HY4730	1
Isle of Coll	NM2055	26
Kentra Moss and Lower Loch Shiel	NM6668	31
Kilconquhar Loch	NO4801	52
Killough Harbour	IJ5436	137
Lake of Menteith	NN5700	37
Langstone Harbour	SU6902	127
Lavan Sands	SH6474	81
Lee Valley GPs	TL3807	115
Lindisfarne	NU1041	61
Loch a` Phuill	NL9541	30
Loch An Eilein	NL9843	29
Loch Bee	NF7743	25
Loch Bhasapoll	NL9746	28
Loch Eye	NH8379	19
Loch Fleet Complex	NH7896	17
Loch Garten	NH9718	39
Loch Gorm	NR2365	33
Loch Gruinart Floods	NR2767	32
Loch Hempriggs	ND3447	14
Loch Ken	NX7168	68
Loch Leven	NO1401	53
Loch Lomond	NS3599	36
Loch Mor: Baleshare	NF7962	24
Loch of Boardhouse	HY2725	4
Loch of Harray	HY2915	7
Loch of Hundland	HY2926	3
Loch of Lintrathen	NO2754	47
Loch of Skaill	HY2418	5
Loch of Skene	NJ7807	44
Loch of Stenness	HY2812	8
Loch of Strathbeg	NK0758	41
Loch of Swannay	HY3128	2
Loch of Wester	ND3259	11
Loch Paible	NF7268	23
Loch Riaghain	NM0347	27
Loch Scarmclate	ND1859	12
Loch Spynie	NJ2366	40
Loch Watten	ND2256	13
Lochs Davan & Kinord	NO4499	45
Lough Foyle	IC5925	143
Loughs Neagh& Beg	IJ0475	142
Lower Derwent Ings	SE6939	72
Lower Lough Erne	IH0960	144
Lower Teviot Valley	NT6725	57
Medway Estuary	TQ8471	109
Melbost Sands, Tong Saltings and Broad Bay (Lewis)	NB4534	15
Mersehead RSPB Reserve	NX9255	63
Mersey Estuary	SJ4578	78
Middle Yare Marshes	TG3504	95
Milldam and Balfour Mains Pools	HY4817	6
Moine Mhor and Add Estuary	NR8293	34
Montrose Basin	NO6958	46
Morecambe Bay	SD4070	71
Nene Washes	TF3300	89
North Norfolk Coast	TF8546	92
North West Solent	SZ3395	130
Orchardton and Auchencairn Bays	NX8151	65
Orwell Estuary	TM2238	100
Ouse Washes	TL5394	90
Outer Ards Shoreline	IJ6660	139
Pagham Harbour	SZ8796	122
Pegwell Bay	TR3563	111
Pitsford Reservoir	SP7669	87
Poole Harbour	SY9988	131
Portsmouth Harbour	SU6204	126
R Clyde: Carstairs to Thankerton	NS9842	55
Ribble Estuary	SD3825	75
R. Avon: Fordingbridge to Ringwood	SU1410	124
R. Avon: Ringwood to Christchurch	SZ1499	128
R. Nith: Keltonbank to Nunholm	NX9774	62
River Tay - Haughs of Kercock	NO1439	48
Rutland Water	SK9307	88
Rye Harbour and Pett Level	TQ9418	114
Severn Estuary	ST5084	119
Severn Hams	SO8426	86
Slains Lochs (Meikle and Sand and Cotehill)	NK0230	42
Solway Estuary	NY1060	67
Somerset Levels	ST4137	120
Southampton Water	SU4507	125
St Benet`s Levels	TG3815	93
Stour Estuary	TM1732	101
Strangford Lough	IJ5460	140
Swale Estuary	TQ9765	110
Taw-Torridge Est	SS4731	134
Tay Estuary	NO4828	49
Tees Estuary	NZ5528	69
Teviot Haughs	NT6925	56
Thames Estuary	TQ7880	108
The Wash	TF5540	91
Tring Reservoirs	SP9113	116
Tweed Estuary	NT9853	59
Upper Lough Erne	IH3131	145
Upper Quoile River	IJ4745	138
Walland Marsh	TQ9824	112
Wigtown Bay	NX4456	66
WWT Caerlaverock	NY0565	64
WWT Martin Mere	SD4214	76
Ythan Estuary	NK0026	43

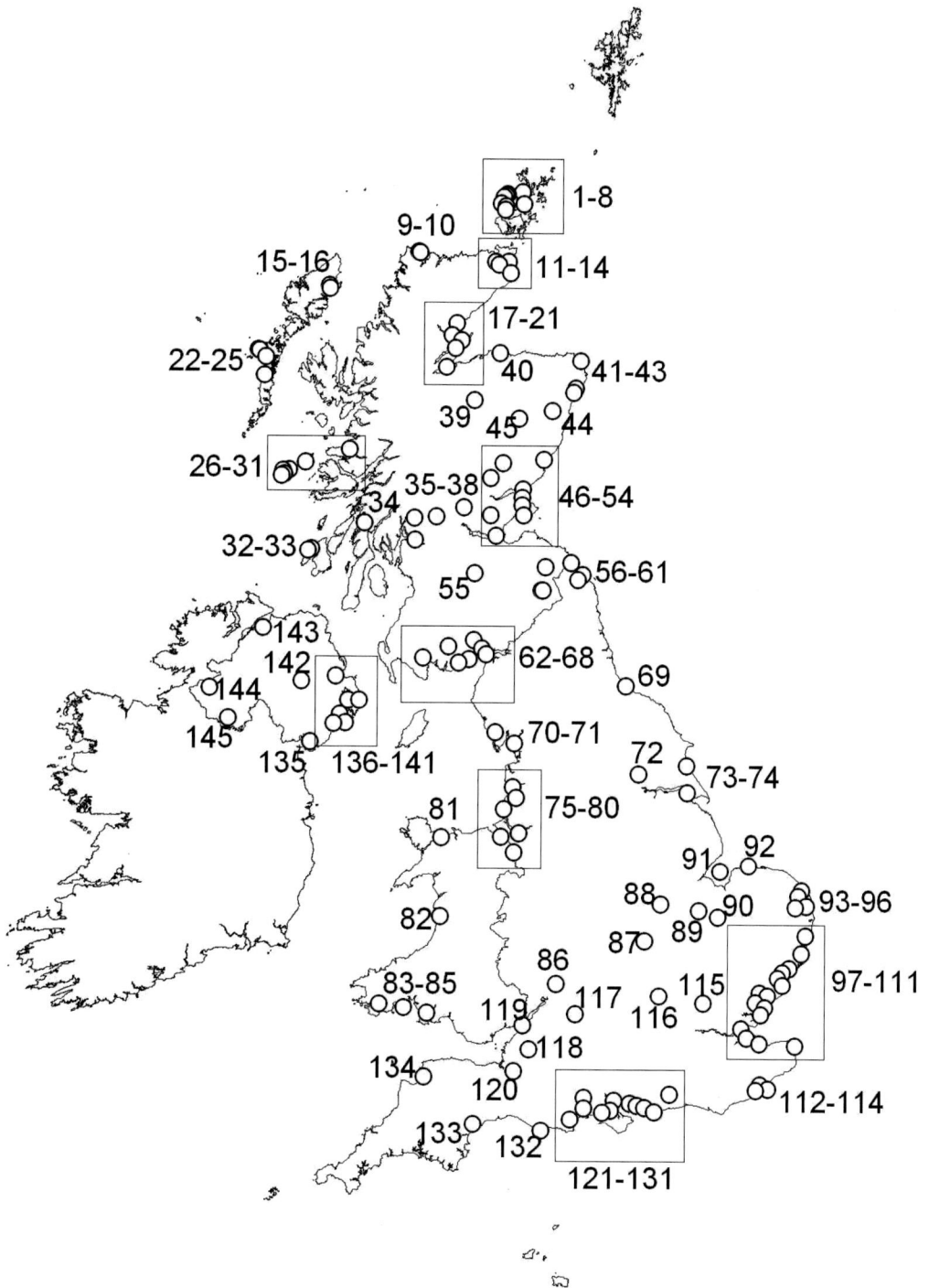

Figure A1. Locations of Core WeBS sites supporting more than 10,000 waterbirds or which support internationally important numbers of one or more waterbird species (see *PRINCIPAL SITES*). Numbers refer to sites listed in Table A2.